GREEK SYMBOLS

Capital	Lower Case	Greek Name	Pronunciation	English Letter
A	α	Alpha	al-fah	a
B	β	Beta	bay-tah	b
Γ	γ	Gamma	gam-ah	g
Δ	δ	Delta	del-tah	d
E	ε	Epsilon	ep-si-lon	e
Z	ζ	Zeta	zat-tah	z
H	η	Eta	ay-tah	h
Θ	θ	Theta	thay-tah	th
I	ι	Iota	eye-oh-tah	i
K	κ	Kappa	cap-ah	k
Λ	λ	Lambda	lamb-da	l
M	μ	Mu	mew	m
N	ν	Nu	new	n
Ξ	ξ	Xi	sah-eye	x
O	o	Omicron	oh-mi-cron	o
Π	π	Pi	pie	p
P	ρ	Rho	roe	r
Σ	σ	Sigma	sig-mah	s
T	τ	Tau	tah-hoe	t
Υ	υ	Upsilon	oop-si-lon	u
Φ	φ	Phi	fah-eye	ph
X	χ	Chi	kigh	ch
Ψ	ψ	Psi	sigh	ps
Ω	ω	Omega	Oh-mega	o

ROMAN NUMERALS

1	I	14	XIV	27	XXVII	150	CL
2	II	15	XV	28	XXVIII	200	CC
3	III	16	XVI	29	XXIX	300	CCC
4	IV	17	XVII	30	XXX	400	CD
5	V	18	XVIII	31	XXXI	500	D
6	VI	19	XIX	40	XL	600	DC
7	VII	20	XX	50	L	700	DCC
8	VIII	21	XXI	60	LX	800	DCCC
9	IX	22	XXII	70	LXX	900	CM
10	X	23	XXIII	80	LXXX	1000	M
11	XI	24	XXIV	90	XC	1600	MDC
12	XII	25	XXV	100	C	1700	MDCC
13	XIII	26	XXVI	101	CI	1900	MCM

USEFUL MATHEMATICAL EQUATIONS

$$\sum_{n=0}^{\infty} \frac{x^n}{n!} = e^x$$

$$\sum_{k=1}^{n} k^3 = 1 + 8 + 27 + \dots + n^3 = \frac{n^2(n+1)^2}{4} = \left[\frac{n(n+1)^2}{2}\right] = \left[\sum_{k=1}^{n} k\right]^2$$

$$\sum_{n=0}^{\infty} \frac{x^n}{n} = \ln\left(\frac{1}{1-x}\right)$$

$$\sum_{x=1}^{\infty} \frac{1}{x} = 1 + \frac{1}{2} + \frac{1}{3} + \dots \text{(does not converge)}$$

$$\sum_{n=0}^{k} x^n = \frac{x^{k+1}-1}{x-1}, \; x \neq 1$$

$$\sum_{m=0}^{k} ma^m = \frac{a}{(1-a)^2}\left[1 - (k+1)a^k + ka^{k+1}\right] = \sum_{m=1}^{k} ma^m$$

$$\sum_{n=1}^{k} x^n = \frac{x - x^{k+1}}{1-x}, \; x \neq 1$$

$$\sum_{k=0}^{n} (1) = n$$

$$\sum_{n=2}^{k} x^n = \frac{x^2 - x^{k+1}}{1-x}, \; x \neq 1$$

$$\sum_{k=0}^{n} \binom{n}{k} = 2^n$$

$$\sum_{n=0}^{\infty} p^n = \frac{1}{1-p}, \text{ if } |p| < 1$$

$$(a+b)^n = \sum_{k=0}^{n} \binom{n}{k} a^k b^{n-k}$$

$$\sum_{n=0}^{\infty} nx^n = \frac{x}{(1-x)^2}, \; x \neq 1$$

$$\prod_{n=1}^{\infty} a_n = e^{\left(\sum_{n=1}^{\infty} \ln(a_n)\right)}$$

$$\sum_{n=0}^{\infty} n^2 x^n = \frac{2x^2}{(1-x)^3} + \frac{x}{(1-x)^2} = \frac{x(1+x)}{(1-x)^3}, \; |x| < 1$$

$$\ln\left(\prod_{n=1}^{\infty} a_n\right) = \sum_{n=1}^{\infty} \ln a_n$$

$$\sum_{n=0}^{\infty} n^3 x^n = \frac{6x^3}{(1-x)^4} + \frac{6x^2}{(1-x)^3} + \frac{x}{(1-x)^2}, \; |x| < 1$$

$$\ln(x) = \sum_{k=1}^{\infty} \frac{1}{k}\left(\frac{x-1}{x}\right)^k, \; x \geq \frac{1}{2}$$

$$\sum_{n=0}^{M} nx^n = \frac{x\left[1 - (M+1)x^M + Mx^{M+1}\right]}{(1-x)^2}, \; |x| < 1$$

$$\lim_{h \to \infty} (1+h)^{1/h} = e$$

$$\sum_{x=0}^{\infty} \binom{r+x-1}{x} u^x = (1-u)^{-r}, \quad \text{if } |u| < 1$$

$$\lim_{n \to \infty} \left(1 + \frac{x}{n}\right)^n = e^{-x}$$

$$\sum_{k=1}^{\infty} (-1)^{k+1} \frac{1}{k} = 1 - \frac{1}{2} + \frac{1}{3} - \frac{1}{4} + \frac{1}{5} - \frac{1}{6} + \dots = \ln 2$$

$$\lim_{n \to \infty} \sum_{k=0}^{n} \frac{e^{-n} n^n}{K!} = \frac{1}{2}$$

$$\sum_{k=1}^{\infty} (-1)^{k+1} \frac{1}{(2k-1)} = 1 - \frac{1}{3} + \frac{1}{5} - \frac{1}{7} + \frac{1}{9} - \dots = \frac{\pi}{4}$$

$$\lim_{k \to \infty} \left(\frac{x^k}{k!}\right) = 0$$

$$\sum_{k=0}^{\infty} (-1)^k x^k = \frac{1}{1+x}, \quad -1 < x < 1$$

$$|x+y| \leq |x| + |y|$$

$$|x - y| \geq |x| - |y|$$

$$\sum_{k=1}^{n} (-1)^k \binom{n}{k} = 1, \quad \text{for } n \geq 2$$

$$\ln(1+x) = \sum_{k=1}^{\infty} (-1)^{k+1}\left(\frac{x^k}{k}\right), \text{ if } -1 < x \leq 1$$

$$\sum_{k=0}^{n} \binom{n}{k}^2 = \binom{2n}{n}$$

$$\Gamma\left(\frac{1}{2}\right) = \sqrt{\pi}$$

$$\sum_{k=1}^{n} k = 1 + 2 + 3 + \dots + n = \frac{n(n+1)}{2}$$

$$\Gamma(\alpha+1) = \alpha\Gamma(\alpha)$$

$$\sum_{k=1}^{n} k^2 = 1 + 4 + 9 + \dots + n^2 = \frac{n(n+1)(2n+1)}{6}$$

$$\Gamma\left(\frac{n}{2}\right) = \frac{\sqrt{\pi}\,(n-1)!}{2^{n-1}\left(\frac{n-1}{2}\right)!}, \quad n \text{ odd}$$

$$\sum_{k=0}^{n-1} k^2 x^k = \frac{(x-1)^2 n^2 x^n - 2(x-1)nx^{n+1} + x^{n+2} - x^2 + x^{n+1} - x}{(x-1)^3}$$

$$\Gamma(n) = \int_0^{\infty} e^{-x} x^{n-1} dx$$

$$\sum_{k=1}^{n} k^3 = 1 + 8 + 27 + \dots + n^3 = \left(\frac{n(n+1)}{2}\right)^2$$

$$\binom{n}{2} = \frac{1}{2}(n^2 - n) = \sum_{k=1}^{n-1} k$$

$$\sum_{k=1}^{n} (2k) = 2 + 4 + 6 + \dots + 2n = n(n-1)$$

$$\binom{n+1}{2} = \binom{n}{2} + n$$

$$\sum_{k=1}^{n} (2k-1) = 1 + 3 + 5 + \dots + (2n-1) = n^2$$

$$2.4.6.8 \dots 2n = \prod_{k=1}^{n} 2k = 2^n n!$$

$$\sum_{k=0}^{\infty} (a+kd)r^k = a + (a+d)r + (a+2d)r^2 + \dots + = \frac{a}{1-r} + \frac{rd}{(1-r)^2}$$

$$1.3.5.7 \dots (2n-1) = \frac{(2n-1)!}{2^{2n-2}(2n-2)!} = \frac{2n-1}{2^{2n-2}}$$

Handbook of Industrial Engineering Equations, Formulas, and Calculations

Industrial Innovation Series

Series Editor
Adedeji B. Badiru
Department of Systems and Engineering Management
Air Force Institute of Technology (AFIT) – Dayton, Ohio

PUBLISHED TITLES

Computational Economic Analysis for Engineering and Industry
Adedeji B. Badiru & Olufemi A. Omitaomu

Conveyors: Applications, Selection, and Integration
Patrick M. McGuire

Global Engineering: Design, Decision Making, and Communication
Carlos Acosta, V. Jorge Leon, Charles Conrad, and Cesar O. Malave

Handbook of Industrial Engineering Equations, Formulas, and Calculations
Adedeji B. Badiru & Olufemi A. Omitaomu

Handbook of Industrial and Systems Engineering
Adedeji B. Badiru

Handbook of Military Industrial Engineering
Adedeji B.Badiru & Marlin U. Thomas

Industrial Project Management: Concepts, Tools, and Techniques
Adedeji B. Badiru, Abidemi Badiru, and Adetokunboh Badiru

Inventory Management: Non-Classical Views
Mohamad Y. Jaber

Knowledge Discovery from Sensor Data
Auroop R. Ganguly, João Gama, Olufemi A. Omitaomu, Mohamed Medhat Gaber, and Ranga Raju Vatsavai

Moving from Project Management to Project Leadership: A Practical Guide to Leading Groups
R. Camper Bull

Social Responsibility: Failure Mode Effects and Analysis
Holly Alison Duckworth & Rosemond Ann Moore

STEP Project Management: Guide for Science, Technology, and Engineering Projects
Adedeji B. Badiru

Systems Thinking: Coping with 21st Century Problems
John Turner Boardman & Brian J. Sauser

Techonomics: The Theory of Industrial Evolution
H. Lee Martin

Triple C Model of Project Management: Communication, Cooperation, Coordination
Adedeji B. Badiru

FORTHCOMING TITLES

Essentials of Engineering Leadership and Innovation
Pamela McCauley-Bush & Lesia L. Crumpton-Young

Industrial Control Systems: Mathematical and Statistical Models and Techniques
Adedeji B. Badiru, Oye Ibidapo-Obe, & Babatunde J. Ayeni

Innovations of Kansei Engineering
Mitsuo Nagamachi & Anitawani Mohd Lokmanr

Kansei/Affective Engineering
Mitsuo Nagamachi

Kansei Engineering - 2 volume set
Mitsuo Nagamachi

Learning Curves: Theory, Models, and Applications
Mohamad Y. Jaber

Modern Construction: Productive and Lean Practices
Lincoln Harding Forbes

Project Management: Systems, Principles, and Applications
Adedeji B. Badiru

Research Project Management
Adedeji B. Badiru

Statistical Techniques for Project Control
Adedeji B. Badiru

Technology Transfer and Commercialization of Environmental Remediation Technology
Mark N. Goltz

Handbook of Industrial Engineering Equations, Formulas, and Calculations

Adedeji B. Badiru
Olufemi A. Omitaomu

CRC Press
Taylor & Francis Group
Boca Raton   London   New York

CRC Press is an imprint of the
Taylor & Francis Group, an **Informa** business

CRC Press
Taylor & Francis Group
6000 Broken Sound Parkway NW, Suite 300
Boca Raton, FL 33487-2742

First issued in paperback 2020

ISBN 13: 978-0-367-57042-2 (pbk)
ISBN 13: 978-1-4200-7627-1 (hbk)

Library of Congress Cataloging-in-Publication Data

Badiru, Adedeji Bodunde, 1952-
  Handbook of industrial engineering equations, formulas, and calculations / authors, Adedeji B. Badiru, Olufemi A. Omitaomu.
      p. cm. -- (Industrial innovation series)
  "A CRC title."
  Includes bibliographical references and index.
  ISBN 978-1-4200-7627-1 (hardcover : alk. paper)
    1. Industrial engineering--Mathematics--Handbooks, manuals, etc. 2. Engineering mathematics--Handbooks, manuals, etc. I. Omitaomu, Olufemi Abayomi. II. Title. III. Series.

T57.B33 2011
620.001'51--dc22   2010028413

Visit the Taylor & Francis Web site at
http://www.taylorandfrancis.com

and the CRC Press Web site at
http://www.crcpress.com

To our families for their unflinching support of our intellectual pursuits.

Contents

2 Basic Mathematical Calculations

3 Statistical Distributions, Methods, and Applications

5 Computations for Economic Analysis

6 Industrial Production Calculations

9 Risk Computations

10 Computations for Project Analysis

11 Product Shape and Geometrical Calculations

12 General Engineering Calculations

Preface

Calculations form the basis for engineering practice. Engineering researchers, instructors, students, and practitioners all need simple guides for engineering calculations. Although several books are available in the market for general engineering calculations, none is available directly for industrial engineering calculations. This book is designed to fill that void. It presents a general collection of mathematical equations that are likely to be encountered in the practice of industrial engineering.

Industrial engineering practitioners do not have to be computational experts; they just have to know where to get the computational resources that they need. This book provides access to computational resources needed by industrial engineers.

Industrial engineering is one of the most versatile and flexible branches of engineering. It has been said that engineers make things, whereas industrial engineers make things better. To make something better requires an understanding of its basic characteristics. The underlying equations and calculations facilitate that understanding. This book consists of several sections, each with a focus on a particular problem area. The book include the following topics:

Basic math calculations
Engineering math calculations
Production engineering calculations
Engineering economics calculations
Ergonomics calculations
Facility layout calculations
Production sequencing and scheduling calculations
Systems engineering calculations
Data engineering calculations
Project engineering calculations
Simulation and statistical equations

The book is unique in the market because it is the first book of its kind to focus exclusively on industrial engineering calculations with a correlation to applications for practice. The book will be of interest to engineering instructors, practicing engineers, upper-level undergraduate engineering students, graduate students, industry engineers, general practitioners, researchers, and consultants. This book will also be a useful review reference for practicing engineers preparing for the Principles and Practice of Engineering (PE) examinations. It is expected to have a long and sustainable tenure in the market in a manner similar to the machinery handbook which has undergone multiple editions. The modular layout and sectioning of the book will permit additions and expansions as new calculations emerge from research and practice of industrial engineering. As this is the first effort at compiling a specific handbook of industrial engineering equations, we expect the compilation to grow and improve over time.

Adedeji B. Badiru
Olufemi A. Omitaomu

Authors

Adedeji Badiru is a professor and head of the Department of Systems and Engineering Management at the Air Force Institute of Technology, Dayton, Ohio. He was previously professor and head of the Department of Industrial and Information Engineering at the University of Tennessee in Knoxville, Tennessee. Prior to that, he was professor of Industrial Engineering and Dean of University College at the University of Oklahoma. He is a registered professional engineer (PE), a certified project management professional (PMP), a fellow of the Institute of Industrial Engineers, and a fellow of the Nigerian Academy of Engineering. He holds BS in Industrial Engineering, MS in Mathematics, and MS in Industrial Engineering from Tennessee Technological University, and PhD in Industrial Engineering from the University of Central Florida. His areas of interest include mathematical modeling, project modeling and analysis, economic analysis, systems engineering, and productivity analysis and improvement. He is the author of several books and technical journal articles. He is the editor of the *Handbook of Industrial and Systems Engineering* (CRC Press, 2006). He is a member of several professional associations including Institute of Industrial Engineers (IIE), Institute of Electrical and Electronics Engineers (IEEE), Society of Manufacturing Engineers (SME), Institute for Operations Research and Management Science (INFORMS), American Society for Engineering Education (ASEE), American Society for Engineering Management (ASEM), New York Academy of Science (NYAS), and Project Management Institute (PMI). He has served as a consultant to several organizations across the world including Russia, Mexico, Taiwan, Nigeria, and Ghana. He has conducted customized training workshops for numerous organizations including Sony, AT&T, Seagate Technology, U.S. Air Force, Oklahoma Gas & Electric, Oklahoma Asphalt Pavement Association, Hitachi, Nigeria National Petroleum Corporation, and ExxonMobil. He has won several awards for his teaching, research, publications, administration, and professional accomplishments. He holds a leadership certificate from the University of Tennessee Leadership Institute. He has served as a technical project reviewer, curriculum reviewer, and proposal reviewer for several organizations including the Third World Network of Scientific Organizations, Italy, National Science Foundation, National Research Council, and the American Council on Education. He is on the editorial and review boards of several technical journals and book publishers. He has also served as an Industrial Development Consultant to the United Nations Development Program. He is also a Program Evaluator (PEV) for ABET (Accreditation Board for Engineering and Technology).

Olufemi Omitaomu is a research scientist in the Computational Sciences and Engineering Division at the Oak Ridge National Laboratory, Tennessee, and Adjunct assistant professor in the Department of Industrial and Information Engineering at the University of Tennessee, Knoxville, Tennessee. He holds BS in Mechanical Engineering from Lagos State University, Nigeria, MS in Mechanical Engineering from the University of Lagos, Nigeria, and PhD in Industrial and Information Engineering from the University of Tennessee. His areas of expertise include data mining and knowledge discovery from sensor data and data streams, complex networks modeling and analysis, and risk assessment in space and time. He has published several papers in prestigious journals and conferences including IEEE Transactions, ASME Journal, ICDM Workshops, and KDD Workshops. He is a co-organizer of the International Workshop on Knowledge Discovery from Sensor Data (SensorKDD). He is a member of Institute of Electrical and Electronics Engineers (IEEE) and Institute for Operations Research and Management Science (INFORMS).

1

Computational Foundations of Industrial Engineering

Mathematical calculations have played a major role in the development and advancement of our modern world. Practically all of our common consumer products have origins that can be traced to mathematical calculations. Equations and calculations provide the basis for the design of physical products and efficiency of services, which are within the direct domains of industrial engineering. In 2003, the U.S. National Academy of Engineering (NAE) published *A Century of Innovation: Twenty Engineering Achievements that Transformed Our Lives*. The book celebrated the top 20 technological advances of the twentieth century that fundamentally changed society. These advances have influenced where and how we live, what we eat, what we do for work or leisure, and how we think about our world and the universe.

Efficacy of Mathematical Modeling

Mathematical equations bring innovations to reality. Mathematical, analytical, theoretical, iconic, and scientific models are used by industrial engineers to represent decision scenarios in order to achieve more efficient decisions. Such models are essential for simulation, optimization, or computational applications that are central to the work of industrial engineers. Even in dynamic environments, robust analytical techniques can still offer insights into system behavior. Many mathematical theories and models of real-world problems have helped scientists and engineers tackle complex tasks. Today, mathematical techniques reach even further into our society with the evolving complexities in global techno-socioeconomic interaction. In addition to making technology more efficient and effective, mathematical techniques help organizations deal with financial, operational, and even marketing challenges.

Industrial Engineering and Computations

Industrial engineering (IE) is about how humans interact with different things—tools, technology, people, and processes. It is about the science, art, technology, and engineering of making things more efficient. To accomplish this, formulas, equations, and calculations are very essential. There is an increasing surge in the demand for analytical reasoning in human endeavors. The analytical approach of IE is used to solve complex and important problems that mankind face. Life itself is about choices. The practice of IE is about making the right choices in a dynamic environment of competing alternatives. Be it in launching a space shuttle or executing project control, decisions must be made in an environment of risk and scarcity. Analytical techniques of IE help solve these problems by formulating both qualitative and quantitative components into an integrated model that can be optimized. IE computational techniques have a wide range of applications including the following:

- Transportation
- Manufacturing
- Engineering
- Technology
- Science
- Mathematics
- Electronics
- Environment
- Medicine

For example, the operational properties of Laplace transform and Z-transform are used to ease computational complexity in applications such as cash flow series analysis, queuing analysis, traffic flow modeling, and material handling moves. Similarly, statistical equations are used for forecasting, geometry formulas are used for facility layout problems, and advanced algebraic computations are used in operations research optimization problems. The diverse capabilities of IE bring unique strengths in strategic management and organizational change and innovation. IE programs are mutually complementary with respect to the discipline's competency in systems engineering by adding the research methods that blend engineering technology with management expertise to implement long-term strategies. Sustainability is essential for IE to help organizations achieve integrated solutions to emerging challenges, to help enhance the interface between technology and human resources, and to help decision-makers focus on systems thinking and process design that improve operational efficiency.

The appreciation and inclusion of qualitative aspects of a problem are what set IE apart from other technical disciplines. In this respect, IE is more human-centric than other engineering disciplines. Human-based decision-making is central to any technical undertaking. A pseudo-code of a typical mathematical formulation might look as follows:

Problem scenario:	production planning
Desired objective:	optimize product, service, or results
Human resource options:	H1, H2, H3, H4
Technology options:	T1, T2, T3
Location options:	L1, L2, L3, L4
Operational constraints:	C1, C2, C3

Using the above problem framework, an organization can pursue cost reduction, profit maximization, or cost-avoidance goals. Superimposed upon the analytical framework are the common decision-making questions making up the *effectiveness* equation presented below:

$$E = W^5H,$$

where the elements are expanded below:

- What
- Who
- Where
- Why
- When
- How

An excellent example of solving complex decision problems using analytical modeling is the *Bridges to Somewhere* puzzle by Toczek (2009). The problem is paraphrased as follows:

> The five residents of Hometown live in houses represented by the letters 'A' through 'E' as shown on the left side of Figure 1.1. The offices where they are working are represented by their matching letters on the island of Worktown. Because a river lies between Hometown and Worktown, the residents are unable to get to work. They have in their budget enough funds to build two bridges that could connect Hometown to Worktown. The locations where these bridges could be built are indicated by the hashed 1 × 3 tiles. The two bridges can only be built in these approved areas. Once the bridges are built, the resident would then be able to commute to work. A commuter will always take the shortest path from home to work and can only travel in up, down, left, or right directions (no diagonals). Each tile represents a 1-km-by-km distance. As an example, if bridge number four were built, resident 'E' would have to travel 10 km to reach his workplace.
>
> Decision required: Which two bridges should be built in order to minimize the total commuting distance of all residents?

If this problem is to be solved by non-seat-of-the-pants approach, then a mathematical representation of the problem scenario must be developed. For this reason, IE and operations research often work hand-in-hand whereby one provides the comprehensive problem space and data requirements, whereas the other provides the analytical engine to solve the model. Equations help to link both sides of the decision-making partnership.

Many times, the qualitative formulation of a multifaceted problem is more difficult than the mathematical solution. A famous quote in this respect is echoed here:

> The mere formulation of a problem is far more essential than its solution, which may be merely a matter of mathematical or experimental skills. To raise new questions, new possibilities, to regard old problems from a new angle, requires creative imagination and marks real advances in science.

Albert Einstein

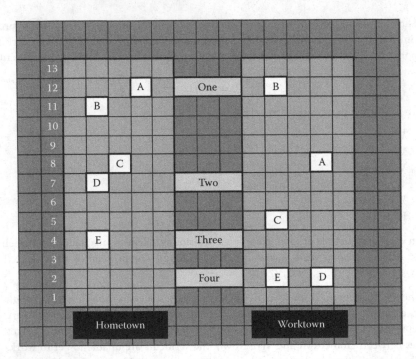

FIGURE 1.1 Bridge proposals connecting Hometown to Worktown. (Adapted from Toczek, J. 2009. Bridges to somewhere. *ORMS Today*, 36(4), 12.)

Einstein's quote is in agreement with the very essence of IE, which embodies versatility and flexibility of practice beyond other engineering disciplines. The versatility of IE is evidenced by the fact that many of those trained as industrial engineers end up working in a variety of professional positions in science, technology, engineering, business, industry, government, and the military. In all of these professional roles, industrial engineers use fundamental calculations to design products and services and also to make business decisions. This book is designed primarily as a reference material for students and practicing industrial engineers who may be unsure about the mathematical identity of the IE profession. Industrial engineers use analytical, computational, and experimental skills to accomplish integrated systems of people, materials, information, equipment, and energy. IE is one unique discipline that factors in human behavioral aspects in the workplace. Issues such as psychology, economics, health, sociology, and technology all have linkages that are best expressed through analytical means. Therein lies the importance of IE calculations.

Many modern, complex computational processes that are most visibly associated with other disciplines do, indeed, have their roots in one form or another of the practice of IE dating back to the Industrial Revolution. Thus, having a *Handbook of Industrial Engineering Equations, Formulas, and Calculations* is a fitting representation of the computationally rich history of the profession. The definition below aptly describes the wide span of IE practice.

Definition and Applications

IE is a broad-based problem-solving discipline that considers all facets of a problem in arriving at a holistic solution. The discipline embraces both the technical and managerial approaches in solving complex problems. Because of their integrated nature, IE solutions are robust and sustainable. Principles of IE are applicable to all areas of human endeavor. The essence of IE is captured in the official definition below:

> *Industrial Engineer*—one who is concerned with the design, installation, and improvement of integrated systems of people, materials, information, equipment, and energy by drawing upon specialized knowledge and skills in the mathematical, physical, and social sciences, together with the principles and methods of engineering analysis and design to specify, predict, and evaluate the results to be obtained from such systems.

The above definition embodies the various aspects of what an industrial engineer does. For several decades, the military, business, and industry have called upon the discipline of IE to achieve program effectiveness and operational efficiencies. IE is versatile, flexible, adaptive, and diverse. It can be seen from the definition that a systems orientation permeates the work of industrial engineers. This is particularly of interest to contemporary organizations, where operations and functions are constituted by linking systems. The major functions and applications of industrial engineers include the following:

- Designing integrated systems of people, technology, process, and methods.
- Modeling operations to optimize cost, quality, and performance trade-offs.
- Developing performance modeling, measurement, and evaluation for systems.
- Developing and maintain quality standards for government, industry, and business.
- Applying production principles to pursue improvements in service organizations.
- Incorporating technology effectively into work processes.
- Developing cost mitigation, avoidance, or containment strategies.
- Improving overall productivity of integrated systems of people, tools, and processes.
- Planning, organizing, scheduling, and controlling programs and projects.
- Organizing teams to improve efficiency and effectiveness of an organization.
- Installing technology to facilitate workflow.
- Enhancing information flow to facilitate smooth operation of systems.
- Coordinating materials and equipment for effective systems performance.
- Designing work systems to eliminate waste and reduce variability.
- Designing jobs (determining the most economic way to perform work).
- Setting performance standards and benchmarks for quality, quantity, and cost.
- Designing and installing facilities to incorporate human factors and ergonomics.

IE takes human as a central focus in the practice of the profession. In this regard, mathematical and statistical calculations are valuable tools in making decisions about people and society. IE activities (traditional and nontraditional), even though they may be called by other names, can be found in a variety of job functions, including the following:

Aircraft design
Maintenance
Engineering science
Mechanical design
Facility layout
City planning
Environmental science

Orientation to STEM

The definition of IE embraces STEM (science, technology, engineering, and mathematics). To this end, the organization of the book covers the following:

- Basic mathematics
  - Geometry
  - Trigonometry
- Advanced mathematics
- Statistics
- Engineering science
- Technology
- Science
- Social science
  - Human performance
  - Behavioral science

IE Catchphrases

As an attempt to convey the versatility, flexibility, and utility of the profession, several IE catchphrases have been proposed over the years. Those presented below represent a sampling of the most commonly cited.

One right way.
Engineers build things, industrial engineers make it better.
The IEs have it.

The official insignia of the Institute of Industrial Engineers says, "A better way."

Span and Utility of IE

IE can be described as the practical application of the combination of engineering fields together with the principles of scientific management. It is the engineering of work processes and the application of engineering methods, practices, and knowledge to production and service enterprises. IE places a strong emphasis on an understanding of workers and their needs in order to increase and improve production and service activities. IE serves as umbrella discipline for subspecialties as depicted as summarized below. Some of the subareas are independent disciplines in their own rights.

Information systems
Simulation
Systems engineering
Human factors
Ergonomics
Quality control
Lean six sigma
Operations research
Manufacturing
Product development
Design

Heritage from Industrial Revolution

IE has a proud heritage with a link that can be traced back to the *industrial revolution*. Although the practice of IE has been in existence for centuries, the work of Frederick Taylor in the early twentieth century was the first formal emergence of the profession. It has been referred to with different names and connotations. Scientific management was one of the early names used to describe what industrial engineers did.

Industry, the root of the profession's name, clearly explains what the profession is about. The dictionary defines industry generally as the ability to produce and deliver goods and services. The "industry" in IE can be viewed as the application of skills and creativity to achieve work objectives. This relates to how human effort is harnessed innovatively to carry out work. Thus, any activity can be defined as "industry" because it generates a product: be it service or physical product. A systems view of IE encompasses all the details and aspects necessary for applying skills and cleverness to produce work efficiently. The academic curriculum of IE continues to change, evolve, and adapt to the changing operating environment of the profession.

It is widely recognized that the occupational discipline that has contributed the most to the development of modern society is *engineering*, through its various segments of focus. Engineers design and build infrastructures that sustain the society. These include roads, residential and commercial buildings, bridges, canals, tunnels, communication systems, healthcare facilities, schools, habitats, transportation systems, and factories. In all of these, the IE process of systems integration facilitates the success of the efforts. In this sense, the scope of IE goes through the levels of activity, task, job, project, program, process, system, enterprise, and society. This handbook of IE calculations presents essential computational tools for the levels embodied by this hierarchy of functions. From the age of horse-drawn carriages and steam engines to the present age of intelligent automobiles and aircraft, the impacts of IE cannot be mistaken, even though the contributions may not be recognized in the context of the conventional work of industry.

Historical Accounts

Going further back in history, several developments helped form the foundation for what later became known as IE. In America, George Washington was said to have been fascinated by the design of farm implements on his farm in Mt. Vermon. He had an English manufacturer send him a plow built according to his specifications that included a mold on which to form new irons when old ones were worn out or would need repairs. This can be described as one of the earliest attempts to create a process of achieving a system of interchangeable parts. Thomas Jefferson invented a wooden moldboard which, when fastened to a plow, minimized the force required to pull the plow at various working depths. This is an example of early agricultural industry innovation. Jefferson also invented a device that allowed a farmer to seed four rows at a time. In pursuit of higher productivity, he invented a horse-drawn threshing machine that did the work of 10 men.

Meanwhile, in Europe, productivity growth, through reductions in manpower, marked the technological innovations of the 1769–1800 Europe. Sir Richard Arkwright developed a practical code of factory disciplines. Matthew Boulton and James Watt developed, in their foundry, a complete and integrated engineering plant to manufacture steam engines. They developed extensive methods of market research, forecasting, plant location planning, machine layout, work flow, machine operating standards, standardization of product components, worker training, division of labor, work study, and other creative approaches to increasing productivity. Charles Babbage, who is credited with the first idea of a computer, documented ideas on scientific methods of managing industry in his book *On the Economy of Machinery and Manufactures* (Babbage, 1833). The book contained ideas on division of labor, paying less for less important tasks, organization charts, and labor relations. These were all forerunners of modern IE.

In the early history of the United States, several efforts emerged to form the future of the IE profession. Eli Whitney used mass production techniques to produce muskets for the U.S. Army. In 1798, Whitney developed the idea of having machines make each musket part so that it could be interchangeable with other similar parts. By 1850, the principle of interchangeable parts was widely adopted. It eventually became the basis for modern mass production for assembly lines. It is believed that Eli Whitney's principle of interchangeable parts contributed significantly to the Union victory during the U.S. Civil War. Thus, the early practice of IE made significant contribution to the military. That heritage has continued until today.

The management's attempts to improve productivity prior to 1880 did not consider the human element as an intrinsic factor. However, from 1880 through the first quarter of the twentieth century, the works of Frederick W. Taylor, Frank and Lillian Gilbreth, and Henry L. Gantt created a long-lasting impact on productivity growth through consideration of the worker and his or her environment.

Frederick Winslow Taylor (1856–1915) was born in the Germantown section of Philadelphia to a well-to-do family. At the age of 18, he entered the labor force, having abandoned his admission to Harvard University due to an impaired vision. He became an apprentice machinist and patternmaker in a local machine shop. In 1878, when he was 22, he went to work at the Midvale Steel Works. The economy was in a depressed state at the time. Frederick was employed as a laborer. His superior intellect was very quickly recognized. He was soon advanced to the positions of time clerk, journeyman, lathe operator, gang boss, and foreman of the machine shop. By the age of 31, he was made chief engineer of the company. He attended night school and earned a degree in mechanical engineering in 1883 from Stevens Institute. As a work leader, Taylor faced the following common questions:

1. Which is the best way to do this job?
2. What should constitute a day's work?

These are still questions faced by industrial engineers of today. Taylor set about the task of finding the proper method for doing a given piece of work, instructing the worker in following the method, maintaining standard conditions surrounding the work so that the task could be properly accomplished, and setting a definite time standard and payment of extra wages for doing the task as specified. Taylor later documented his industry management techniques in his book, *The Principles of Scientific Management* (Taylor, 1911).

The work of Frank and Lillian Gilbreth agree with that of Frederick Taylor. In 1895, on his first day on the job as a bricklayer, Frank Gilbreth noticed that the worker assigned to teach him how to lay brick did his work in three different ways. The bricklayer was insulted when Frank tried to tell him of his work inconsistencies—when training someone on the job, when performing the job himself, and when speeding up. Frank thought it was essential to find one best way to do work. Many of Frank Gilbreth's ideas were similar to Taylor's ideas. However, Gilbreth outlined procedures for analyzing each step of workflow. Gilbreth made it possible to apply science more precisely in the analysis and design of the workplace. Developing *therbligs*, which is a moniker for Gilbreth spelled backwards, as elemental predetermined time units, Frank and Lillian Gilbreth were able to analyze the motions of a worker in performing most factory operations in a maximum of 18 steps. Working as a team, they developed techniques that later became known as work design, methods improvement, work simplification, value engineering, and optimization. Lillian (1878–1972) brought to the engineering profession the concern for human relations. The foundation for establishing the profession of IE was originated by Frederick Taylor and Frank and Lillian Gilbreth.

The work of Henry Gantt (1861–1919) advanced the management movement from an industrial management perspective. He expanded the scope of managing industrial operations. His concepts emphasized the unique needs of the worker by recommending the following considerations for managing work:

1. Define his task, after a careful study.
2. Teach him how to do it.
3. Provide an incentive in terms of adequate pay or reduced hours.
4. Provide an incentive to surpass it.

Henry Gantt's major contribution is the Gantt Chart, which went beyond the works of Frederick Taylor or the Gilbreths. The Gantt Chart related every activity in the plant to the factor of time. This was a revolutionary concept for the time. It led to better production planning control and better production control. This involved visualizing the plant as a whole, like one big system made up of interrelated subsystems.

Chronology of Applications

The major chronological events marking the origin and applications of IE are summarized below.

1440   Venetian ships were reconditioned and refitted on an assembly line.

1474   Venetian Senate passed the first patent law and other industrial laws.

1568   Jacques Besson published illustrated book on iron machinery as a replacement for wooden machines.

1622   William Oughtred invented the slide rule.

1722   Rene de Reaunur published the first handbook on iron technology.

1733   John Kay patented the flying shuttle for textile manufacture—a landmark in textile mass production.

1747   Jean Rodolphe Perronet established the first engineering school.

1765   Watt invented the separate condenser, which made the steam engine the power source.

1770   James Hargreaves patented his "Spinning Jenny." Jesse Ramsden devised a practical screw-cutting lathe.

1774   John Wilkinson built the first horizontal boring machine.

1775   Richard Arkwright patented a mechanized mill in which raw cotton is worked into thread.

1776   James Watt built the first successful steam engine, which became a practical power source.

1776   Adam Smith discussed the division of labor in *The Wealth of Nations*.

1785   Edmund Cartwright patented a power loom.

1793   Eli Whitney invented the "cotton gin" to separate cotton from its seeds.

1797   Robert Owen used modern labor and personnel management techniques in a spinning plant in the New Lanark Mills in Manchester, UK.

1798   Eli Whitney designed muskets with interchangeable parts.

1801   Joseph Marie Jacquard designed automatic control for pattern-weaving looms using punched cards.

1802   "Health and Morals Apprentices Act" in Britain aimed at improving standards for young factory workers.

Marc Isambard Brunel, Samuel Benton, and Henry Maudsey designed an integrated series of 43 machines to mass produce pulley blocks for ships.

1818   Institution of Civil Engineers founded in Britain.

1824   The repeal of the Combination Act in Britain legalized trade unions.

1829   Mathematician Charles Babbage designed "analytical engine," a forerunner of the modern digital computer.

1831   Charles Babbage published *On the Economy of Machines and Manufactures* (1839).

1832   The Sadler Report exposed the exploitation of workers and the brutality practiced within factories.

1833   Factory law enacted in the United Kingdom. The Factory Act regulated British children's working hours.

A general Trades Union was formed in New York.

1835	Andrew Ure published *Philosophy of Manufacturers.*
	Samuel Morse invented the telegraph.
1845	Friederich Engels published *Condition of the Working Classes in England.*
1847	Factory Act in Britain reduced the working hours of women and children to 10 h per day.
	George Stephenson founded the Institution of Mechanical Engineers.
1856	Henry Bessemer revolutionized the steel industry through a novel design for a converter.
1869	Transcontinental railroad completed in the United States.
1871	British Trade Unions legalized by Act of Parliament.
1876	Alexander Graham Bell invented a usable telephone.
1877	Thomas Edison invented the phonograph.
1878	Frederick W. Taylor joined Midvale Steel Company.
1880	American Society of Mechanical Engineers (ASME) was organized.
1881	Frederick Taylor began time study experiments.
1885	Frank B. Gilbreth began motion study research.
1886	Henry R. Towne presented the paper *The Engineer as Economist.*
	American Federation of Labor (AFL) was organized.
	Vilfredo Pareto published *Course in Political Economy.*
	Charles M. Hall and Paul L. Herault independently invented an inexpensive method of making aluminum.
1888	Nikola Tesla invented the alternating current induction motor, enabling electricity to takeover from steam as the main provider of power for industrial machines.
	Dr. Herman Hollerith invented the electric tabulator machine, the first successful data-processing machine.
1890	Sherman Anti-Trust Act was enacted in the United States.
1892	Gilbreth completed the motion study of bricklaying.
1893	Taylor began work as a consulting engineer.
1895	Taylor presented the paper *A Piece-Rate System* to ASME.
1898	Taylor began time study at Bethlehem Steel.
	Taylor and Maunsel White developed process for heat-treating high-speed tool steels.
1899	Carl G. Barth invented a slide rule for calculating metal cutting speed as part of Taylor system of management.
1901	American national standards were established.
	Yawata Steel began operation in Japan.
1903	Taylor presented the paper *Shop Management* to ASME.
	H.L. Gantt developed the "Gantt Chart."
	Hugo Diemers wrote *Factory Organization and Administration.*
	Ford Motor Company was established.
1904	Harrington Emerson implemented Santa Fe Railroad improvement.
	Thorstein B. Veblen: *The Theory of Business Enterprise.*
1906	Taylor established metal-cutting theory for machine tools.

	Vilfredo Pareto: *Manual of Political Economy.*
1907	Gilbreth used time study for construction.
1908	Model T Ford built.
	Pennsylvania State College introduced the first university course in IE.
1911	Taylor published *The Principles of Scientific Management.*
	Gilbreth published *Motion Study.*
	Factory laws enacted in Japan.
1912	Harrington Emerson published *The Twelve Principles of Efficiency.*
	Frank and Lillian Gilbreth presented the concept of "therbligs."
	Yokokawa translated into Japanese Taylor's *Shop Management and the Principles of Scientific Management.*
1913	Henry Ford established a plant at Highland Park, MI, which utilized the principles of uniformity and interchangeability of parts, and of the moving assembly line by means of conveyor belt.
	Hugo Munstenberg published *Psychology of Industrial Efficiency.*
1914	World War I.
	Clarence B. Thompson edited *Scientific Management*, a collection of articles on Taylor's system of management.
1915	Taylor's system used at Niigata Engineering's Kamata plant in Japan.
	Robert Hoxie published *Scientific Management and Labour.*
1916	Lillian Gilbreth published *The Psychology of Management.*
	Taylor Society established in the United States.
1917	The Gilbreths published *Applied Motion Study.*
	The Society of Industrial Engineers was formed in the United States.
1918	Mary P. Follet published *The New State: Group Organization, the Solution of Popular Government.*
1919	Henry L. Gantt published *Organization for Work.*
1920	Merrick Hathaway presented the paper *Time Study as a Basis for Rate Setting.*
	General Electric established divisional organization.
	Karel Capek: *Rossum's Universal Robots.*
	This play coined the word "robot."
1921	The Gilbreths introduced process-analysis symbols to ASME.
1922	Toyoda Sakiichi's automatic loom developed.
	Henry Ford published *My Life and Work.*
1924	The Gilbreths announced the results of micromotion study using therbligs.
	Elton Mayo conducted illumination experiments at Western Electric.
1926	Henry Ford published *Today and Tomorrow.*
1927	Elton Mayo and others began relay-assembly test room study at the Hawthorne plant.
1929	Great Depression.
	International Scientific Management Conference held in France.
1930	Hathaway: *Machining and Standard Times.*
	Allan H. Mogensen discussed 11 principles for work simplification in *Work Simplification.*

Henry Ford published *Moving Forward.*

1931   Dr. Walter Shewhart published *Economic Control of the Quality of Manufactured Product.*

1932   Aldous Huxley published *Brave New World*, the satire which prophesied a horrifying future ruled by industry.

1934   General Electric performed micro-motion studies.

1936   The word "automation" was first used by D.S. Harder of General Motors. It was used to signify the use of transfer machines which carry parts automatically from one machine to the next, thereby linking the tools into an integrated production line.

Charlie Chaplin produced *Modern Times*, a film showing an assembly line worker driven insane by routine and unrelenting pressure of his job.

1937   Ralph M. Barnes published *Motion and Time Study.*

1941   R.L. Morrow: *Ratio Delay Study*, an article in *Mechanical Engineering* journal.

Fritz J. Roethlisberger: *Management and Morale.*

1943   ASME work standardization committee published glossary of IE terms.

1945   Marvin E. Mundel devised "memo-motion" study, a form of work measurement using time-lapse photography.

Joseph H. Quick devised work factors (WF) method.

1945   Shigeo Shingo presented concept of production as a network of processes and operations and identified lot delays as source of delay between processes, at a technical meeting of the Japan Management Association.

1946   The first all-electronic digital computer ENIAC (Electronic Numerical Integrator and Computer) was built at Pennsylvania University.

The first fully automatic system of assembly was applied at the Ford Motor Plant.

1947   American mathematician, Norbert Wiener: *Cybernetics.*

1948   H.B. Maynard and others introduced methods time measurement (MTM) method.

Larry T. Miles developed value analysis (VA) at General Electric.

Shigeo Shingo announced process-based machine layout.

American Institute of Industrial Engineers was formed.

1950   Marvin E. Mundel: *Motion and Time Study, Improving Productivity.*

1951   Inductive statistical quality control was introduced to Japan from the United States.

1952   Role and sampling study of IE conducted at ASME.

1953   B.F. Skinner: *Science of Human Behaviour.*

1956   New definition of IE was presented at the American Institute of Industrial Engineering Convention.

1957   Chris Argyris: *Personality and Organization.*

Herbert A. Simon: *Organizations.*

R.L. Morrow: *Motion and Time Study.*

1957   Shigeo Shingo introduced scientific thinking mechanism (STM) for improvements.

	The Treaty of Rome established the European Economic Community.
1960	Douglas M. McGregor: *The Human Side of Enterprise.*
1961	Rensis Lickert: *New Patterns of Management.*
1961	Shigeo Shingo devised ZQC (source inspection and poka-yoke) systems.
1961	Texas Instruments patented the silicon chip integrated circuit.
1963	H.B. Maynard: *Industrial Engineering Handbook.*
	Gerald Nadler: *Work Design.*
1964	Abraham Maslow: *Motivation and Personality.*
1965	Transistors were fitted into miniaturized "integrated circuits."
1966	Frederick Hertzberg: *Work and the Nature of Man.*
1968	Roethlisberger: *Man in Organization.*
	U.S. Department of Defense: *Principles and Applications of Value Engineering.*
1969	Shigeo Shingo developed single-minute exchange of dies (SMED).
	Shigeo Shingo introduced preautomation.
	Wickham Skinner: *Manufacturing—Missing Link in Corporate Strategy*; Harvard Business Review.
1971	Taiichi Ohno completed the Toyota production system.
1971	Intel Corporation developed the micro-processor chip.
1973	First annual Systems Engineering Conference of AIIE.
1975	Shigeo Shingo extolled NSP-SS (nonstock production) system.
	Joseph Orlicky: "MRP: Material Requirements Planning."
1976	IBM marketed the first personal computer.
1980	Matsushita Electric used Mikuni method for washing machine production.
	Shigeo Shingo: *Study of the Toyota Production System from an Industrial Engineering Viewpoint.*
1981	Oliver Wight: *Manufacturing Resource Planning: MRP II.*
1982	Gavriel Salvendy: *Handbook of Industrial Engineering.*
1984	Shigeo Shingo: *A Revolution in Manufacturing: The SMED System.*
2009	Adedeji Badiru and Marlin Thomas: Publication of *Handbook of Military Industrial Engineering.*
2010	Adedeji Badiru and Olufemi Omitaomu: Publication of *Handbook of Industrial Engineering Equations, Formulas, and Calculations.*

As can be seen from the historical details above, industry engineering has undergone progressive transformation over the past several decades.

Importance of IE Calculations

Calculations form the basis of engineering practice. Engineering researchers, instructors, students, and practitioners all need simple guides for engineering calculations. Although several books are available in the market for general engineering calculations, none is available directly for IE calculations. This book is designed to fill that void.

For decades, industrial engineers have been dedicated for finding ways to work more effectively and for more productivity. This quest for continuous improvement has its basis on calculations. In general, calculations form the basis for engineering analysis, design, and implementation. In particular, calculations form the basis of IE practice. Industrial engineers use calculations everywhere and for everything, even in areas not traditionally recognized as being within the IE discipline. Because the term *industry* defines production, sale, delivery, and utilization of goods and services, the discipline of *industrial engineering* has wide and direct applications in all human endeavors. Thus, the discipline is the most diverse and flexible of the engineering disciplines.

The span of IE calculations covers diverse areas including the following:

Research. The process of gathering information, discovering inherent knowledge, creating new knowledge, and disseminating what is found using calculations at every stage of the endeavor.

Operations analysis. The process of using number sense and skills, relationships, and calculations to determine how work works.

Facility design. The process of identifying, classifying, and analyzing dimensional objects, understanding their properties, and using that knowledge to design the layout of facilities.

Measurement. The process of making accurate measurements and calculations; representing patterns and relationships using tables, graphs, and symbols to draw inferences about the work environment.

Data analysis and probability. The process of organizing and interpreting results through data collection and calculations to answer questions, solve problems, show relationships, and make predictions.

Mathematical modeling. The process of applying problem-solving skills, reasoning, and mathematical calculations to represent decision objects and draw conclusions.

Life sciences. The process of understanding the structure and function of living systems, particularly humans, and how they interact with the work environment.

Physical sciences. The process of understanding physical systems, concepts, and properties of matter, energy, forces, and motion to improve human work.

Social sciences and human factors. The process of identifying both similarities and differences in the traditions and cultures of various groups of people and their relationships with the work environment.

Science and technology. The process of understanding the relationship between science and technology to design and construct devices to perform and improve human work.

Scientific enquiry and management. The process of using scientific approach to ask questions, conduct investigations, collect data, analyze data, extract information, and disseminate information to manage and enhance work.

Economic analysis. The process of understanding how to make decisions on the basis of economic factors and multidimensional calculations.

Ergonomics. The process of using applied science of equipment design for the workplace with the purpose of maximizing productivity by reducing operator fatigue and discomfort.

Importance of Calculations Guide

Engineering researchers, instructors, students, and practitioners all need simple guides for engineering calculations. Although several books are available in the market for general engineering calculations, none is available directly for IE calculations. This book is designed to fill that void. IE practitioners do not have to be computational experts; they just have to know where to get the computational resources that they need. This book provides access to computational resources needed by industrial engineers.

Being the most versatile and flexible branches of engineering, it has been said that "engineers make things, whereas industrial engineers make things better." To make something better requires an understanding of its basic characteristics. The underlying equations and calculations facilitate that understanding. This book consists of several sections, each with a focus on a particular specialization area of IE. Topics covered in the book include the following:

Basic math calculations
Engineering math calculations
Product shape geometrical calculations
Production engineering calculations
Engineering economics calculations
Ergonomics calculations
Facility layout calculations
Production sequencing and scheduling calculations
Systems engineering calculations
Data engineering calculations
Project engineering calculations
Simulation and statistical equations

The book is unique in the market because it is the first of its kind to focus exclusively on IE calculations with a correlation to applications for practice. The book will be of interest to engineering instructors, practicing engineers, upper-level undergraduate engineering students, graduate students, industry engineers, general practitioners, researchers, and consultants. This book will also be a useful review reference for practicing engineers preparing for the principles and practice of engineering (PE) examinations. The modular layout and sectioning of the book will permit additions and expansions as new calculation emerge from research and practice of IE.

IE practitioners do not have to be computational experts; they just have to know where to get the computational resources that they need. This book provides access to computational resources needed by industrial engineers.

IE is one of the most versatile and flexible branches of engineering. It has been said that engineers make things, whereas industrial engineers make things better. To make something better requires an understanding of its basic characteristics. The underlying equations and calculations facilitate that understanding. The sections that follow illustrate one area of the fundamental analytical computations that form the foundation of IE and its complementary field of operations research.

Basic Queuing Equations

Our organizational and socioeconomic systems often require that we "wait" for a service or a resource. This waiting process is achieved through the process of queuing. Bank tellers, gasoline service stations, post offices, grocery check-out lanes, ATM machines, food service counters, and benefits offices are common examples of where we experience queues. Not all queues involve humans. In many cases, the objects and subjects in a queue are themselves service elements. Thus, we can have an automated service waiting on another automated service. Electronic communication systems often consist of this type of queuing relationship. There can also be different mixtures or combinations of humans and automated services in a composite queuing system. In each case, the desire of industrial engineers is to make the queue more efficient. This is accomplished through computational analyses. Because the goal of IE is centered on improvement, if we improve the queuing process, we can make significant contributions to both organizational and technical processes.

In terms of a definition, a queuing system consists of one or more servers that provide some sort of service to arriving customers. Customers who arrive to find all servers busy generally enter one or more lines in front of the servers. This process of arrival–service–departure constitutes the *queuing system*. Each queuing system is characterized by three components:

- The arrival process
- The service mechanism
- The queue discipline

Arrivals into the queuing system may originate from one or several sources referred to as the *calling population*. The calling population (source of inputs) can be limited or unlimited. The arrival process consists of describing how customers arrive to the system. This requires knowing the time in-between the arrivals of successive customers. From the inter-arrival times, the arrival frequency can be computed. So, if the average inter-arrival time is denoted as λ, then the arrival frequency is $1/\lambda$.

The *service mechanism* of a queuing system is specified by the number of servers with each server having its/his/her own queue or a common queue. Servers are usually denoted by the letter s. Also specified is the probability distribution of customer service times. If we let S_i denote the service time of the ith customer, then the mean service time of a customer can be denoted by $E(S)$. Thus, the service rate of a server is denoted as $\mu = 1/E(S)$. The variables, λ and μ, are the key elements of queuing equations.

Queue discipline of a queuing system refers to the rule that governs how a server chooses the next customer from the queue (if any) when the server completes the service of the current customer. Commonly used queue disciplines are:

- FIFO: first-in, first-out basis
- LIFO: last-in, first-out
- SIRO: service in random order
- GD: general queue discipline

- Priority: customers are served in order of their importance on the basis of their service requirements or some other set of criteria.

Banks are notorious for tweaking their queuing systems, service mechanisms, and queue discipline based on prevailing needs and desired performance outcomes. The performance of a queue can be assessed based on several *performance measures*. Some of the common elements of queue performance measures are:

- Waiting time in the system for the ith customer, W_i
- Delay in the queue for the ith customer, D_i
- Number of customers in the queue, $Q(t)$
- The number of customers in the queue system at time t, $L(t)$
- The probability that a delay will occur
- The probability that the total delay will be greater than some predetermined value
- The probability that all service facilities will be idle
- The expected idle time of the total facility
- The probability of balking (un-served departures) due to insufficient waiting accommodation

Note that

$$W_i = D_i + S_i$$

$$L(t) = Q(t) + \text{number of customers being served at } t.$$

If a queue is not chaotic, it is expected to reach steady-state conditions that are expressed as

- Steady-state average delay, d
- Steady-state average waiting time in the system, w
- Steady-state time average number in queue, Q
- Steady-state time average number in the system, L

$$d = \lim_{n \to \infty} \frac{\sum_{i=1}^{n} D_i}{n},$$

$$w = \lim_{n \to \infty} \frac{\sum_{i=1}^{n} W_i}{n},$$

$$Q = \lim_{T \to \infty} \frac{1}{T} \int_{0}^{T} Q(t) \, dt,$$

$$L = \lim_{T \to \infty} \frac{1}{T} \int_{0}^{T} L(t) \, dt,$$

$$Q = \lambda d, \quad L = \lambda w, \quad w = d + E(S).$$

The following short-hand representation of a queue embeds the inherent notation associated with the particular queue:

$$[A/B/s] : \{d/e/f\}$$

where A is the probability distribution of the arrivals, B the probability distribution of the departures, s the number of servers or channels, d the capacity of the queue(s), e the size of the calling population, and f the queue ranking or ordering rule.

A summary of the standard notation for queuing short-hand representation is presented below:

M random arrival, service rate, or departure distribution that is a Poisson process
E Erlang distribution
G general distribution
GI general independent distribution
D deterministic service rate (constant rate of serving customers)

Using the above, we have the following example:

$$[M/M/1] : \{\infty/\infty/\text{FCFS}\},$$

which represents a queue system where the arrivals and departures follow a Poisson distribution, with a single server, infinite queue length, infinite calling population, and the queue discipline of First Come, First Served (FCFS). This is the simplest queue system that can be studied mathematically. It is often simply referred to as the $M/M/1$ queue. The following notation and equations apply:

λ arrival rate
μ service rate
ρ system utilization factor (traffic intensity) = fraction of time that servers are busy $(= \lambda/\mu)$
s number of servers
M random arrival and random service rate (Poisson)
D deterministic service rate
L the average number of customers in the queuing system
L_q the average queue length (customers waiting in line)
L_s average number of customers in service
W average time a customer spends in the system
W_q average time a customer spends waiting in line
W_s average time a customer spends in service
P_n probability that exactly n customers are in the system

For any queuing system in steady-state condition, the following relationships hold:

$$L = \lambda W$$
$$L_q = \lambda W_q$$
$$L_s = \lambda W_s$$

The common cause of not achieving a steady state in a queuing system is where the arrival rate is at least as large as the maximum rate at which customers can be served.

For steady state to exist, we must have $\rho < 1$.

$$\frac{\lambda}{\mu} = \rho,$$

$$L = (1 - \rho)\frac{\rho}{(1 - \rho)^2}$$

$$= \frac{\rho}{1 - \rho}$$

$$= \frac{\lambda}{\mu - \lambda},$$

$$L_q = L - L_s$$

$$= \frac{\rho}{1 - \rho} - \rho$$

$$= \frac{\rho^2}{1 - \rho}$$

$$= \frac{\lambda^2}{\mu(\mu - \lambda)}.$$

If the mean service time is constant for every customer (i.e., $1/\mu$), then

$$W = W_q + \frac{1}{\mu}$$

The queuing formula $L = \lambda W$ is very general and can be applied to many situations that do not, on the surface, appear to be queuing problems. For example, it can be applied to inventory and stocking problems, whereby an average number of units is present in the system (L), new units arrive into the system at a certain rate (λ), and the units spend an average time in the system (W). For example, consider how a fast food restaurant stocks, depletes, and replenishes hamburger meat patties in the course of a week.

$M/D/1$ case (random arrival, deterministic service, and one service channel)

Expected average queue length, $L_q = (2\rho - \rho^2)/2(1 - \rho)$
Expected average total time in the system, $W = (2 - \rho)/2\mu(1 - \rho)$
Expected average waiting time, $W_q = \rho/2\mu(1 - \rho)$

$M/M/1$ case (random arrival, random service, and one service channel)

The probability of having zero vehicles in the systems, $P_0 = 1 - \rho$
The probability of having n vehicles in the systems, $P_n = \rho^n P_0$
Expected average queue length, $L_q = \rho/(1 - \rho)$
Expected average total time in system, $W = \rho/\lambda(1 - \rho)$
Expected average waiting time, $W_q = W - 1/\mu$

It can be seen from the above expressions that different values of parameters of the queuing system will generally have different consequences. For example, limiting the queue length could result in the following:

- Average idle time might increase;
- Average queue length will decrease;

- Average waiting time will decrease;
- A portion of the customers will be lost.

Queuing Birth–Death Processes

A queuing *birth–death process* is a continuous-time stochastic process for which the system's state at any time is a nonnegative integer. We use birth–death processes to answer questions about several different types of queuing systems. We define the number of people present in any queuing system at time t to be the *state* of the queuing systems at time t and define the following:

$\pi_j =$ steady-state probability of state j
$P_{ij}(t) =$ transient behavior of the queuing system before the steady state is reached

Laws of Motion of Queuing Birth and Death

Queuing Birth–Death Law 1

With probability $\lambda_j \Delta t + o(\Delta t)$, a birth occurs between time t and time $t + \Delta t$. A birth increases the system state by 1, to $j + 1$. The variable λ_j is called the *birth rate* in state j. In most cases, a birth is simply an arrival.

Queuing Birth–Death Law 2

With probability $\mu_j \Delta t + o(\Delta t)$, a death occurs between time t and time $t + \Delta t$. A death decreases the system state by 1, to $j - 1$. The variable μ_j is the death rate in state j. In most queuing systems, a death is a service completion. Note that $\mu_0 = 0$ must hold, otherwise a negative state could occur.

Queuing Birth–Death Law 3

Births and deaths are independent of each other.

The primary approach here is to relate $P_{ij}(t + \Delta t)$ to $P_{ij}(t)$ for small Δt. Thus, we have

$$\pi_{j-1}\lambda_{j-1} + \pi_{j+1}\mu_{j+1} = \pi_j(\lambda_j + \mu_j), \quad j = 1, 2, \ldots,$$
$$\pi_1\mu_1 = \pi_0\lambda_0.$$

The above equations are called the *flow balance equations* or *conservation of flow equations* for a birth–death process. It should be cautioned that mathematical computations of queuing systems, such as many mathematical analyses, must be tempered by reason, real-life logic, and practical heuristics. Mathematically, formulations are often too restrictive to be able to model real-world scenarios accurately. This is because the underlying assumptions may not match real-world situations. The complexity of production lines with product-specific characteristics cannot be handled with mathematical models alone. Hybrid computations and analyses offer more pragmatic approaches that allow us to better visualize, design, simulate, analyze, and optimize queuing systems. In queuing analysis, mathematical models often assume infinite numbers of customers, infinite queue capacity, or no bounds on inter-arrival or service times. These are clearly not the case in most practical operational situations, although lack of statistical significance may allow us to ignore some of the real-life restrictions. In some cases, the mathematical solution may be intractable, overly complex, or lack sufficient data to be useful for practical decisions. This is where the human element of IE decision analysis is most needed. Simulation and empirical or experimental tools are used by industrial engineers for the special cases.

Data Types for Computational Analysis

Computational models are sensitive to the type of data that form the basis for the models. Data collection, measurement, and analysis are essential for developing robust computational models. Each data set must be measurable on an appropriate scale. The common data types and scales are presented below.

Nominal Scale

Nominal scale is the lowest level of measurement scales. It classifies items into categories. The categories are mutually exclusive and collectively exhaustive. That is, the categories do not overlap and they cover all possible categories of the characteristics being observed. For example, in the analysis of the critical path in a project network, each job is classified as either critical or not critical. Gender, type of industry, job classification, and color are some examples of measurements on a nominal scale.

Ordinal Scale

Ordinal scale is distinguished from a nominal scale by the property of order among the categories. An example is the process of prioritizing project tasks for resource allocation. We know that first is above second, but we do not know how far above. Similarly, we know that better is preferred to good, but we do not know by how much. In quality control, the ABC classification of items based on the Pareto distribution is an example of a measurement on an ordinal scale.

Interval Scale

Interval scale is distinguished from an ordinal scale by having equal intervals between the units of measure. The assignment of priority ratings to project objectives on a scale of 0 to 10 is an example of a measurement on an interval scale. Even though an objective may have a priority rating of zero, it does not mean that the objective has absolutely no significance to the project team. Similarly, the scoring of zero on an examination does not imply that a student knows absolutely nothing about the materials covered by the examination. Temperature is a good example of an item that is measured on an interval scale. Even though there is a zero point on the temperature scale, it is an arbitrary relative measure. Other examples of interval scale are IQ measurements and aptitude ratings.

Ration Scale

Ratio scale has the same properties of an interval scale, but with a true zero point. For example, an estimate of zero time unit for the duration of a task is a ratio scale measurement. Other examples of items measured on a ratio scale are cost, time, volume, length, height, weight, and inventory level. Many of the items measured in a project management environment will be on a ratio scale.

Cardinal Scale

Cardinal scale is a counting scale that refers to the number of elements in a collection. This is used for comparison purposes. The cardinality of a collection of items is the number of items contained in the collection. For example, in a productivity incentive program, the cardinality of production cycles can be used to determine which division is rewarded with bonuses.

References

Babbage, C. 1833. *On the Economy of Machinery and Manufactures*, 3rd Edition. London: Charles Knight & Pall Kast Publishers.

Badiru, A. B., ed. 2006. *Handbook of Industrial & Systems Engineering*. Boca Raton, FL: Taylor & Francis/CRC Press.

Taylor, F. W. 1911. *The Principles of Scientific Management*, Harper Bros, New York, NY.

Hillier, F. S. and Lieberman, G. J. 1974. *Operations Research*, 2nd edn. San Francisco, CA: Holden-Day.

Taha, H. A. 1982. *Operations Research: An Introduction*, 3rd edn. New York, NY: MacMillan

Toczek, J. 2009. Bridges to somewhere, *ORMS Today*, 36(4), 12.

U.S. National Academy of Engineering 2003. *A Century of Innovation: Twenty Engineering Achievements that Transformed Our Lives*. Washington, D.C.: U.S. National Academy of Engineering.

Whitehouse, G. E. and Wechsler, B. L. 1976. *Applied Operations Research: A Survey*. New York, NY:Wiley.

2

Basic Mathematical Calculations

Quadratic Equation

The roots of the quadratic equation

$$ax^2 + bx + c = 0 \tag{2.1}$$

are given by the quadratic formula

$$x = \frac{-b \pm \sqrt{b^2 - 4ac}}{2a}.$$

The roots are complex if $b^2 - 4ac < 0$, the roots are real if $b^2 - 4ac > 0$, and the roots are real and repeated if $b^2 - 4ac = 0$.

Dividing both sides of Equation 2.1 by a, where $a \neq 0$, we obtain

$$x^2 + \frac{b}{a}x + \frac{c}{a} = 0.$$

Note that the solution of $ax^2 + bx + c = 0$ is $x = -c/b$ if $a = 0$.

Rewrite Equation 2.1 as

$$\left(x + \frac{b}{2a}\right)^2 - \frac{b^2}{4a^2} + \frac{c}{a} = 0,$$

$$\left(x + \frac{b}{2a}\right)^2 = \frac{b^2}{4a^2} - \frac{c}{a} = \frac{b^2 - 4ac}{4a^2},$$

$$x + \frac{b}{2a} = \pm\sqrt{\frac{b^2 - 4ac}{4a^2}} = \pm\frac{\sqrt{b^2 - 4ac}}{2a},$$

$$x = -\frac{b}{2a} \pm \sqrt{\frac{b^2 - 4ac}{4a^2}},$$

$$x = \frac{-b \pm \sqrt{b^2 - 4ac}}{2a}.$$

Overall Mean

The mean for n number of elements is generally given by the following equation:

$$\bar{x} = \frac{n_1\bar{x}_1 + n_2\bar{x}_2 + n_3\bar{x}_3 + \cdots + n_k\bar{x}_k}{n_1 + n_2 + n_3 + \cdots + n_k} = \frac{\sum n\bar{x}}{\sum n}.$$

Chebyshev's Theorem

For any positive constant k, the probability that a random variable will take on a value within k standard deviations of the mean is at least

$$1 - \frac{1}{k^2}.$$

Permutations

A permutation of m elements from a set of n elements is any arrangement, without repetition, of the m elements. The total number of all the possible permutations of n distinct objects taken m times is

$$P(n, m) = \frac{n!}{(n - m)!}, \quad n \geq m.$$

Example

Find the number of ways a president, a vice president, a secretary, and a treasurer can be chosen from a committee of eight members.

SOLUTION

$$P(n, m) = \frac{n!}{(n - m)!} = P(8, 4) = \frac{8!}{(8 - 4)!} = \frac{8 \cdot 7 \cdot 6 \cdot 5 \cdot 4 \cdot 3 \cdot 2 \cdot 1}{4 \cdot 3 \cdot 2 \cdot 1} = 1680.$$

There are 1680 ways of choosing the four officials from the committee of eight members.

Combinations

The number of combinations of n distinct elements taken is given by

$$C(n, m) = \frac{n!}{m!(n - m)!}, \quad n \geq m.$$

Example

How many poker hands of five cards can be dealt from a standard deck of 52 cards?

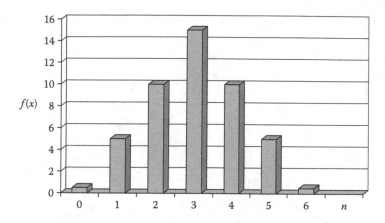

FIGURE 2.1 Probability distribution plot.

SOLUTION

Note: The order in which the five cards can be dealt with is not important.

$$C(n, m) = \frac{n!}{m!(n - m)!} = C(52, 5) = \frac{52!}{5!(52 - 5)!}$$

$$= \frac{52!}{5!47!} = \frac{52 \cdot 51 \cdot 50 \cdot 49 \cdot 48}{5 \cdot 4 \cdot 3 \cdot 2 \cdot 1} = 2{,}598{,}963.$$

Failure

If a trial results in any of n equally likely ways and s is the number of successful ways, then the probability of failure is

$$q = 1 - p = \frac{n - s}{n}.$$

Probability Distribution

An example of probability distribution is shown in the histogram plot in Figure 2.1. A probability density function can be inferred from the probability distribution as shown by the example in Figure 2.2.

Probability

$$P(X \leq x) = F(x) = \int_{-\infty}^{x} f(x)\, dx.$$

Distribution Function

Figure 2.3 shows the general profile of the cumulative probability function of the probability density function in Figure 2.2.

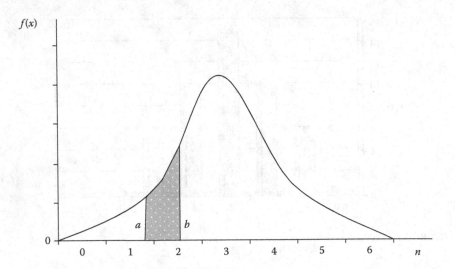

FIGURE 2.2   Example of probability density function on plot.

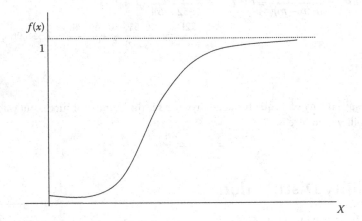

FIGURE 2.3   Cumulative probability plot.

Expected Value

The expected value μ is defined by

$$\mu = \sum x f(x)$$

provided this sum converges absolutely.

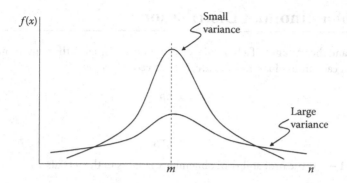

FIGURE 2.4 Graphical illustration of variance.

Variance

The variance (which is the square of the standard deviation, i.e., σ^2) is defined as the average of the squared differences from the mean, that is,

$$\sigma^2 = \sum (x - \mu)^2 f(x) \quad \text{or} \quad \sigma^2 = \int_{-\infty}^{\infty} (x - \mu)^2 f(x)\, dx.$$

Figure 2.4 illustrates the distribution spread conveyed by the variance measure.

Binomial Distribution

A binomial random variable is the number of successes x in n repeated trials of a binomial experiment. The probability distribution of a binomial random variable is called a binomial distribution (also known as a Bernoulli distribution) and is given by

$$f(x) = {}^n c_x p^x (1 - p)^{n-x}.$$

Poisson Distribution

The Poisson distribution is a mathematical rule that assigns probabilities to the number occurrences. The distribution function for the Poisson distribution is given by

$$f(x) = \frac{(np)^x e^{-np}}{x!}.$$

Mean of a Binomial Distribution

The mean and the variance of a binomial distribution are equal to the sums of means and variances of each individual trial and are, respectively, given by

$$\mu = np$$

and

$$\sigma^2 = npq,$$

where $q = 1 - p$ is the probability of obtaining x failures in the n trials.

Normal Distribution

A normal distribution in a variate x with mean μ and variance σ^2 is a statistical distribution with probability density function

$$f(x) = \frac{1}{\sigma\sqrt{2\pi}}e^{-(x-\mu)^2/2\sigma^2}$$

on the domain $x \in (-\infty, \infty)$.

Cumulative Distribution Function

Let X be a numerical random variable. It is completely described by the probability for a realization of the variable to be less than x for any x. This probability is denoted by

$$F(x) = P(X \le x) = \frac{1}{\sigma\sqrt{2\pi}}\int_{-\infty}^{x} e^{-(x-\mu)^2/2\sigma^2}\,dx.$$

Here $F(x)$ is called the cumulative distribution function of the variable X.

Population Mean

Population mean is defined as the mean of a numerical set that includes all the numbers within the entire group and is given by

$$\mu_{\bar{x}} = \mu.$$

Standard Error of the Mean

The standard error of the mean ($\sigma_{\bar{x}}$) is the standard deviation of the sample mean estimate of a population mean. It can also be viewed as the standard deviation of the error in the

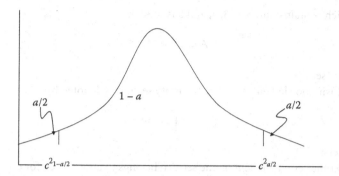

FIGURE 2.5 Chi-squared distribution.

sample mean relative to the true mean, since the sample mean is an unbiased estimator. It is usually estimated by the sample estimate of the population standard deviation (sample standard deviation) divided by the square root of the sample size (assuming statistical independence of the values in the sample):

$$\sigma_{\bar{x}} = \frac{\sigma}{\sqrt{n}}.$$

t-Distribution

t-Distribution is a type of probability distribution that is theoretical and resembles a normal distribution. A *t*-distribution differs from the normal distribution by its degrees of freedom. The higher the degrees of freedom, the closer that distribution will resemble a standard normal distribution with a mean of 0 and a standard deviation of 1. It is also known as the "Student's *t*-distribution" and is given by

$$\bar{x} - t_{\alpha/2}\left(\frac{s}{\sqrt{n}}\right) \leq \mu \leq \bar{x} + t_{\alpha/2}\left(\frac{s}{\sqrt{n}}\right),$$

where $\bar{x}$ is the sample mean, μ the population mean, and s the sample standard deviation.

Chi-Squared Distribution

The chi-squared distribution is shown graphically in Figure 2.5.

$$\frac{(n-1)s^2}{\chi^2_{\alpha/2}} \leq \sigma^2 \leq \frac{(n-1)s^2}{\chi^2_{1-\alpha/2}}.$$

Definition of Set and Notation

A set is a collection of objects called elements. In mathematics, we write a set by putting its elements between the curly brackets { }.

Set A which contains numbers 3, 4, and 5 is written as

$$A = \{3, 4, 5\}.$$

(a) Empty set
A set with no elements is called an empty set and is denoted by

$$\{\,\} = \Phi.$$

(b) Subset
Sometimes every element of one set also belongs to another set, for example,

$$A = \{3, 4, 5\} \quad \text{and} \quad B = \{1, 2, 3, 4, 5, 6, 7\}.$$

Here set A is the subset of set B because every element of set A is also an element of set B, and hence it is written as

$$A \subseteq B.$$

(c) Set equality
Sets A and B are equal if and only if they have exactly the same elements, and the equality is written as
$$A = B.$$

(d) Set union
The union of set A and set B is the set of all elements that belong to either A or B or both, and is written as

$$A \cup B = \{x \mid x \in A \text{ or } x \in B \text{ or both}\}.$$

Set Terms and Symbols

{}	set braces
$\in$	is an element of
$\notin$	is not an element of
$\subseteq$	is a subset of
$\not\subseteq$	is not a subset of
A'	complement of set A
$\cap$	set intersection
$\cup$	set union

Venn Diagrams

Venn diagrams are used to visually illustrate relationships between sets. Examples are shown in Figure 2.6.

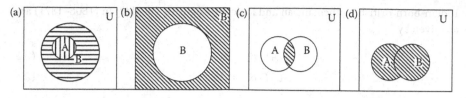

FIGURE 2.6 Venn diagram examples.

These Venn diagrams illustrate the following statements:

(a) Set A is the subset of set B ($A \subset B$).
(b) Set B' is the complement of B.
(c) Two sets A and B with their intersection $A \cap B$.
(d) Two sets A and B with their union $A \cup B$.

Operations on Sets

If A, B, and C are arbitrary subsets of universal set U, then the following rules govern the operations on sets:

(1) Commutative law for union
$$A \cup B = B \cup A.$$

(2) Commutative law for intersection
$$A \cap B = B \cap A.$$

(3) Associative law for union
$$A \cup (B \cup C) = (A \cup B) \cup C.$$

(4) Associative law for intersection
$$A \cap (B \cap C) = (A \cap B) \cap C.$$

(5) Distributive law for union
$$A \cup (B \cap C) = (A \cup B) \cap (A \cup C).$$

(6) Distributive law for intersection
$$A \cap (B \cap C) = (A \cap B) \cup (A \cap C).$$

De Morgan's Laws

In set theory, De Morgan's laws relate the three basic set operations to each other: the union, the intersection, and the complement. De Morgan's laws are named after the

Indian-born British mathematician and logician Augustus De Morgan (1806–1871) and are given by

$$(A \cup B)' = A' \cap B', \tag{2.2}$$

$$(A \cap B)' = A' \cup B'. \tag{2.3}$$

The complement of the union of two sets is equal to the intersection of their complements (Equation 2.2). The complement of the intersection of two sets is equal to the union of their complements (Equation 2.3).

Counting the Elements in a Set

The number of the elements in a finite set is determined by simply counting the elements in the set.

If A and B are disjoint sets, then

$$n(A \cup B) = n(A) + n(B).$$

In general, A and B need not to be disjoint, so

$$n(A \cup B) = n(A) + n(B) - n(A \cap B),$$

where n is the number of the elements in a set.

Permutations

A permutation of m elements from a set of n elements is any arrangement, without repetition, of the m elements. The total number of all the possible permutations of n distinct objects taken m times is

$$P(n, m) = \frac{n!}{(n-m)!}, \quad n \geq m.$$

Example

Find the number of ways a president, a vice president, a secretary, and a treasurer can be chosen from a committee of eight members.

SOLUTION

$$P(n, m) = \frac{n!}{(n-m)!} = P(8, 4) = \frac{8!}{(8-4)!} = \frac{8 \cdot 7 \cdot 6 \cdot 5 \cdot 4 \cdot 3 \cdot 2 \cdot 1}{4 \cdot 3 \cdot 2 \cdot 1} = 1680.$$

There are 1680 ways of choosing the four officials from the committee of eight members.

Combinations

The number of combinations of n distinct elements taken is given by

$$C(n, m) = \frac{n!}{m!(n-m)!}, \quad n \geq m.$$

Example

How many poker hands of five cards can be dealt from a standard deck of 52 cards?

SOLUTION

Note: The order in which the five cards can be dealt with is not important.

$$C(n, m) = \frac{n!}{m!(n-m)!} = C(52, 5) = \frac{52!}{5!(52-5)!}$$

$$= \frac{52!}{5!47!} = \frac{52 \cdot 51 \cdot 50 \cdot 49 \cdot 48}{5 \cdot 4 \cdot 3 \cdot 2 \cdot 1} = 2{,}598{,}963.$$

Probability Terminology

A number of specialized terms are used in the study of probability.

Experiment: An experiment is an activity or occurrence with an observable result.
Outcome: The result of the experiment.
Sample point: An outcome of an experiment.
Event: An event is a set of outcomes (a subset of the sample space) to which a probability is assigned.

Basic Probability Principles

Consider a random sampling process in which all the outcomes solely depend on chance, that is, each outcome is equally likely to happen. If S is a uniform sample space and the collection of desired outcomes is E, then the probability of the desired outcomes is

$$P(E) = \frac{n(E)}{n(S)},$$

where $n(E)$ is the number of favorable outcomes in E and $n(S)$ the number of possible outcomes in S.

As E is a subset of S,

$$0 \leq n(E) \leq n(S),$$

then the probability of the desired outcome is

$$0 \leq P(E) \leq 1.$$

Random Variable

A random variable is a rule that assigns a number to each outcome of a chance experiment.

1. A coin is tossed six times. The random variable X is the number of tails that are noted. X can only take the values $1, 2, \ldots, 6$, so X is a discrete random variable.
2. A light bulb is burned until it burns out. The random variable Y is its lifetime in hours. Y can take any positive real value, so Y is a continuous random variable.

Mean Value $\hat{x}$ or Expected Value μ

The mean value or expected value of a random variable indicates its average or central value. It is a useful summary value of the variable's distribution.

1. If a random variable X is a discrete mean value, then

$$\hat{x} = x_1 p_1 + x_2 p_2 + \cdots + x_n p_n = \sum_{i=1}^{n} x_1 p_1,$$

where p_i is the probability densities.

2. If X is a continuous random variable with probability density function $f(x)$, then the expected value of X is

$$\mu = E(X) = \int_{-\infty}^{+\infty} x f(x)\, dx,$$

where $f(x)$ is the probability densities.

Series Expansions

(a) **Expansions of Common Functions**

A series expansion is a representation of a particular function as a sum of powers in one of its variables, or by a sum of powers of another (usually elementary) function $f(x)$. Here are series expansions (some Maclaurin, some Laurent, and some Puiseux) for a number of common functions:

$$e = 1 + \frac{1}{1!} + \frac{1}{2!} + \frac{1}{3!} + \cdots,$$

$$e^x = 1 + x + \frac{x^2}{2!} + \frac{x^3}{3!} + \cdots,$$

$$a^x = 1 + x \ln a + \frac{(x \ln a)^2}{2!} + \frac{(x \ln a)^3}{3!} + \cdots,$$

$$e^{-x^2} = 1 - x^2 + \frac{x^4}{2!} - \frac{x^6}{3!} + \frac{x^8}{4!} - \cdots,$$

$$\ln x = (x-1) - \frac{1}{2}(x-1)^2 + \frac{1}{3}(x-1)^3 - \cdots, \quad 0 < x \le 2,$$

$$\ln x = \frac{x-1}{x} + \frac{1}{2}\left(\frac{x-1}{x}\right)^2 + \frac{1}{3}\left(\frac{x-1}{x}\right)^3 + \cdots, \quad x > \frac{1}{2},$$

$$\ln x = 2\left[\frac{x-1}{x+1} + \frac{1}{3}\left(\frac{x-1}{x+1}\right)^3 + \frac{1}{5}\left(\frac{x-1}{x+1}\right)^5 + \cdots\right], \quad x > 0,$$

$$\ln(1+x) = x - \frac{x^2}{2} + \frac{x^3}{3} - \frac{x^4}{4} + \cdots, \quad |x| \le 1,$$

$$\ln(a+x) = \ln a + 2\left[\frac{x}{2a+x} + \frac{1}{3}\left(\frac{x}{2a+x}\right)^3 + \frac{1}{5}\left(\frac{x}{2a+x}\right)^5 + \cdots\right],$$

$$a > 0, \; -a < x < +\infty,$$

$$\ln\left(\frac{1+x}{1-x}\right) = 2\left(x + \frac{x^3}{3} + \frac{x^5}{5} + \frac{x^7}{7} + \cdots\right), \quad x^2 < 1,$$

$$\ln\left(\frac{1+x}{1-x}\right) = 2\left[\frac{1}{x} + \frac{1}{3}\left(\frac{1}{x}\right)^3 + \frac{1}{5}\left(\frac{1}{x}\right)^5 + \left(\frac{1}{x}\right)^7 + \cdots\right], \quad x^2 > 1,$$

$$\ln\left(\frac{1+x}{x}\right) = 2\left[\frac{1}{2x+1} + \frac{1}{3(2x+1)^3} + \frac{1}{5(2x+1)^5} + \cdots\right], \quad x > 0,$$

$$\sin x = x - \frac{x^3}{3!} + \frac{x^5}{5!} - \frac{x^7}{7!} + \cdots,$$

$$\cos x = 1 - \frac{x^2}{2!} + \frac{x^4}{4!} - \frac{x^6}{6!} + \cdots,$$

$$\tan x = x + \frac{x^3}{3} + \frac{2x^5}{15} + \frac{17x^7}{315} + \frac{62x^9}{2835} + \cdots, \quad x^2 < \frac{\pi^2}{4},$$

$$\sin^{-1} x = x + \frac{x^3}{6} + \frac{1}{2}\cdot\frac{3}{4}\cdot\frac{x^3}{5} + \frac{1}{2}\cdot\frac{3}{4}\cdot\frac{5}{6}\cdot\frac{x^7}{7} + \cdots, \quad x^2 < 1,$$

$$\tan^{-1} x = x - \frac{1}{3}x^3 + \frac{1}{5}x^5 - \frac{1}{7}x^7 + \cdots, \quad x^2 < 1,$$

$$\tan^{-1} x = \frac{\pi}{2} - \frac{1}{x} + \frac{1}{3x^3} - \frac{1}{5x^5} + \cdots, \quad x^2 > 1,$$

$$\sinh x = x + \frac{x^3}{3!} + \frac{x^5}{5!} + \frac{x^7}{7!} + \cdots,$$

$$\cosh x = 1 + \frac{x^2}{2!} + \frac{x^4}{4!} + \frac{x^6}{6!} + \cdots,$$

$$\tanh x = x - \frac{x^3}{3} + \frac{2x^5}{15} - \frac{17x^7}{315} + \cdots,$$

$$\sinh^{-1} x = x - \frac{1}{2} \cdot \frac{x^3}{3} + \frac{1 \cdot 3}{2 \cdot 4} \cdot \frac{x^5}{5} - \frac{1 \cdot 3 \cdot 5}{2 \cdot 4 \cdot 6} \cdot \frac{x^7}{7} + \cdots, \quad x^2 < 1,$$

$$\sinh^{-1} x = \ln 2x + \frac{1}{2} \cdot \frac{1}{2x^2} - \frac{1 \cdot 3}{2 \cdot 4} \cdot \frac{1}{4x^4} + \frac{1 \cdot 3 \cdot 5}{2 \cdot 4 \cdot 6} \cdot \frac{1}{6x^6} - \cdots, \quad x > 1,$$

$$\cosh^{-1} x = \ln 2x - \frac{1}{2} \cdot \frac{1}{2x^2} - \frac{1 \cdot 3}{2 \cdot 4} \cdot \frac{1}{4x^4} - \frac{1 \cdot 3 \cdot 5}{2 \cdot 4 \cdot 6} \cdot \frac{1}{6x^6} - \cdots,$$

$$\tanh^{-1} x = x + \frac{x^3}{3} + \frac{x^5}{5} + \frac{x^7}{7} + \cdots, \quad x^2 < 1.$$

(b) **Binomial Theorem**

The binomial theorem describes the algebraic expansion of powers of a binomial. According to this theorem, it is possible to expand the power $(a + x)^n$ into a sum involving terms of the form $ba^c x^d$, where the coefficient of each term is a positive integer and the sum of the exponents of a and x in each term is n. For example

$$(a + x)^n = a^n + na^{n-1}x + \frac{n(n-1)}{2!}a^{n-2}x^2 + \frac{n(n-1)(n-2)}{3!}a^{n-3}x^3 + \cdots,$$

$$x^2 < a^2.$$

(c) **Taylor Series Expansion**

A function $f(x)$ may be expanded about $x = a$ if the function is continuous, and its derivatives exist and are finite at $x = a$.

$$f(x) = f(a) + f'(a)\frac{(x-a)}{1!} + f''(a)\frac{(x-a)^2}{2!} + f'''(a)\frac{(x-a)^3}{3!} + \cdots$$

$$+ f^{n-1}(a)\frac{(x-a)^{n-1}}{(n-1)!} + R_n.$$

(d) **Maclaurin Series Expansion**

The Maclaurin series expansion is a special case of the Taylor series expansion for $a = 0$.

$$f(x) = f(0) + f'(0)\frac{x}{1!} + f''(0)\frac{x^2}{2!} + f'''(0)\frac{x^3}{3!} + \cdots + f^{(n-1)}(0)\frac{x^{n-1}}{(n-1)!} + R_n.$$

(e) **Arithmetic Progression**

The sum to n terms of the arithmetic progression

$$S = a + (a + d) + (a + 2d) + \cdots + [a + (n-1)d]$$

is (in terms of the last number l)

$$S = \frac{n}{2}(a + l),$$

where $l = a + (n-1)d$.

(f) **Geometric Progression**

The sum of the geometric progression to n terms is

$$S = a + ar + ar^2 + \cdots + ar^{n-1} = a\left(\frac{1-r^n}{1-r}\right).$$

(g) **Sterling's Formula for Factorials**

Sterling's formula gives an approximate value for the factorial function $n!$ The approximation can most simply be derived for n, an integer, by approximating the sum over the terms of the factorial with an integral, so that

$$n! \approx \sqrt{2\pi}\, n^{n+1/2} e^{-n}.$$

Mathematical Signs and Symbols

There are numerous mathematical symbols that can be used in mathematics. Here is the list of some of the most commonly used math symbols and their definition.

$\pm\ (\mp)$	plus or minus (minus or plus)
:	divided by, ratio sign
::	proportional sign
$<$	less than
$>$	greater than
$\cong$	approximately equals, congruent
$\sim$	similar to
$\equiv$	equivalent to
$\neq$	not equal to
$\doteq$	approaches, is approximately equal to
$\propto$	varies as
∞	infinity
$\therefore$	therefore
$\sqrt{\ }$	square root
$\sqrt[3]{\ }$	cube root
$\sqrt[n]{\ }$	nth root
$\angle$	angle
$\perp$	perpendicular to
$\parallel$	parallel to
$|x|$	numerical value of x
log or $\log_{10}$	common logarithm or Briggsian logarithm
$\log_e$ or ln	natural logarithm or hyperbolic logarithm or Napierian logarithm
e	base (2.718) of natural system of logarithms
$a°$	an angle a degrees

a'	a prime, an angle a minutes
a''	a double prime, an angle a seconds, a second
sin	sine
cos	cosine
tan	tangent
ctn or cot	cotangent
sec	secant
csc	cosecant
vers	versed sine
covers	coversed sine
exsec	exsecant
$\sin^{-1}$	antisine or angle whose sine is
sinh	hyperbolic sine
cosh	hyperbolic cosine
tanh	hyperbolic tangent
$\sinh^{-1}$	antihyperbolic sine or angle whose hyperbolic sine is
$f(x)$ or $\phi(x)$	function of x
Δx	increment of x
$\sum$	summation of
dx	differential of x
dy/dx or y'	derivative of y with respect to x
d^2y/dx^2 or y''	second derivative of y with respect to x
d^ny/dx^n	nth derivative of y with respect to x
$\partial y/\partial x$	partial derivative of y with respect to x
$\partial^n y/\partial x^n$	nth partial derivative of y with respect to x
$\partial^n y/\partial x \partial y$	nth partial derivative with respect to x and y
$\int$	integral of
$\int_a^b$	integral between the limits a and b
$\dot{y}$	first derivative of y with respect to time
$\ddot{y}$	second derivative of y with respect to time
Δ or ∇^2	the Laplacian $\left(\dfrac{\partial^2}{\partial x^2} + \dfrac{\partial^2}{\partial y^2} + \dfrac{\partial^2}{\partial x^2}\right)$
δ	sign of a variation
ξ	sign of integration around a closed path

Greek Alphabets

Alpha $= A,\ \alpha = A, a$
Beta $= B,\ \beta = B, b$
Gamma $= \Gamma,\ \gamma = G, g$
Delta $= \Delta,\ \delta = D, d$

Epsilon　$= E,\ \varepsilon = E, e$
Zeta　　$= Z,\ \zeta = Z, z$
Eta　　$= H,\ \eta = E, e$
Theta　$= \Theta,\ \theta = \text{Th,th}$
Iota　　$= I,\ \iota = I, i$
Kappa　$= K,\ \kappa = K, k$
Lambda　$= \Lambda,\ \lambda = L, 1$
Mu　　$= M,\ \mu = M, m$
Nu　　$= N,\ \nu = N, n$
Xi　　$= \Xi,\ \xi = X, x$
Omicron $= O,\ o = O, o$
Pi　　$= \Pi,\ \pi = P, p$
Rho　　$= P,\ p = R, r$
Sigma　$= \Sigma,\ \sigma = S, s$
Tau　　$= T,\ \tau = T, t$
Upsilon　$= T,\ \upsilon = U, u$
Phi　　$= \Phi,\ \phi = \text{Ph,ph}$
Chi　　$= X,\ \chi = \text{Ch,ch}$
Psi　　$= \Psi,\ \psi = \text{Ps,ps}$
Omega　$= \Omega,\ \omega = O, o$

Algebra

Laws of Algebraic Operations

Many operations on real numbers are based on the commutative, associative, and distributive laws. The effective use of these laws is important. These laws will be stated in written form as well as algebraic form, where letters or symbols are used to represent an unknown number.

(a)　Commutative law: $a + b = b + a$, $ab = ba$;
(b)　Associative law: $a + (b + c) = (a + b) + c$, $a(bc) = (ab)c$;
(c)　Distributive law: $c(a + b) = ca + cb$.

Special Products and Factors

$$(x + y)^2 = x^2 + 2xy + y^2,$$
$$(x - y)^2 = x^2 - 2xy + y^2,$$
$$(x + y)^3 = x^3 + 3x^2y + 3xy^2 + y^3,$$
$$(x - y)^3 = x^3 - 3x^2y + 3xy^2 - y^3,$$
$$(x + y)^4 = x^4 + 4x^3y + 6x^2y^2 + 4xy^3 + y^4,$$

$$(x-y)^4 = x^4 - 4x^3y + 6x^2y^2 - 4xy^3 + y^4,$$
$$(x+y)^5 = x^5 + 5x^4y + 10x^3y^2 + 10x^2y^3 + 5xy^4 + y^5,$$
$$(x-y)^5 = x^5 - 5x^4y + 10x^3y^2 - 10x^2y^3 + 5xy^4 - y^5,$$
$$(x+y)^6 = x^6 + 6x^5y + 15x^4y^2 + 20x^3y^3 + 15x^2y^4 + 6xy^5 + y^6,$$
$$(x-y)^6 = x^6 - 6x^5y + 15x^4y^2 - 20x^3y^3 + 15x^2y^4 - 6xy^5 + y^6.$$

The above results are special cases of the binomial formula.

$$x^2 - y^2 = (x-y)(x+y),$$
$$x^3 - y^3 = (x-y)(x^2 + xy + y^2),$$
$$x^3 + y^3 = (x+y)(x^2 - xy + y^2),$$
$$x^4 - y^4 = (x-y)(x+y)(x^2 + y^2),$$
$$x^5 - y^5 = (x-y)(x^4 + x^3y + x^2y^2 + xy^3 + y^4),$$
$$x^5 + y^5 = (x+y)(x^4 - x^3y + x^2y^2 - xy^3 + y^4),$$
$$x^6 - y^6 = (x-y)(x+y)(x^2 + xy + y^2)(x^2 - xy + y^2),$$
$$x^4 + x^2y^2 + y^4 = (x^2 + xy + y^2)(x^2 - xy + y^2),$$
$$x^4 + 4y^4 = (x^2 + 2xy + 2y^2)(x^2 - 2xy + 2y^2).$$

Some generalization of the above are given by the following results where n is a positive integer:

$$x^{2n+1} - y^{2n+1} = (x-y)(x^{2n} + x^{2n-1}y + x^{2n-2}y^2 + \cdots + y^{2n})$$

$$= (x-y)\left(x^2 - 2xy\cos\frac{2\pi}{2n+1} + y^2\right)\left(x^2 - 2xy\cos\frac{4\pi}{2n+1} + y^2\right)\cdots$$

$$\times \left(x^2 - 2xy\cos\frac{2n\pi}{2n+1} + y^2\right),$$

$$x^{2n+1} + y^{2n+1} = (x+y)(x^{2n} - x^{2n-1}y + x^{2n-2}y^2 - \cdots + y^{2n})$$

$$= (x+y)\left(x^2 + 2xy\cos\frac{2\pi}{2n+1} + y^2\right)\left(x^2 + 2xy\cos\frac{4\pi}{2n+1} + y^2\right)\cdots$$

$$\times \left(x^2 + 2xy\cos\frac{2n\pi}{2n+1} + y^2\right),$$

$$x^{2n} - y^{2n} = (x-y)(x+y)(x^{n-1} + x^{n-2}y + x^{n-3}y^2 + \cdots)$$
$$\times (x^{n-1} - x^{n-2}y + x^{n-3}y^2 - \cdots)$$

$$= (x-y)(x+y)\left(x^2 - 2xy\cos\frac{\pi}{n} + y^2\right)\left(x^2 - 2xy\cos\frac{2\pi}{n} + y^2\right)\cdots$$

$$\times \left(x^2 - 2xy\cos\frac{(n-1)\pi}{n} + y^2\right),$$

$$x^{2n} + y^{2n} = \left(x^2 + 2xy\cos\frac{\pi}{2n} + y^2\right)\left(x^2 + 2xy\cos\frac{3\pi}{2n} + y^2\right)\cdots$$

$$\times \left(x^2 + 2xy\cos\frac{(2n-1)\pi}{2n} + y^2\right).$$

Powers and Roots

Powers or exponents are used to write down the multiplication of a number with itself, over and over again. Roots are the opposite of powers.

$$a^x \times a^y = a^{x+y},$$

$$a^0 = 1 \quad \text{if } a \neq 0,$$

$$(ab)^x = a^x b^x,$$

$$\frac{a^x}{a^y} = a^{x-y},$$

$$a^{-x} = \frac{1}{a^x},$$

$$\left(\frac{a}{b}\right)^x = \frac{a^x}{b^x},$$

$$(a^x)^y = a^{xy},$$

$$a^{1/x} = \sqrt[x]{a},$$

$$\sqrt[x]{ab} = \sqrt[x]{a}\,\sqrt[x]{b},$$

$$\sqrt[x]{\sqrt[y]{a}} = \sqrt[xy]{a},$$

$$a^{x/y} = \sqrt[y]{a^x},$$

$$\sqrt[x]{\frac{a}{b}} = \frac{\sqrt[x]{a}}{\sqrt[x]{b}}.$$

Proportion

If

$$\frac{a}{b} = \frac{c}{d},$$

then

$$\frac{a+b}{b} = \frac{c+d}{d},$$

$$\frac{a-b}{b} = \frac{c-d}{d},$$

and

$$\frac{a-b}{a+b} = \frac{c-d}{c+d}.$$

Sum of Arithmetic Progression to *n* Terms

An arithmetic progression or arithmetic sequence is a sequence of numbers such that the difference of any two successive members of the sequence is a constant. If the initial term of an arithmetic progression is a and the common difference of successive members is d, then the sum of the members of a finite arithmetic progression is given by

$$a + (a+d) + (a+2d) + \cdots + (a+(n-1)d) = na + \frac{1}{2}n(n-1)d = \frac{n}{2}(a+l),$$

where $l = a + (n-1)d$.

Sum of Geometric Progression to *n* Terms

We can show that the sum of geometric progression is equivalent to

$$s_n = a + ar + ar^2 + \cdots + ar^{n-1} = \frac{a(1-r^n)}{1-r},$$

$$\lim_{n \to \infty} 8_n = a!(1-r), \quad -1 < r < 1.$$

Arithmetic Mean of *n* Quantities, *A*

Arithmetic mean of the given numbers is defined as

$$A = \frac{a_1 + a_2 + \cdots + a_n}{n}.$$

Geometric Mean of *n* Quantities, *G*

The geometric mean of n quantities is given by

$$G = (a_1 a_2 \cdots a_n)^{1/n}, \quad a_k > 0, \quad k = 1, 2, \ldots, n.$$

Harmonic Mean of *n* Quantities, *H*

The harmonic mean is one of the three Pythagorean means. For all data sets containing at least one pair of nonequal values, the harmonic mean is always the least of the three means, while the arithmetic mean is always the greatest of the three and the geometric mean is always in between. The harmonic mean of n numbers is given by the formula

$$\frac{1}{H} = \frac{1}{n}\left(\frac{1}{a_1} + \frac{1}{a_2} + \cdots + \frac{1}{a_n}\right), \quad a_k > 0, \quad k = 1, 2, \ldots, n.$$

Generalized Mean

A generalized mean, also known as power mean or Hölder mean (named after Otto Hölder), is an abstraction of the Pythagorean means including arithmetic, geometric, and harmonic means and is given by

$$M(t) = \left(\frac{1}{n} \sum_{k=1}^{n} a_k^l \right)^{1/t},$$

$$M(t) = 0, \quad t < 0, \text{ some } a_k \text{ zero,}$$

$$\lim_{t \to \infty} M(t) = \max, \quad a_1, a_2, \ldots, a_n = \max \, a,$$

$$\lim_{t \to -\infty} M(t) = \min, \quad a_1, a_2, \ldots, a_n = \min \, a,$$

$$\lim_{t \to 0} M(t) = G,$$

$$M(1) = A,$$

$$M(-1) = H.$$

Solution of Quadratic Equations

Given $az^2 + bz + c = 0$,

$$z_{1,2} = - \left(\frac{b}{2a} \right) \pm \frac{1}{2a} q^{1/2}, \quad q = b^2 - 4ac,$$

$$z_1 + z_2 = -\frac{b}{a}, \quad z_1 z_2 = \frac{c}{a}.$$

The two roots are real if $q > 0$, equal if $q = 0$, and are complex conjugates if $q < 0$.

Solution of Cubic Equations

Given $z^2 + a_2 z^2 + a_1 z + a_0 = 0$, let

$$q = \frac{1}{3} a_1 - \frac{1}{9} a_2^2,$$

$$r = \frac{1}{6} (a_1 a_2 - 3a_0) - \frac{1}{27} a_2^3.$$

There are one real root and a pair of complex conjugate roots if $q^3 + r^2 > 0$, all roots are real and at least two are equal if $q^3 + r^2 = 0$, and all roots are real (irreducible case) if $q^3 + r^2 < 0$.

Let

$$s_1 = [r + (q^3 + r^2)^{1/2}]^{1/2}$$

and

$$s_2 = [r - (q^3 + r^2)^{1/2}]^{1/2},$$

then

$$z_1 = (s_1 + s_2) - \frac{a_2}{3},$$

$$z_2 = -\frac{1}{2}(s_1 + s_2) - \frac{a_2}{3} + \frac{i\sqrt{3}}{2}(s_1 - s_2),$$

and

$$z_3 = -\frac{1}{2}(s_1 + s_2) - \frac{a_2}{3} - \frac{i\sqrt{3}}{2}(s_1 - s_2).$$

If z_1, z_2, and z_3 are the roots of the cubic equation, then

$$z_1 + z_2 + z_3 = -a_2,$$

$$z_1 z_2 + z_1 z_3 + z_2 z_3 = a_1,$$

and

$$z_1 z_2 z_3 = a_0.$$

Trigonometric Solution of the Cubic Equation

The form $x^3 + ax + b = 0$ with $ab \neq 0$ can always be solved by transforming it to the trigonometric identity

$$4 \cos^3 \theta - 3 \cos \theta - \cos(3\theta) \equiv 0.$$

Let $x = m \cos \theta$, then

$$x^3 + ax + b = m^3 \cos^3 \theta + am \cos \theta + b = 4 \cos^3 \theta - 3 \cos \theta - \cos(3\theta) \equiv 0.$$

Hence

$$\frac{4}{m^3} = -\frac{3}{am} = \frac{-\cos(3\theta)}{b},$$

from which we obtain

$$m = 2\sqrt{-\frac{a}{3}}, \quad \cos(3\theta) = \frac{3b}{am}.$$

Any solution θ_1 which satisfies $\cos(3\theta) = 3b/am$ will also have the solutions

$$\theta_1 + \frac{2\pi}{3} \quad \text{and} \quad \theta_1 + \frac{4\pi}{3}.$$

Therefore the roots of the cubic equation $x^3 + ax + b = 0$ are

$$2\sqrt{-\frac{a}{3}} \cos \theta_1,$$

$$2\sqrt{-\frac{a}{3}} \cos \left(\theta_1 + \frac{2\pi}{3} \right),$$

$$2\sqrt{-\frac{a}{3}} \cos \left(\theta_1 + \frac{4\pi}{3} \right).$$

Solution of Quadratic Equations

Given $z^4 + a_3 z^3 + a_2 z^2 + a_1 z + a_0 = 0$, find the real root u_1 of the cubic equation

$$u^3 - a_2 u^2 + (a_1 a_3 - 4a_0)u - (a_1^2 + a_0 a_3^2 - 4a_0 a_2) = 0$$

and determine the four roots of the quartic as solutions of the two quadratic equations

$$v^2 + \left[\frac{a_3}{2} \mp \left(\frac{a_3^2}{4} + u_1 - a_2\right)^{1/2}\right] v + \frac{u_1}{2} \mp \left[\left(\frac{u_1}{2}\right)^2 - a_0\right]^{1/2} = 0.$$

If all roots of the cubic equation are real, use the value of u_1 which gives real coefficients in the quadratic equation and select signs such that

$$z^4 + a_3 z^3 + a_2 z^3 + a_1 z + a_0 = (z^2 + p_1 z + q_1)(z^2 + p_2 z + q_2),$$

then

$$p_1 + p_2 = a_3, \quad p_1 p_2 + q_1 + q_2 = a_2, \quad p_1 q_2 + p_2 q_1 = a_1, \quad q_1 q_2 = a_0.$$

If z_1, z_2, z_3, and z_4 are the roots, then

$$\sum z_t = -a_3, \quad \sum z_t z_j z_k = -a_1,$$

$$\sum z_t z_j = a_2, \quad z_1 z_2 z_3 z_4 = a_0.$$

Partial Fractions

This section applies only to rational algebraic fractions with numerators of lower degree than denominators. Improper fractions can be reduced to proper fractions by long division.

Every fraction may be expressed as the sum of component fractions whose denominators are factors of the denominator of the original fraction.

Let $N(x)$ be the numerator, then a polynomial of the form

$$N(x) = n_0 + n_1 x + n_2 x^2 + \cdots + n_1 x^1.$$

Nonrepeated Linear Factors

$$\frac{N(x)}{(x-a)G(x)} = \frac{A}{x-a} + \frac{F(x)}{G(x)},$$

$$A = \left[\frac{N(x)}{G(x)}\right]_{x=a}.$$

$F(x)$ is determined by methods discussed in the following sections.

Repeated Linear Factors

$$\frac{N(x)}{x^m G(x)} = \frac{A_0}{x^m} + \frac{A_1}{x^{m-1}} + \cdots + \frac{A_{m-1}}{x} + \frac{F(x)}{G(x)},$$

$$N(x) = n_0 + n_1 x + n_2 x^2 + n_3 x^3 + \cdots,$$

$$F(x) = f_0 + f_1 x + f_2 x^2 + \cdots,$$

$$G(x) = g_0 + g_1 x + g_2 x^2 + \cdots,$$

$$A_0 = \frac{n_0}{g_0},$$

$$A_1 = \frac{n_1 - A_0 g_1}{g_0},$$

$$A_2 = \frac{n_2 - A_0 g_2 - A_1 g_1}{g_0}.$$

General terms

$$A_0 = \frac{n_0}{g_0}, \quad A_k = \frac{1}{g_0}\left[n_k - \sum_{t=0}^{k-1} A_t g_k - t \right] k \geq 1,$$

$$m^* = 1 \begin{cases} f_0 = n_1 - A_0 g_1, \\ f_1 = n_2 - A_0 g_2, \\ f_1 = n_{j+1} - A_0 g_{t+1}, \end{cases}$$

$$m = 2 \begin{cases} f_0 = n_2 - A_0 g_2 - A_1 g_1, \\ f_1 = n_3 - A_0 g_3 - A_1 g_2, \\ f_1 = n_{j+2} - [A_0 g_{1+2} + A_1 g_1 + 1], \end{cases}$$

$$m = 3 \begin{cases} f_0 = n_3 - A_0 g_3 - A_1 g_2 - A_2 g_1, \\ f_1 = n_3 - A_0 g_4 - A_1 g_3 - A_2 g_2, \\ f_1 = n_{j+3} - [A_0 g_{j+3} + A_1 g_{j+2} + A_2 g_{j+1}]. \end{cases}$$

Any m:

$$f_1 = n_{m+1} - \sum_{i=0}^{m-1} A_1 g_{m+j-1},$$

$$\frac{N(x)}{(x-a)^m G(x)} = \frac{A_0}{(x-a)^m} + \frac{A_1}{(x-a)^{m-1}} + \cdots + \frac{A_{m-1}}{(x-a)} + \frac{F(x)}{G(x)}.$$

Change to form $N'(y)/y^m G'(y)$ by substituting $x = y + a$. Resolve into partial fractions in terms of y as described above. Then express in terms of x by substituting $y = x - a$.

Repeated Linear Factors

Alternative method of determining coefficients:

$$\frac{N(x)}{(x-a)^m G(x)} = \frac{A_0}{(x-a)^m} + \cdots + \frac{A_k}{(x-a)^{m-k}} + \cdots + \frac{A_{m-1}}{x-a} + \frac{F(x)}{G(x)},$$

$$A_k = \frac{1}{k!}\left\{ D_x^k \left[\frac{N(x)}{G(x)} \right] \right\}_{x-G},$$

where D_x^k is the differentiating operator, and the derivative of zero order is defined as

$$D_x^0 u = u.$$

Factors of Higher Degree

Factors of higher degree have the corresponding numerators indicated.

$$\frac{N(x)}{(x^2 + h_1 x + h_0)G(x)} = \frac{a_1 x + a_0}{x^2 + h_1 x + h_0} + \frac{F(x)}{G(x)},$$

$$\frac{N(x)}{(x^2 + h_1 x + h_0)^2 G(x)} = \frac{a_1 x + a_0}{(x^2 + h_1 x + h_0)^2} + \frac{b_1 x + b_0}{(x^2 + h_1 x + h_0)} + \frac{F(x)}{G(x)},$$

$$\frac{N(x)}{(x^3 + h_2 x^2 + h_1 x + h_0)G(x)} = \frac{a_2 x^2 + a_1 x + a_0}{x^3 + h_2 x^2 + h_1 x + h_0} + \frac{F(x)}{G(x)},$$

and so on.

Problems of this type are determined first by solving for the coefficients due to linear factors as shown above and then determining the remaining coefficients by the general methods given below.

Geometry

Mensuration formulas are used for measuring angles and distances in geometry. Examples are presented below.

Triangles

Let K be the area, r the radius of the inscribed circle, and R the radius of the circumscribed circle.

Right Triangle

$$A + B = C = 90°,$$

$$c^2 = a^2 + b^2 \text{ (Pythagorean relations),}$$

$$a = \sqrt{(c+b)(c-b)},$$

$$K = \frac{1}{2}ab,$$

$$r = \frac{ab}{a+b+c}, \quad R = \frac{1}{2}c,$$

$$h = \frac{ab}{c}, \quad m = \frac{b^2}{c}, \quad n = \frac{a^2}{c}.$$

Equilateral Triangle

$$A = B = C = 60°,$$

$$K = \frac{1}{4}a^2\sqrt{3},$$

$$r = \frac{1}{6}a\sqrt{3}, \quad R = \frac{1}{3}a\sqrt{3},$$

$$h = \frac{1}{2}a\sqrt{3}.$$

General Triangle

Let $s = \frac{1}{2}(a+b+c)$, h_c be the length of the altitude on side c, t_c the length of the bisector of angle C, m_c the length of the median to side c.

$$A + B + C = 180°,$$

$$c^2 = a^2 + b^2 - 2ab\cos C \text{ (law of cosines),}$$

$$K = \frac{1}{2}h_c c = \frac{1}{2}ab\sin C$$

$$= \frac{c^2 \sin A \sin B}{2\sin C}$$

$$= rs = \frac{abc}{4R}$$

$$= \sqrt{s(s-a)(s-b)(s-c)} \quad \text{(Heron's formula),}$$

$$r = c\sin\frac{A}{2}\sin\frac{B}{2}\sec\frac{C}{2} = \frac{ab\sin C}{2s} = (s-c)\tan\frac{C}{2}$$

$$= \sqrt{\frac{(s-a)(s-b)(s-c)}{s}} = \frac{K}{s} = 4R\sin\frac{A}{2}\sin\frac{B}{2}\sin\frac{C}{2},$$

$$R = \frac{c}{2\sin C} = \frac{abc}{4\sqrt{s(s-a)(s-b)(s-c)}} = \frac{abc}{4K},$$

$$h_c = a \sin B = b \sin A = \frac{2K}{c},$$

$$t_c = \frac{2ab}{a+b} \cos \frac{C}{2} = \sqrt{ab \left\{ 1 - \frac{c^2}{(a+b)^2} \right\}}$$

$$m_c = \sqrt{\frac{a^2}{2} + \frac{b^2}{2} - \frac{c^2}{4}}.$$

Menelaus' Theorem

A necessary and sufficient condition for points D, E, and F on the respective side lines BC, CA, and AB of a triangle ABC to be collinear is that

$$BD \cdot CE \cdot AF = -DC \cdot EA \cdot FB,$$

where all segments in the formula are directed segments.

Ceva's Theorem

A necessary and sufficient condition for AD, BE, and CF, where D, E, and F are points on the respective side lines BC, CA, and AB of a triangle ABC, to be concurrent is that

$$BD \cdot CE \cdot AF = +DC \cdot EA \cdot FB,$$

where all segments in the formula are directed segments.

Quadrilaterals

Let K be the area and p and q are diagonals.

Rectangle

$$A = B = C = D = 90°,$$

$$K = ab, \quad p = \sqrt{a^2 + b^2}.$$

Parallelogram

$$A = C, \quad B = D, \quad A + B = 180°,$$

$$K = bh = ab \sin A = ab \sin B,$$

$$h = a \sin A = a \sin B,$$

$$p = \sqrt{a^2 + b^2 - 2ab \cos A},$$

$$q = \sqrt{a^2 + b^2 - 2ab \cos B} = \sqrt{a^2 + b^2 + 2ab \cos A}.$$

Rhombus

$$p^2 + q^2 = 4a^2,$$
$$K = \frac{1}{2}pq.$$

Trapezoid

$$m = \frac{1}{2}(a+b),$$
$$K = \frac{1}{2}(a+b)h = mh.$$

General Quadrilateral

Let

$$s = \frac{1}{2}(a+b+c+d).$$

$$K = \frac{1}{2}pq\sin\theta$$

$$= \frac{1}{4}(b^2 + d^2 - a^2 - c^2)\tan\theta$$

$$= \frac{1}{4}\sqrt{4p^2q^2 - (b^2 + d^2 - a^2 - c^2)^2} \text{ (Bretschneider's formula)}$$

$$= \sqrt{(s-a)(s-b)(s-c)(s-d) - abcd\cos^2\left(\frac{A+B}{2}\right)}.$$

Theorem

The diagonals of a quadrilateral with consecutive sides a, b, c, and d are perpendicular if and only if $a^2 + c^2 = b^2 + d^2$.

Regular Polygon of n Sides Each of Length b

$$\text{Area} = \frac{1}{4}nb^2\cot\frac{\pi}{n} = \frac{1}{4}nb^2\frac{\cos(\pi/n)}{\sin(\pi/n)},$$

with the perimeter nb.

Circle of radius r

$$\text{Area} = \pi r^2,$$

with perimeter $2\pi r$.

Regular Polygon of *n* sides inscribed in a Circle of Radius *r*

$$\text{Area} = \frac{1}{2}nr^2 \sin \frac{2\pi}{n} = \frac{1}{2}nr^2 \sin \frac{360°}{n},$$

$$\text{Perimeter} = 2nr \sin \frac{\pi}{n} = 2nr \sin \frac{180°}{n}.$$

Regular Polygon of *n* Sides Circumscribing a Circle of Radius *r*

$$\text{Area} = nr^2 \tan \frac{\pi}{n} = nr^2 \tan \frac{180°}{n},$$

$$\text{Perimeter} = 2nr \tan \frac{\pi}{n} = 2nr \tan \frac{180°}{n}.$$

Cyclic Quadrilateral

Let R be the radius of the circumscribed circle.

$$A + C = B + D = 180°,$$

$$K = \sqrt{(s-a)(s-b)(s-c)(s-d)} = \frac{\sqrt{(ac+bd)(ad+bc)(ab+cd)}}{4R},$$

$$p = \sqrt{\frac{(ac+bd)(ab+cd)}{ad+bc}},$$

$$q = \sqrt{\frac{(ac+bd)(ad+bc)}{ab+cd}},$$

$$R = \frac{1}{2}\sqrt{\frac{(ac+bd)(ad+bc)(ab+cd)}{(s-a)(s-b)(s-c)(s-d)}},$$

$$\sin \theta = \frac{2K}{ac+bd}.$$

Prolemy's Theorem

A convex quadrilateral with consecutive sides a, b, c, d and diagonals p and q is cyclic if and only if $ac + bd = pq$.

Cyclic-Inscriptable Quadrilateral

Let r be the radius of the inscribed circle, R the radius of the circumscribed circle, and m the distance between the centers of the inscribed and the circumscribed circles. Then

$$A + C = B + D = 180°,$$

$$a + c = b + d,$$

$$K = \sqrt{abcd},$$

$$\frac{1}{(R-m)^2} + \frac{1}{(R+m)^2} = \frac{1}{r^2},$$

$$r = \frac{\sqrt{abcd}}{s},$$

$$R = \frac{1}{2}\sqrt{\frac{(ac+bd)(ad+bc)(ab+cd)}{abcd}}.$$

Sector of Circle of Radius r

$$\text{Area} = \frac{1}{2}r^2\theta \quad (\theta \text{ in radians}),$$

$$\text{Arc length } s = r\theta.$$

Radius of a Circle Inscribed in a Triangle of Sides a, b, and c

$$r = \frac{\sqrt{8(8-a)(8-b)(8-c)}}{8},$$

where $s = \frac{1}{2}(a+b+c)$ is the semiperimeter.

Radius of a Circle Circumscribing a Triangle of Sides a, b, and c

$$R = \frac{abc}{4\sqrt{8(8-a)(8-b)(8-c)}},$$

where $s = \frac{1}{2}(a+b+c)$ is the semiperimeter.

Segment of a Circle of Radius r

$$\text{Area of shaded part} = \frac{1}{2}r^2(\theta - \sin\theta).$$

Ellipse of Semimajor Axis a and Semiminor Axis b

$$\text{Area} = \pi ab,$$

$$\text{Perimeter} = 4a \int_0^{\pi/2} \sqrt{1 - k^2 \sin^2 \theta} \, d\theta$$

$$= 2\pi \sqrt{\frac{1}{2}(a^2 + b^2)} \quad \text{(approximately)},$$

where $k = \sqrt{a^2 - b^2}/a$.

Segment of a Parabola

$$\text{Area} = \frac{2}{8} ab,$$

$$\text{Arc length } ABC = \frac{1}{2} \sqrt{b^2 + 16a^2} + \frac{b^2}{8a} \ln \left(\frac{4a + \sqrt{b^2 + 16a^2}}{b} \right).$$

Planar Areas by Approximation

Divide the planar area K into n strips by equidistant parallel chords of lengths $y_0, y_1, y_2, \ldots, y_n$ (where y_0 and/or y_n may be zero) and let h denote the common distance between the chords. Then, approximately we have the following rules.

Trapezoidal Rule:

$$K = h \left(\frac{1}{2} y_0 + y_1 + y_2 + \cdots + y_{n-1} + \frac{1}{2} y_n \right).$$

Durand's Rule:

$$K = h \left(\frac{4}{10} y_0 + \frac{11}{10} y_1 + y_2 + y_3 + \cdots + y_{n-2} + \frac{11}{10} y_{n-1} + \frac{4}{10} y_n \right).$$

Simpson's Rule (n even):

$$K = \frac{1}{3} h (y_0 + 4y_1 + 2y_2 + 4y_3 + 2y_4 + \cdots + 2y_{n-2} + 4y_{n-1} + y_n).$$

Weddle's Rule ($n = 6$):

$$K = \frac{3}{10} h (y_0 + 5y_1 + y_2 + 6y_3 + y_4 + 5y_5 + y_6).$$

Solids Bounded By Planes

In the following, S is the lateral surface, T the total surface, and V the volume.

Cube

Let a be the length of each edge.

$$T = 6a^2, \text{diagonal of face} = a\sqrt{2},$$

$$V = a^3, \text{diagonal of cube} = a\sqrt{3}.$$

Rectangular Parallelepiped (or Box)

Let a, b, and c be the lengths of its edges.

$$T = 2(ab + bc + ca), \quad V = abc,$$

$$\text{diagonal} = \sqrt{a^2 + b^2 + c^2}.$$

Prism

A prism is a solid figure with a uniform cross-section. The surface area of any prism equals the sum of the areas of its faces, which include the floor, roof, and walls. Because the floor and the roof of a prism have the same shape, the surface area can always be found as follows:

$$S = (\text{perimeter of right section}) \times (\text{lateral edge}).$$

We determine the volume of the prism by

$$V = (\text{area of right section}) \times (\text{lateral edge})$$
$$= (\text{area of base}) \times (\text{altitude})$$

Truncated Triangular Prism

A truncated triangular prism is equivalent to the sum of three pyramids whose common base is the base of the prism and whose vertices are the three vertices of the upper base and its volume is given by

$$V = (\text{area of right section}) \times \frac{1}{3}(\text{sum of the three lateral edges}).$$

Pyramid

A pyramid is a building where the outer surfaces are triangular and converge at a point. The base of a pyramid can be trilateral, quadrilateral, or any polygon shape, meaning that a pyramid has at least three outer surfaces (at least four faces including the base). The square pyramid, with square base and four triangular outer surfaces, is a common version.

The surface area of a pyramid is given by

$$S \text{ of regular pyramid} = \frac{1}{2}(\text{perimeter of base}) \times (\text{slant height})$$

and its volume is given by

$$V = \frac{1}{3}(\text{area of base}) \times (\text{altitude}).$$

Frustum of a Pyramid

In geometry, a frustum is the portion of a solid (normally a cone or pyramid) which lies between two parallel planes cutting it. If a pyramid is cut through by a plane parallel to its base portion of the pyramid between that plane, then the base is called frustum of the pyramid.

Let B_1 be the area of lower base, B_2 the area of upper base, and h the altitude, then the surface area and the volume are, respectively, given by

$$S \text{ of regular figure} = \frac{1}{2}(\text{sum of perimeters of base}) \times (\text{slant height})$$

and

$$V = \frac{1}{3}h(B_1 + B_2 + \sqrt{B_1 B_2}).$$

Prismatoid

A prismatoid is a polyhedron having for bases two polygons in parallel planes, and for lateral faces triangles or trapezoids with one side lying in one base, and the opposite vertex or side lying in the other base, of the polyhedron. Let B_1 be the area of the lower base, M the area of the midsection, B_2 the area of the upper base, and h the altitude. Then the volume is given by

$$V = \frac{1}{6}h(B_1 + 4M + B_2) \quad \text{(the prismoidal formula)}.$$

Note: As cubes, rectangular parallelepipeds, prisms, pyramids, and frustums of pyramids are all examples of prismatoids, the formula for the volume of a prismatoid subsumes most of the above volume formulae.

Regular Polyhedra

Let v be the number of vertices, e the number of edges, f the number of faces, α each dihedral angle, a the length of each edge, r the radius of the inscribed sphere, R the radius of the circumscribed sphere, A the area of each face, T the total area, and V the volume, then

$$v - e + f = 2,$$
$$T = fA,$$
$$V = \frac{1}{3}rfA = \frac{1}{3}rT.$$

Name	Nature of Surface	T	V
Tetrahedron	4 equilateral triangles	$1.73205a^2$	$0.11785a^3$
Hexahedron (cube)	6 squares	$6.00000a^2$	$1.00000a^3$
Octahedron	8 equilateral triangles	$3.46410a^2$	$0.47140a^3$
Dodecahedron	12 regular pentagons	$20.64573a^2$	$7.66312a^3$
Icosahedron	20 equilateral triangles	$8.66025a^2$	$2.18169a^2$

Name	v	e	f	α	a	r
Tetrahedron	4	6	4	70°32′	1.633R	0.333R
Hexahedron	8	12	6	90°	1.155R	0.577R
Octahedron	6	12	8	190°28′	1.414R	0.577R
Dodecahedron	20	30	12	116°34′	0.714R	0.795R
Icosahedron	12	30	20	138°11′	1.051R	0.795R

Name	A	r	R	V
Tetrahedron	$\frac{1}{4}a^2\sqrt{3}$	$\frac{1}{12}a\sqrt{6}$	$\frac{1}{4}a\sqrt{6}$	$\frac{1}{12}a^3\sqrt{2}$
Hexahedron (cube)	a^2	$\frac{1}{2}a$	$\frac{1}{2}a\sqrt{3}$	a^3
Octahedron	$\frac{1}{4}a^2\sqrt{3}$	$\frac{1}{6}a\sqrt{6}$	$\frac{1}{2}a\sqrt{2}$	$\frac{1}{3}a^3\sqrt{2}$
Dodecahedron	$\frac{1}{4}a^2\sqrt{25+10\sqrt{5}}$	$\frac{1}{20}a\sqrt{250+110\sqrt{5}}$	$\frac{1}{4}a(\sqrt{15}+\sqrt{3})$	$\frac{1}{4}a^3(15+7\sqrt{5})$
Icosahedron	$\frac{1}{4}a^2\sqrt{3}$	$\frac{1}{12}a\sqrt{42+18\sqrt{5}}$	$\frac{1}{4}a\sqrt{10+2\sqrt{5}}$	$\frac{5}{12}a^3(3+\sqrt{5})$

Sphere of Radius *r*

A sphere is a perfectly round geometrical object in three-dimensional space. A perfect sphere is completely symmetrical around its center, with all points on the surface lying the same distance *r* from the center point. This distance *r* is known as the radius of the sphere. The volume and surface area of a sphere are, respectively, given as

$$\text{Volume} = \frac{3}{4}\pi r^3$$

and

$$\text{Surface area} = 4\pi r^2.$$

Right Circular Cylinder of Radius *r* and Height *h*

A cylinder is one of the most basic curvilinear geometric shapes, the surface formed by the points at a fixed distance from a given straight line, the axis of the cylinder. The surface area and the volume of a cylinder have been derived from the following formulae:

$$\text{Volume} = \pi r^2 h$$

and

$$\text{Lateral surface area} = 2\pi rh.$$

Circular Cylinder of Radius *r* and Slant Height ℓ

$$\text{Volume} = \pi r^2 h = \pi r^2 \ell \sin \theta,$$

$$\text{Lateral surface area} = p\ell.$$

Cylinder of Cross-Sectional Area *A* and Slant Height ℓ

$$\text{Volume} = Ah = A\ell \sin \theta,$$

$$\text{Lateral surface area} = p\ell.$$

Right Circular Cone of Radius *r* and Height *h*

A cone is a three-dimensional geometric shape that tapers smoothly from a flat, usually circular base to a point called the apex or vertex. More precisely, it is the solid figure bounded by a plane base and the surface (called the lateral surface) formed by the locus of all straight-line segments joining the apex to the perimeter of the base. Both the volume and the surface area of a cone are given, respectively, by

$$\text{Volume} = \frac{1}{3}\pi r^2 h$$

and

$$\text{Lateral surface area} = \pi r \sqrt{r^2 + h^2} = \pi r l.$$

Spherical Cap of Radius *r* and Height *h*

$$\text{Volume (shaded in figure)} = \frac{1}{3}\pi h^2 (3r - h)$$

and

$$\text{Surface area} = 2\pi r h.$$

Frustum of a Right Circular Cone of Radii *a* and *b* and Height *h*

$$\text{Volume} = \frac{1}{3}\pi h(a^2 + ab + b^2),$$

$$\text{Lateral surface area} = \pi(a + b)\sqrt{h^2 + (b - a)^2}$$

$$= \pi(a + b)l.$$

Zone and Segment of Two Bases

$$S = 2\pi Rh = \pi Dh,$$

$$V = \frac{1}{6}\pi h(3a^2 + 3b^2 + h^2).$$

Lune

A lune is a plane figure bounded by two circular arcs of unequal radii, that is, a crescent. Its surface area is given by

$$S = 2R^3\theta,$$

where θ is in radians.

Spherical Sector

A spherical sector is a solid of revolution enclosed by two radii from the center of a sphere. The volume of a spherical sector is given by

$$V = \frac{2}{3}\pi R^2 h = \frac{1}{6}\pi D^2 h.$$

Spherical Triangle and Polygon

Let A, B, and C be the angles (in radians) of the triangle. Let θ be the sum of the angles (in radians) of a spherical polygon on n sides. Then the surface area is given by

$$S = (A + B + C - \pi)R^2$$

or

$$S = [\theta - (n-2)\pi]R^2.$$

Spheroids

Ellipsoid

Let a, b, and c be the lengths of the semi-axes, then its volume is

$$V = \frac{4}{3}\pi abc.$$

Oblate Spheroid

An oblate spheroid is formed by the rotation of an ellipse about its minor axis. Let a and b be the major and minor semi-axes, respectively, and $\in$ the eccentricity of the revolving ellipse, then

$$S = 2\pi a^2 + \pi \frac{b^2}{\in}\log_e\frac{1+\in}{1-\in},$$

$$V = \frac{4}{3}\pi a^2 b.$$

Prolate Spheroid

A prolate spheroid is formed by the rotation of an ellipse about its major axis. Let a and b be the major and minor semi-axes, respectively, and ϵ the eccentricity of the revolving ellipse, then

$$S = 2\pi b^2 + 2\pi \frac{ab}{\epsilon} \sin^{-1} \epsilon,$$

$$V = \frac{3}{4} \pi a b^2.$$

Circular Torus

A circular torus is formed by the rotation of a circle about an axis in the plane of the circle and not cutting the circle. Let r be the radius of the revolving circle and let R be the distance of its center from the axis of rotation, then

$$S = 4\pi^2 Rr,$$

$$V = 2\pi^2 Rr^2.$$

Formulas from Plane Analytic Geometry

Distance d between Two Points

If $P_1(x_1, y_1)$ and $P_2(x_2, y_2)$ are any two points on a line, then the distance d between these two points is given by

$$d = \sqrt{(x_2 - x_1)^2 + (y_2 - y_1)^2}.$$

Slope m of Line Joining Two Points

If $P_1(x_1, y_1)$ and $P_2(x_2, y_2)$ are any two points on a line, then the slope m of a line is expressed by the following formula:

$$m = \frac{y_2 - y_1}{x_2 - x_1} = \tan \theta.$$

Equation of a Line Joining Two Points

Let the line pass through $P_1(x_1, y_1)$ and $P_2(x_2, y_2)$, the two given points. Therefore, the equation of the line is given as

$$\frac{y - y_1}{x - x_1} = \frac{y_2 - y_1}{x_2 - x_1} = m \quad \text{or} \quad y - y_1 = m(x - x_1),$$

$$y = mx + b,$$

where $b = y_1 - mx_1 = (x_2 y_1 - x_1 y_2)/(x_2 - x_1)$ is the intercept on the y-axis, that is, the y-intercept.

Equation of a Line in Terms of x-intercept $a \neq 0$ and y-intercept $b \neq 0$

$$\frac{x}{a} + \frac{y}{b} = 1.$$

Normal Form for Equation of a Line

The equation of the line in the intercept form is given by

$$x \cos \alpha + y \sin \alpha = p,$$

where p is the perpendicular distance from origin O to the line and α the angle of inclination of perpendicular with the positive x-axis.

General Equation of a Line

A common form of a linear equation in the two variables x and y is

$$Ax + By + C = 0,$$

where at least one of A and B is not zero.

Distance From a Point (x_1, y_1) to the Line $Ax + By + C = 0$

The distance from a point (x_1, y_1) to the line is given by

$$\frac{Ax_1 + By_1 + C}{\pm\sqrt{A^2 + B^2}},$$

where the sign is chosen so that the distance is nonnegative.

Angle ψ between Two Lines Having Slopes m_1 and m_2

If θ is the angle between two lines, then

$$\tan \psi = \frac{m_2 - m_1}{1 + m_1 m_2},$$

where m_1 and m_2 are the slopes of the two lines and are finite.

The lines are parallel or coincident if and only if $m_1 = m_2$ and are perpendicular if and only if $m_2 = -1/m_1$.

Area of a Triangle with Vertices

The area of a triangle having vertices (x_1, y_1), (x_2, y_2), and (x_3, y_3) is the absolute value of the determinant and is given by

$$\text{Area} = \pm \frac{1}{2} \begin{vmatrix} x_1 & y_1 & 1 \\ x_2 & y_2 & 1 \\ x_3 & y_3 & 1 \end{vmatrix}$$

$$= \pm \frac{1}{2}(x_1 y_2 + y_1 x_3 + y_3 x_2 - y_2 x_3 - y_1 x_2 - x_1 y_3),$$

where the sign is chosen so that the area is nonnegative.

If the area is zero, then all the points lie on a line.

Transformation of Coordinates Involving Pure Translation

$$\begin{cases} x = x' + x_0 \\ y = y' + y_0 \end{cases} \quad \text{or} \quad \begin{cases} x' = x + x_0, \\ y' = y + y_0, \end{cases}$$

where x, y are old coordinates (i.e., coordinates relative to xy system), (x', y') are new coordinates (relative to $x'y'$ system), and (x_0, y_0) are the coordinates of the new origin O' relative to the old xy-coordinate system.

Transformation of Coordinates Involving Pure Rotation

$$\begin{cases} x = x' \cos\alpha - y' \sin\alpha \\ y = x' \sin\alpha + y' \cos\alpha \end{cases} \quad \text{or} \quad \begin{cases} x' = x \cos\alpha + y \sin\alpha, \\ y' = y \cos\alpha - x \sin\alpha, \end{cases}$$

where the origins of the old $[xy]$- and new $[x'y']$-coordinate systems are the same but the x'-axis makes an angle α with the positive x-axis.

Transformation of Coordinates Involving Translation and Rotation

$$\begin{cases} x = x' \cos\alpha - y' \sin\alpha + x_0 \\ y = x' \sin\alpha + y' \cos\alpha + y_0 \end{cases}$$

or

$$\begin{cases} x' = (x - x_0)\cos\alpha + (y - y_0)\sin\alpha, \\ y' = (y - y_0)\cos\alpha - (x - x_0)\sin\alpha, \end{cases}$$

where the new origin O' of $x'y'$-coordinate system has coordinates (x_0, y_0) relative to the old xy-coordinate system and the x'-axis makes an angle α with the positive x-axis.

Polar Coordinates (r, θ)

A point P can be located by rectangular coordinates (x, y) or polar coordinates (r, θ). The transformation between these coordinates is

$$\begin{cases} x = r\cos\theta \\ y = r\sin\theta \end{cases} \quad \text{or} \quad \begin{cases} r = \sqrt{x^2 + y^2}, \\ \theta = \tan^{-1}\left(\dfrac{y}{x}\right). \end{cases}$$

Plane Curves

A plane curve is a curve in a Euclidian plane and is given by

$$(x^2 + y^2)^2 = ax^2 y,$$

$$r = a\sin\theta\cos^2\theta.$$

Catenary, Hyperbolic Cosine

The curve described by a uniform, flexible chain hanging under the influence of gravity is called the *catenary*. It is a hyperbolic cosine curve, and its slope varies as the hyperbolic sine. The equation of a catenary in Cartesian co-ordinates has the form

$$y = \frac{a}{2}(e^{x/e} + e^{-x/e}) = a\cosh\frac{x}{a}.$$

Cardioid

A cardioid is the curve traced by a point on the edge of a circular wheel that is rolling around a fixed wheel of the same size. The resulting curve is roughly heart-shaped, with a cusp at the place where the point touches the fixed wheel. It is given by the following parametric equations:

$$(x^2 + y^2 - ax)^2 = a^2(x^2 + y^2),$$

$$r = a(\cos\theta + 1),$$

or

$$r = a(\cos\theta - 1),$$

$$P'A = AP = a.$$

Circle

A circle is a simple shape of Euclidean geometry consisting of those points in a plane which are equidistant from a given point called the center. The circle with center $(0,0)$ and radius a has the equation

$$x^2 + y^2 = a^2,$$

$$r = a.$$

Cassinian Curves

Think about a magnetic field created by n parallel threads with the same current. The magnetic field lines in a plane orthogonal to the threads are Cassinian curves for which the foci are the intersections of the threads with the plane. The Cassinian curve has been studied by Serret (in 1843), and the curve is named after Cassini. It is described by the following equations:

$$x^2 + y^2 = 2ax,$$

$$r = 2a \cos \theta,$$

$$x^2 + y^2 = ax + by,$$

$$r = a \cos \theta + b \sin \theta.$$

Cotangent Curve

$$y = \cot x.$$

Cubical Parabola

$$y = ax^3, \quad a > 0,$$

$$r^2 = \frac{1}{a} \sec^2 \theta \tan \theta, \quad a > 0.$$

Cosecant Curve

$$y = \csc x.$$

Cosine Curve

$$y = \cos x.$$

Ellipse

$$\frac{x^2}{a^2} + \frac{y^2}{b^2} = 1,$$

$$x = a \cos \phi, \quad y = b \sin \phi.$$

Gamma Function

$$\Gamma(n) = \int_0^\infty x^{n-1} e^{-x} \, dx, \quad n > 0,$$

$$\Gamma(n) = \frac{\Gamma(n+1)}{n}, \quad 0 > n \neq -1, -2, -3, \ldots.$$

Hyperbolic Functions

$$\sinh x = \frac{e^x - e^{-x}}{2}, \quad \operatorname{csch} x = \frac{2}{e^x - e^{-x}},$$

$$\cosh x = \frac{e^x - e^{-x}}{2}, \quad \operatorname{csch} x = \frac{2}{e^x - e^{-x}},$$

$$\tanh x = \frac{e^x - e^{-x}}{e^x + e^{-x}}, \quad \coth x = \frac{e^x + e^{-x}}{e^x - e^{-x}}.$$

Inverse Cosine Curve

$$y = \arccos x.$$

Inverse Sine Curve

$$y = \arcsin x.$$

Inverse Tangent Curve

$$y = \arctan x.$$

Logarithmic Curve

$$y = \log_a x.$$

Parabola

$$y = x^2.$$

Cubical Parabola

$$y = x^3.$$

Tangent Curve

$$y = \tan x.$$

Ellipsoid:

$$\frac{x^2}{a^2} + \frac{y^2}{b^2} + \frac{z^2}{c^2} = 1.$$

Elliptic Cone:

$$\frac{x^2}{a^2} + \frac{y^2}{b^2} - \frac{z^2}{c^2} = 0.$$

Elliptic Cylinder:

$$\frac{x^2}{a^2} + \frac{y^2}{b^2} = 1.$$

Hyperboloid of One Sheet:

$$\frac{x^2}{a^2} + \frac{y^2}{b^2} - \frac{z^2}{c^2} = 1.$$

Elliptic Paraboloid:

$$\frac{x^2}{a^2} + \frac{y^2}{b^2} = cz.$$

Hyperboloid of Two Sheets:

$$\frac{z^2}{c^2} - \frac{x^2}{a^2} - \frac{y^2}{b^2} = 1.$$

Hyperbolic Paraboloid:

$$\frac{x^2}{a^2} - \frac{y^2}{b^2} = cz.$$

Sphere:

$$x^2 + y^2 + z^2 = a^2.$$

Distance d between Two Points

The distance d between two points $P_1(x_1, y_1, z_1)$ and $P_2(x_2, y_2, z_2)$ is given by

$$d = \sqrt{(x_2 - x_1)^2 + (y_2 - y_1)^2 + (z_2 - z_1)^2}.$$

Equations of a Line Joining:
Equations of a line joining $P_1(x_1, y_1, z_1)$ and $P_2(x_2, y_2, z_2)$ in standard form are

$$\frac{x - x_1}{x_2 - x_1} = \frac{y - y_1}{y_2 - y_1} = \frac{z - z_1}{z_2 - z_1}$$

or

$$\frac{x - x_1}{l} = \frac{y - y_1}{m} = \frac{z - z_1}{n}.$$

Equations of a Line Joining:
Equations of a line joining $P_1(x_1, y_1, z_1)$ and $P_2(x_2, y_2, z_2)$ in parametric form are

$$x = x_1 + lt, \quad y = y_1 + mt, \quad z = z_1 + nt.$$

Angle ϕ between Two Lines with Direction Cosines l_1, m_1, n_1 and l_2, m_2, n_2:

$$\cos\phi = l_1 l_2 + m_1 m_2 + n_1 n_2.$$

General Equation of a Plane

$$Ax + By + Cz + D = 0,$$

where A, B, C, and D are constants.

Equation of a Plane Passing Through Points:
The equation of a plane passing through points (x_1, y_1, z_1), (x_2, y_2, z_2), and (x_3, y_3, z_3) is given by

$$\begin{vmatrix} x - x_1 & y - y_1 & z - z_1 \\ x_2 - x_1 & y_2 - y_1 & z_2 - z_1 \\ x_3 - x_1 & y_3 - y_1 & z_3 - z_1 \end{vmatrix} = 0$$

or

$$\begin{vmatrix} y_2 - y_1 & z_2 - z_1 \\ y_3 - y_1 & z_3 - z_1 \end{vmatrix} (x - x_1) + \begin{vmatrix} z_2 - z_1 & x_2 - x_1 \\ z_3 - z_1 & x_3 - x_1 \end{vmatrix} (y - y_1)$$

$$+ \begin{vmatrix} x_2 - x_1 & y_2 - y_1 \\ x_3 - x_1 & y_3 - y_1 \end{vmatrix} (z - z_1) = 0.$$

Equation of a Plane in Intercept Form:
Equation of a plane in intercept form is given by

$$\frac{x}{a} + \frac{y}{b} + \frac{z}{c} = 1,$$

where a, b, and c are the intercepts on the x-, y-, and z-axes, respectively.

Equations of a Line through (x_0, y_0, z_0) and Perpendicular to Plane:

$$Ax + By + Cz + D = 0,$$

$$\frac{x - x_0}{A} = \frac{y - y_0}{B} = \frac{z - z_0}{C},$$

or

$$x = x_0 + At, \quad y = y_0 + Bt, \quad z = z_0 + Ct.$$

Distance from Point (x, y, z) to Plane $Ax + By + D = 0$:

$$\frac{Ax_0 + By_0 + Cz_0 + D}{\pm\sqrt{A^2 + B^2 + C^2}},$$

where the sign is chosen so that the distance is nonnegative.

Normal form for Equation of Plane:

$$x \cos \alpha + y \cos \beta + z \cos \gamma = p,$$

where p is the perpendicular distance from O to plane at P and α, β, γ are the angles between OP and positive x-, y-, and z-axes.

Transformation of Coordinates Involving Pure Translation:

$$
\begin{aligned}
x &= x' + x_0 & x' &= x + x_0, \\
y &= y' + y_0 \quad \text{or} \quad & y' &= y + y_0, \\
z &= z' + z_0 & z' &= z + z_0,
\end{aligned}
$$

where (x, y, z) are old coordinates (i.e., coordinates relative to xyz system), (x', y', z') are new coordinates [relative to (x', y', z') system], and (x_0, y_0, z_0) are the coordinates of the new origin O' relative to the old xyz coordinate system.

Transformation of Coordinates Involving Pure Rotation:

$$
\begin{aligned}
x &= l_1 x' + l_2 y' + l_3 z' & x' &= l_1 x + m_1 y + n_1 z, \\
y &= m_1 x' + m_2 y' + m_3 z' \quad \text{or} \quad & y' &= l_2 x + m_2 y + n_3 z, \\
z &= n_1 x' + n_2 y' + n_3 z' & z' &= l_3 x + m_3 y + n_3 z,
\end{aligned}
$$

where the origins of the xyz and x', y', z' systems are the same and $l_1, m_1, n_1; l_2, m_2, n_2; l_3, m_3, n_3$ are the direction cosines of the x', y', z'-axes relative to the x, y, z-axes, respectively.

Transformation of Coordinates Involving Translation and Rotation:

$$
\begin{aligned}
x &= l_1 x' + l_2 y' + l_3 z' + x_0, \\
y &= m_1 x' + m_2 y' + m_3 z' + y_0, \\
z &= n_1 x' + n_2 y' + n_3 z' + z_0,
\end{aligned}
$$

or

$$
\begin{aligned}
x' &= l_1(x - x_0) + m_1(y - y_0) + n_1(z - z_0), \\
y' &= l_2(x - x_0) + m_2(y - y_0) + n_2(z - z_0), \\
z' &= l_3(x - x_0) + m_3(y - y_0) + n_3(z - z_0),
\end{aligned}
$$

where the origin O' of the $x'y'z'$ system has coordinates (x_0, y_0, z_0) relative to the xyz system and $l_1, m_1, n_1; l_2, m_2, n_2; l_3, m_3, n_3$ are the direction cosines of the $x'y'z'$-axes relative to the x, y, z-axes, respectively.

Cylindrical Coordinates (r, θ, z):

A point P can be located by cylindrical coordinates (r, θ, z) as well as rectangular coordinates (x, y, z). The transformation between these coordinates is

$$x = r \cos \theta,$$
$$y = r \sin \theta, \quad \text{or}$$
$$z = z,$$

$$r = \sqrt{x^2 + y^2},$$
$$\theta = \tan^{-1}\left(\frac{y}{x}\right),$$
$$z = z.$$

Spherical Coordinates (r, θ, ϕ):

A point P can be located by cylindrical coordinates (r, θ, ϕ) as well as rectangular coordinates (x, y, z). The transformation between these coordinates is

$$x = r \cos \theta \cos \phi,$$
$$y = r \sin \theta \sin \phi,$$
$$z = r \cos \theta,$$

or

$$r = \sqrt{x^2 + y^2 + z^2},$$
$$\phi = \tan^{-1}\left(\frac{y}{x}\right),$$
$$\theta = \cos^{-1}\left(\frac{z}{\sqrt{x^2 + y^2 + z^2}}\right).$$

Equation of a Sphere in Rectangular Coordinates:

$$(x - x_0)^2 + (y - y_0)^2 + (z - z_0)^2 = R^2,$$

where the sphere has center (x_0, y_0, z_0) and radius R.

Equation of a Sphere in Cylindrical Coordinates:

$$r^2 - 2r_0 r \cos(\theta - \theta_0) + r_0^2 + (z - z_0)^2 = R^2,$$

where the sphere has center (r_0, θ_0, z_0) in cylindrical coordinates and radius R. If the center is at the origin, the equation is

$$r^2 + z^2 = R^2.$$

Equation of a Sphere in Spherical Coordinates:

$$r^2 + r_0^2 - 2r_0 r \sin \theta \sin \theta_0 \cos(\phi - \phi_0) = R^2,$$

where the sphere has center (r_0, θ_0, ϕ_0) in spherical coordinates and radius R. If the center is at the origin, the equation is

$$r = R.$$

Logarithmic Identities

Following is a list of identities that are useful when dealing with logarithms. All of these are valid for all positive real numbers, except that the base of a logarithm may never be 1.

$$\ln(z_1 z_2) = \ln z_1 + \ln z_2,$$

$$\ln(z_1 z_2) = \ln z_1 + \ln z_2 \quad (-\pi < \arg z_1 + \arg z_2 \leq \pi),$$

$$\ln\left(\frac{z_1}{z_2}\right) = \ln z_1 - \ln z_2,$$

$$\ln\left(\frac{z_1}{z_2}\right) = \ln z_1 - \ln z_2 \quad (-\pi < \arg z_1 - \arg z_2 \leq \pi),$$

$$\ln z^n = n \ln z \quad (n \text{ integer}),$$

$$\ln z^n = n \ln z \quad (n \text{ integer}, \ -\pi < n \arg z \leq \pi).$$

Special Values

$$\ln 1 = 0,$$

$$\ln 0 = -\infty,$$

$$\ln(-1) = \pi i,$$

$$\ln(\pm i) = \pm\frac{1}{2}\pi i,$$

$\ln e = 1$, then e is the real number such that

$$\int_1^e \frac{dt}{t} = 1,$$

$$e = \lim_{n \to \infty} \left(1 + \frac{1}{n}\right)^n = 2.7182818284\ldots.$$

Logarithms to General Base

$$\log_a z = \frac{\ln z}{\ln a},$$

$$\log_a z = \frac{\log_b z}{\log_b a},$$

$$\log_a b = \frac{1}{\log_b a},$$

$$\log_e z = \ln z,$$

$$\log_{10} z = \frac{\ln z}{\ln 10} = \log_{10} e \ln z = (0.4342944819\ldots)\ln z,$$

$$\ln z = \ln 10 \log_{10} z = (2.3025850929\ldots)\log_{10} z,$$

$$\left(\begin{array}{l} \log_e x = \ln x, \text{ called natural, Napierian, or hyperbolic logarithms;} \\ \log_{10} x, \text{ called common or Briggs logarithms} \end{array}\right).$$

Series Expansions

A series expansion is a representation of a particular function as a sum of powers in one of its variables, or by a sum of powers of another (usually elementary) function $f(x)$.

$$\ln(1+z) = z - \frac{1}{2}z^2 + \frac{1}{3}z^3 - \cdots, \quad |z| \leq 1 \text{ and } z \neq -1,$$

$$\ln z = \left(\frac{z-1}{z}\right) + \frac{1}{2}\left(\frac{z-1}{z}\right)^2 + \frac{1}{3}\left(\frac{z-1}{z}\right)^3 + \cdots, \quad \text{Re}\, z \geq \frac{1}{2},$$

$$\ln z = (z-1) - \frac{1}{2}(z-1)^2 + \frac{1}{3}(z-1)^3 - \cdots, \quad |z-1| \leq 1, \ z \neq 0,$$

$$\ln z = 2\left[\left(\frac{z-1}{z+1}\right) + \frac{1}{3}\left(\frac{z-1}{z+1}\right)^3 + \frac{1}{5}\left(\frac{z-1}{z+1}\right)^5 + \cdots\right], \quad \text{Re}\, z \geq 0, \ z \neq 0,$$

$$\ln\left(\frac{z+1}{z-1}\right) = 2\left(\frac{1}{z} + \frac{1}{3z^3} + \frac{1}{5z^5} + \cdots\right), \quad |z| \geq 1, \ z \neq \pm 1,$$

$$\ln(z+a) = \ln a + 2\left[\left(\frac{z}{2a+z}\right) + \frac{1}{3}\left(\frac{z}{2a+z}\right)^3 + \frac{1}{5}\left(\frac{z}{2a+z}\right)^5 + \cdots\right],$$

$$a > 0, \ \text{Re}\, z \geq -a \neq z.$$

Limiting Values

$$\lim_{x \to \infty} x^{-\alpha} \ln x = 0, \quad \alpha \text{ constant}, \text{Re}\, \alpha > 0,$$

$$\lim_{x \to 0} x^{\alpha} \ln x = 0, \quad \alpha \text{ constant}, \text{Re}\, \alpha > 0,$$

$$\lim_{m \to \infty} \left(\sum_{k=1}^{m} \frac{1}{k} - \ln m\right) = \gamma, \quad \text{Euler's constant} = 0.5772156649\ldots.$$

Inequalities

$$\frac{x}{1+x} < \ln(1+x) < x, \quad x > -1, \ x \neq 0,$$

$$x < -\ln(1-x) < \frac{x}{1+x}, \quad x < 1, \ x \neq 0,$$

$$|\ln(1-x)| < \frac{3x}{2}, \quad 0 < x \leq 0.5828,$$

$$\ln x \leq x - 1, \quad x > 0,$$

$$\ln x \leq n(x^{1/n} - 1) \text{ for any positive } n, \quad x > 0,$$

$$|\ln(1-z)| \leq -\ln(1-|z|) \quad |z| < 1.$$

Polynomial Approximations

For $\dfrac{1}{\sqrt{10}} \leq x \leq \sqrt{10}$:

$$\log_{10} x = a_1 t + a_3 t^3 + \varepsilon(x), \quad t = \frac{x-1}{x+1},$$

$$|\varepsilon(x)| \leq 6 \times 10^{-4},$$

$$a_1 = 0.86304, \quad a_3 = 0.36415,$$

$$\log_{10} x = a_1 t + a_3 t^3 + a_5 t^5 + a_7 t^7 + a_9 t^9 + \varepsilon(x),$$

$$t = \frac{x-1}{x+1},$$

$$|\varepsilon(x)| \leq 10^{-7},$$

$$a_1 = 0.868591718,$$

$$a_3 = 0.289335524,$$

$$a_5 = 0.177522071,$$

$$a_7 = 0.094376476,$$

$$a_9 = 0.191337714,$$

For $0 \leq x \leq 1$:

$$\ln(1+x) = a_1 x + a_2 x^2 + a_3 x^3 + a_4 x^4 + a_5 x^5 + \varepsilon(x),$$

$$|\varepsilon(x)| \leq 1 \times 10^{-5},$$

$$a_1 = 0.99949556,$$

$$a_2 = 0.49190896,$$

$$a_3 = 0.28947478,$$

$$a_4 = 0.13606275,$$

$$a_5 = 0.03215845,$$

$$\ln(1+x) = a_1 x + a_2 x^2 + a_3 x^3 + a_4 x^4 + a_5 x^5 + a_6 x^6 + a_7 x^7 + a_8 x^8 + \varepsilon(x),$$

$$|\varepsilon(x)| \leq 3 \times 10^{-8},$$

$$a_1 = 0.9999964239,$$

$$a_2 = -0.4998741238,$$

$$a_3 = 0.3317990258,$$

$$a_4 = -0.2407338084,$$

$$a_5 = 0.1676540711,$$

$$a_6 = -0.0953293897,$$

$$a_7 = 0.0360884937,$$

$$a_8 = -0.0064535442.$$

Exponential Function Series Expansion

The exponential function e^x can be defined, in a variety of equivalent ways, as an infinite series. In particular, it may be defined by the following power series:

$$e^z = \exp z = 1 + \frac{z}{1!} + \frac{z^2}{2!} + \frac{z^3}{3!} + \cdots, \quad z = x + iy.$$

Fundamental Properties

$$\ln(\exp z) = z + 2k\pi i \quad (k \text{ any integer}),$$

$$\ln(\exp z) = z, \quad -\pi < \oint z \leq \pi,$$

$$\exp(\ln z) = \exp(\ln\ z) = z,$$

$$\frac{d}{dz} \exp z = \exp z.$$

Definition of General Powers

$$\text{If } N = a^z, \text{ then } z = \log_a N,$$

$$a^z = \exp(z \ln a).$$

$$\text{If } a = |a| \exp(i \arg a), \quad -\pi < \arg a \leq \pi,$$

$$|a^z| = |a|^x e^{-v \arg a},$$

$$\arg(a^z) = y \ln |a| + x \arg a,$$

$$\ln a^z = z \ln a \quad \text{for one of the values of } \ln a^z,$$

$$\ln a^x = x \ln a \quad (a \text{ real and positive}),$$

$$|e^z| = e^x,$$

$$\arg(e^z) = y,$$

$$a^{z_1} a^{z_2} = a^{z_1 + z_2},$$

$$a^z b^z = (ab)^z \quad (-\pi < \arg a + \arg b \leq \pi).$$

Logarithmic and Exponential Functions

Periodic Property

$$e^{z+2\pi ki} = e^z \quad (k \text{ any integer}),$$

$$e^x < \frac{1}{1-x}, \quad x < 1,$$

$$\frac{1}{1-x} < (1-e^{-x}) < x, \quad x > -1,$$

$$x < (e^x - 1) < \frac{1}{1-x}, \quad x < 1,$$

$$1 + x > e^{(x/(1+x))}, \quad x > -1,$$

$$e^x > 1 + \frac{x^n}{n!}, \quad n > 0, x > 0,$$

$$e^x > \left(1 + \frac{x}{y}\right)^y > \frac{xy}{e^{x+y}}, \quad x > 0, y > 0,$$

$$e^{-x} < 1 - \frac{x}{2}, \quad 0 < x \le 1.5936,$$

$$\frac{1}{4}|z| < |e^z - 1| < \frac{7}{4}|z|, \quad 0 < |z| < 1,$$

$$|e^z - 1| \le e^{|z|} - 1 \le |z|e^{|z|} \quad (\text{all } z)$$

$$e^{2a \arctan(1/2)} = 1 + \frac{2a}{z-a+} \frac{a^2+1}{3z+} \frac{a^2+4}{5z+} \frac{a^2+9}{7z+} \cdots.$$

Polynomial Approximations

For $0 \le x \le \ln 2 = 0.693\ldots$:

$$e^{-x} = 1 + a_1 x + a_2 x^2 + \varepsilon(x),$$

$$|\varepsilon(x)| \le 3 \times 10^{-3},$$

$$a_1 = -0.9664,$$

$$a_2 = 0.3536.$$

For $0 \le x \le \ln 2$:

$$e^{-x} = 1 + a_1 x + a_2 x^2 + a_3 x^3 + a_4 x^4 + \varepsilon(x),$$

$$|\varepsilon(x)| \le 3 \times 10^{-5},$$

$$a_1 = -0.9998684,$$

$$a_2 = 0.4982926,$$

$$a_3 = -0.1595332,$$

$$a_4 = 0.0293641.$$

$$e^{-x} = 1 + a_1 x + a_2 x^2 + a_3 x^3 + a_4 x^4 + a_5 x^5 + a_6 x^6 + a_7 x^7 + \varepsilon(x),$$

$$|\varepsilon(x)| \leq 2 \times 10^{-10},$$

$$a_1 = -0.9999999995,$$

$$a_2 = 0.4999999206,$$

$$a_3 = -0.1666653019,$$

$$a_4 = 0.0416573475,$$

$$a_5 = -0.0083013598,$$

$$a_6 = 0.0013298820,$$

$$a_7 = -0.0001413161.$$

For $0 \leq x \leq 1$:

$$10^x = (1 + a_1 x + a_2 x^2 + a_3 x^3 + a_4 x^4)^2 + \varepsilon(x),$$

$$|\varepsilon(x)| \leq 7 \times 10^{-4},$$

$$a_1 = 1.1499196,$$

$$a_2 = 0.6774323,$$

$$a_3 = 0.2080030,$$

$$a_4 = 0.1268089.$$

$$10^x = (1 + a_1 x + a_2 x^2 + a_3 x^3 + a_4 x^4 + a_5 x^5 + a_6 x^6 + a_7 x^7)^2 + \varepsilon(x),$$

$$|\varepsilon(x)| \leq 5 \times 10^{-8},$$

$$a_1 = 1.15129277603,$$

$$a_2 = 0.66273088429,$$

$$a_3 = 0.25439357484,$$

$$a_4 = 0.07295173666,$$

$$a_5 = 0.01742111988,$$

$$a_6 = 0.00255491796,$$

$$a_7 = 0.00093264267.$$

Formulae to find out the surface area and volume of solids are given below. The surface area and the volume of a cylinder are given, respectively, by

$$\text{surface area of a cylinder} = 2\pi r h + 2\pi r^2,$$

and

$$\text{volume of a cylinder} = \pi r^2 h.$$

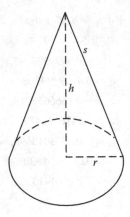

FIGURE 2.7 Surface area of a cone.

The surface area (see Figure 2.7) and volume of a cone are given, respectively, by

$$\text{surface area of a cone} = \pi r^2 + \pi rs,$$

and

$$\text{volume of a cone} = \frac{\pi r^2 h}{3}.$$

The volume of a pyramid is determined by the following formula (see Figure 2.8):

$$\text{Volume of a pyramid} = \frac{Bh}{3},$$

where B is the area of the base.

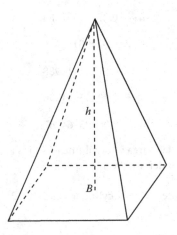

FIGURE 2.8 Volume of a pyramid.

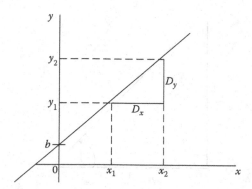

FIGURE 2.9   Equation of a straight line.

Slopes

Let us consider the equation of a straight line

$$y - y_1 = m(x - x_1),$$

where the slope m (=rise/run) is given by

$$\frac{\Delta y}{\Delta x} = \frac{y_2 - y_1}{x_2 - x_1}$$

or

$$y = mx + b,$$

where m is the slope and b is the y-intercept. See Figure 2.9 for the graphical representation of a straight line.

Trigonometric Ratios

The relationships between the angles and the sides of a triangle are expressed in terms of trigonometric ratios. From Figure 2.10, the ratios are defined as follows:

$$\tan \theta = \frac{\sin \theta}{\cos \theta},$$

$$\sin^2 \theta + \cos^2 \theta = 1,$$

$$1 + \tan^2 \theta = \sec^2 \theta,$$

$$1 + \cot^2 \theta = \csc^2 \theta,$$

$$\cos^2 \theta - \sin^2 \theta = \cos 2\theta,$$

$$\sin 45° = \frac{1}{\sqrt{2}},$$

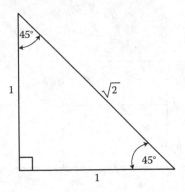

FIGURE 2.10 Triangle equations.

$$\cos 45° = \frac{1}{\sqrt{2}},$$

$$\tan 45° = 1,$$

$$\sin(A + B) = \sin A \cos B + \cos A \sin B,$$

$$\sin(A - B) = \sin A \cos B - \cos A \sin B,$$

$$\cos(A + B) = \cos A \cos B - \sin A \sin B,$$

$$\cos(A - B) = \cos A \cos B + \sin A \sin B,$$

$$\tan(A + B) = \frac{\tan A + \tan B}{1 - \tan A \tan B},$$

$$\tan(A - B) = \frac{\tan A - \tan B}{1 + \tan A \tan B},$$

For a right triangle, there relationships are expressed by the following equations, as shown in Figure 2.11:

$$\sin \theta = \frac{y}{r} \left(\frac{\text{opposite}}{\text{hypotenuse}} \right) = \frac{1}{\csc \theta},$$

$$\cos \theta = \frac{x}{r} \left(\frac{\text{adjacent}}{\text{hypotenuse}} \right) = \frac{1}{\sec \theta},$$

$$\tan \theta = \frac{y}{x} \left(\frac{\text{opposite}}{\text{adjacent}} \right) = \frac{1}{\cot \theta},$$

In a triangle, there are some angles or sub-divided angles within it which relates to each other in some way. The concept of their relationship is simple, as shown in Figure 2.12.

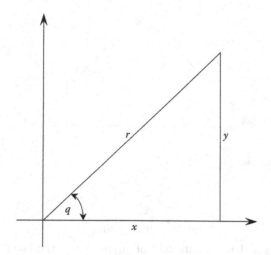

FIGURE 2.11 Right triangle calculations.

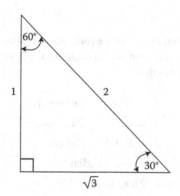

FIGURE 2.12 Angle relationships.

The relationships derived from Figure 2.12 are given below.

$$\sin 30° = \frac{1}{2}, \quad \sin 60° = \frac{\sqrt{3}}{2},$$

$$\cos 30° = \frac{\sqrt{3}}{2}, \quad \cos 60° = \frac{1}{2},$$

$$\tan 30° = \frac{1}{\sqrt{3}}, \quad \tan 60° = \sqrt{3}.$$

Sine Law

The law of sines (also known as the sines law, sine formula, or sine rule) is an equation relating the lengths of the sides of an arbitrary triangle to the sines of its angle. According

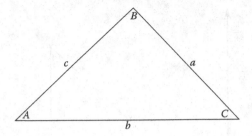

FIGURE 2.13 Cosine law.

to the law

$$\frac{a}{\sin A} = \frac{b}{\sin B} = \frac{c}{\sin C},$$

where a, b, and c are the lengths of the sides of a triangle, and A, B, and C are the opposite angles.

Cosine Law

The law of cosines (also known as the cosine formula or cosine rule) is a statement about a general triangle that relates the lengths of its sides to the cosine of one of its angles (for details, see Figure 2.13). It states that

$$a^2 = b^2 + c^2 - 2bc \cos A,$$
$$b^2 = a^2 + c^2 - 2ac \cos B,$$
$$c^2 = a^2 + b^2 - 2ab \cos C,$$
$$\theta = 1 \text{ radian},$$
$$2\pi \text{ radians} = 360°.$$

Algebra

Expanding

$$a(b + c) = ab + ac,$$
$$(a + b)^2 = a^2 + 2ab + b^2,$$
$$(a - b)^2 = a^2 - 2ab + b^2,$$
$$(a + b)(c + d) = ac + ad + bc + bd,$$
$$(a + b)^3 = a^3 + 3a^2b + 3ab^2 + b^3,$$
$$(a - b)^3 = a^3 - 3a^2b + 3ab^2 - b^3.$$

Factoring

$$a^2 - b^2 = (a+b)(a-b),$$

$$a^2 + 2ab + b^2 = (a+b)^2,$$

$$a^3 + b^3 = (a+b)(a^2 - ab + b^2),$$

$$a^3 b - ab = ab(a+1)(a-1),$$

$$a^2 - 2ab + b^2 = (a-b)^2,$$

$$a^3 - b^3 = (a-b)(a^2 + ab + b^2).$$

Roots of a Quadratic Equation

The solution for a quadratic equation $ax^2 + bx + c = 0$ is given by

$$x = \frac{-b \pm \sqrt{b^2 - 4ac}}{2a}.$$

Law of Exponents

$$a^r \cdot a^s = a^{r+s},$$

$$\frac{a^p a^q}{a^r} = a^{p+q-r},$$

$$\frac{a^r}{a^s} = a^{r-s},$$

$$(a^r)^s = a^{rs},$$

$$(ab)^r = a^r b^r,$$

$$\left(\frac{a}{b}\right)^r = \frac{a^r}{b^r}, \quad b \neq 0,$$

$$a^0 = 1, \quad a \neq 0,$$

$$a^{-r} = \frac{1}{a^r}, \quad a \neq 0,$$

$$a^{r/s} = \sqrt[s]{a^r}, \quad a^{1/2} = \sqrt{a}, \quad a^{1/3} = \sqrt[3]{a}.$$

Logarithms

Example

$$\log(xy) = \log x + \log y, \quad \log\left(\frac{x}{y}\right) = \log x - \log y,$$

$$\log x^r = r \log x,$$

$$\log x = n \leftrightarrow x = 10^n \text{ (common log)}, \quad \pi \simeq 3.14159265,$$

$$\log_a x = n \leftrightarrow x = a^n \text{ (log to the base } a\text{)}, \quad e \simeq 2.71828183,$$

$$\ln x = n \leftrightarrow x = e^n \text{ (natural log)}.$$

Tables 2.1 through 2.7 present a collection of reference materials relevant to the contents in this chapter.

TABLE 2.1 Basic Laplace Transforms

$f(t)$	$L\{f(t)\} = F(s)$
1	$\dfrac{1}{s}$
$t^n, \, n = 1, 2, 3, \ldots$	$\dfrac{n!}{s^{n+1}}$
e^{at}	$\dfrac{1}{s-a}$
$\sin kt$	$\dfrac{k}{s^2+k^2}$
$\cos kt$	$\dfrac{s}{s^2+k^2}$
$\sinh kt$	$\dfrac{k}{s^2-k^2}$
$\cosh kt$	$\dfrac{s}{s^2-k^2}$

TABLE 2.2 Operational Properties of Transforms

$e^{at}f(t)$	$F(s-a)$
$f(t-a)U(t-a), a > 0$	$e^{-as}F(s)$
$t^n f(t), n = 1, 2, 3, \ldots$	$(-1)^n \dfrac{d^n}{ds^n}F(s)$
$f^n(t), n = 1, 2, 3, \ldots$	$s^n F(s) - s^{n-1}f(0) - \cdots - f^{n-1}(0)$
$\displaystyle\int_0^t f(\tau)\,d\tau$	$\dfrac{F(s)}{s}$
$\displaystyle\int_0^t f(\tau)g(t-\tau)\,d\tau$	$F(s)G(s)$

TABLE 2.3 Transforms of Functional Products

$t^n e^{at}, \quad n = 1, 2, 3, \ldots$	$\dfrac{n!}{(s-a)^{n+1}}$
$e^{at} \sin kt$	$\dfrac{k}{(s-a)^2 + k^2}$
$e^{at} \cos kt$	$\dfrac{s-a}{(s-a)^2 + k^2}$
$t \sin kt$	$\dfrac{2ks}{(s^2 + k^2)^2}$
$t \cos kt$	$\dfrac{s^2 - k^2}{(s^2 + k^2)^2}$
$\sin kt - kt \cos kt$	$\dfrac{2k^3}{(s^2 + k^2)^2}$
$\sin kt + kt \cos kt$	$\dfrac{2ks^2}{(s^2 + k^2)^2}$

TABLE 2.4 Units of Measurement

English system

1 foot (ft)	= 12 inches (in); 1' = 12"
1 yard (yd)	= 3 feet
1 mile (mi)	= 1760 yards
1 sq. foot	= 144 sq. inches
1 sq. yard	= 9 sq. feet
1 acre	= 4840 sq. yards = 43,560 ft^2
1 sq. mile	= 640 acres

Metric system

millimeter (mm)	0.001 m
centimeter (cm)	0.01 m
decimeter (dm)	0.1 m
meter (m)	1 m
decameter (dam)	10 m
hectometer (hm)	100 m
kilometer (km)	1000 m

Note: Prefixes also apply to L (liter) and g (gram).

TABLE 2.5 Common Units of Measurement

Unit of measurement	Notation	Description
meter	m	length
hectare	ha	area
tonne	t	mass
kilogram	kg	mass
nautical mile	M	distance (navigation)
knot	kn	speed (navigation)
liter	L	volume or capacity
second	s	time

continued

TABLE 2.5 Common Units of Measurement (*Continued*)

Unit of measurement	Notation	Description
hertz	Hz	frequency
candela	cd	luminous intensity
degree Celsius	°C	temperature
kelvin	K	thermodynamic temperature
pascal	Pa	pressure, stress
joule	J	energy, work
newton	N	force
watt	W	power, radiant flux
ampere	A	electric current
volt	V	electric potential
ohm	Ω	electric resistance
coulomb	C	electric charge

TABLE 2.6 Values of Trig Ratio

θ	0	$\pi/2$	π	$3\pi/2$	2π
$\sin\theta$	0	1	0	-1	0
$\cos\theta$	1	0	-1	0	1
$\tan\theta$	0	∞	0	$-\infty$	0

TABLE 2.7 Physical Science Equations

$$D = \frac{m}{V} \qquad\qquad p = \frac{W}{t}$$

$$d = vt \qquad\qquad \text{K.E.} = \frac{1}{2}mv^2$$

$$a = \frac{v_f - v_i}{t} \qquad\qquad F_e = \frac{kQ_1 Q_2}{d^2}$$

$$d = v_i t + \frac{1}{2}at^2 \qquad\qquad V = \frac{W}{Q}$$

$$F = ma \qquad\qquad l = \frac{Q}{t}$$

$$F_g = \frac{Gm_1 m_2}{d^2} \qquad\qquad W = Vlt$$

$$p = mv \qquad\qquad p = Vl$$

$$W = Fd \qquad\qquad H = cm\,\Delta T$$

Note: D is the density (g/cm^3 = kg/m^3), m is the mass (kg), V is the volume, d is the distance (m), v is the velocity (m/s), t is the time (s), a is the acceleration (m/s^2), v_f is the final velocity (m/s), v_i is the initial velocity (m/s), F_g is the force of gravity (N), G is the universal gravitational constant ($G = 6.67 \times 10^{-11}$ N m^2/kg^2), m_1 and m_2 are the masses of the two objects (kg), p is the momentum (kg m/s), W is the work or electrical energy (J), P is the power (W), K.E. is the kinetic energy (J), F_e is the electric force (N), k Coulomb's constant ($k = 9 \times 10^9$ N m^2/C^2), Q, Q_1, and Q_2 are electric charges (C), V is the electric potential difference (V), l is the electric current (A), H is the heat energy (J), δT is the change in temperature (°C), and c is the specific heat (J/kg °C).

3

Statistical Distributions, Methods, and Applications

Industrial engineering uses statistical distributions and methods extensively in design and process improvement applications. The most common distributions are summarized in Table 3.1.

Discrete Distributions

Probability mass function, $p(x)$

Mean, μ

Variance, σ^2

Coefficient of skewness, β_1

Coefficient of kurtosis, β_2

Moment-generating function, $M(t)$

Characteristic function, $\phi(t)$

Probability-generating function, $P(t)$

Bernoulli Distribution

$$p(x) = p^x q^{x-1}, \quad x = 0, \ 10 \leq p \leq 1, \ q = 1 - p,$$

$$\mu = p, \quad \sigma^2 = pq, \quad \beta_1 = \frac{1 - 2p}{\sqrt{pq}}, \quad \beta_2 = 3 + \frac{1 - 6pq}{pq},$$

$$M(t) = q + pe^t, \quad \phi(t) = q + pe^{it}, \quad P(t) = q + pt.$$

Beta Binomial Distribution

$$p(x) = \frac{1}{n+1} \frac{B(a+x, b+n-x)}{B(x+1, n-x+1)B(a,b)}, \quad x = 0, 1, 2, \ldots, n, \ a > 0, \ b > 0,$$

$$\mu = \frac{na}{a+b}, \quad \sigma^2 = \frac{nab(a+b+n)}{(a+b)^2(a+b+1)},$$

where $B(a, b)$ is the Beta function.

TABLE 3.1 Summary of Common Statistical Distributions

Distribution of Random Variable x	Functional Form	Parameters	Mean	Variance	Range
Binomial	$P_x(k) = \dfrac{n!}{k!(n-k)!}p^k(1-p)^{n-k}$	n, p	np	$np(1-p)$	$0,1,2,\ldots,n$
Poisson	$P_x(k) = \dfrac{\lambda^k e^{-\lambda}}{k!}$	λ	λ	λ	$0,1,2,\ldots$
Geometric	$P_x(k) = p(1-p)^{k-1}$	p	$1/p$	$\dfrac{1-p}{p^2}$	$1,2,\ldots$
Exponential	$f_x(y) = \dfrac{1}{\theta}e^{-y/\theta}$	θ	θ	θ_2	$(0,\infty)$
Gamma	$f_x(y) = \dfrac{1}{\Gamma(\alpha)\beta^\alpha}y^{(\alpha-1)}e^{-y/\beta}$	α, β	$\alpha\beta$	$\alpha\beta^2$	$(0,\infty)$
Beta	$f_x(y) = \dfrac{\Gamma(\alpha+\beta)}{\Gamma(\alpha)\Gamma(\beta)}y^{(\alpha-1)}(1-y)^{(\beta-1)}$	α, β	$\dfrac{\alpha}{\alpha+\beta}$	$\dfrac{\alpha\beta}{(\alpha+\beta)^2(\alpha+\beta+1)}$	$(0,1)$
Normal	$f_x(y) = \dfrac{1}{\sqrt{2\pi}\sigma}e^{-(y-\mu)^2/2\sigma^2}$	μ, σ	μ	σ^2	$(-\infty,\infty)$
Student's t	$f_x(y) = \dfrac{1}{\sqrt{\pi v}}\dfrac{\Gamma((v+1)/2)}{\Gamma(v/2)}\left(1+\dfrac{y^2}{v}\right)^{-(v+1)/2}$	v	0 for $v>1$	$\dfrac{v}{v-2}$ for $v>2$	$(-\infty,\infty)$
Chi square	$f_x(y) = \dfrac{1}{2^{v/2}\Gamma(v/2)}y^{(v-2)/2}\,e^{-y/2}$	v	v	$2v$	$(0,\infty)$
F	$f_x(y) = \dfrac{\Gamma((v_1+v_2)/2)v_1^{v_1/2}v_2^{v_2/2}}{\Gamma(v_1/2)\Gamma(v_2/2)}\dfrac{(y)^{(v_1/2)-1}}{(v_2+v_1y)^{(v_1+v_2)/2}}$	v_1, v_2	$\dfrac{v_2}{v_2-2}$ for $v2>2$	$\dfrac{v_2^2(2v_2+2v_1-4)}{v_1(v_2-2)^2(v_2-4)}$ for $v2>4$	$(0,\infty)$

Beta Pascal Distribution

$$p(x) = \frac{\Gamma(x)\Gamma(\nu)\Gamma(\rho+\nu)\Gamma(\nu+x-(\rho+r))}{\Gamma(r)\Gamma(x-r+1)\Gamma(\rho)\Gamma(\nu-\rho)\Gamma(\nu+x)}, \quad x = r, r+1, \ldots, \nu > p > 0,$$

$$\mu = r\left(\frac{\nu-1}{\rho-1}\right), \quad \rho > 1, \quad \sigma^2 = r(r+\rho-1)\frac{(\nu-1)(\nu-\rho)}{(\rho-1)^2(\rho-2)}, \quad \rho > 2.$$

Binomial Distribution

$$p(x) = \binom{n}{x}p^x q^{n-x}, \quad x = 0, 1, 2, \ldots, n, \quad 0 \le p \le 1, \quad q = 1-p,$$

$$\mu = np, \quad \sigma^2 = npq, \quad \beta_1 = \frac{1-2p}{\sqrt{npq}}, \quad \beta_2 = 3 + \frac{1-6pq}{npq},$$

$$M(t) = (q + pe^t)^n, \quad \phi(t) = (q + pe^{it})^n, \quad P(t) = (q + pt)^n.$$

Discrete Weibull Distribution

$$p(x) = (1-p)^{x^\beta} - (1-p)^{(x+1)^\beta}, \quad x = 0, 1, \ldots, \quad 0 \le p \le 1, \quad \beta > 0.$$

Geometric Distribution

$$p(x) = pq^{1-x}, \quad x = 0, 1, 2, \ldots, \quad 0 \le p \le 1, \quad q = 1-p,$$

$$\mu = \frac{1}{p}, \quad \sigma^2 = \frac{q}{p^2}, \quad \beta_1 = \frac{2-p}{\sqrt{q}}, \quad \beta_2 = \frac{p^2+6q}{q},$$

$$M(t) = \frac{p}{1-qe^t}, \quad \phi(t) = \frac{p}{1-qe^{it}}, \quad P(t) = \frac{p}{1-qt}.$$

Hypergeometric Distribution

$$p(x) = \frac{(M)(N-M)}{\underset{x}{}\underset{n-x}{}}{\displaystyle\binom{N}{n}}, \quad x = 0, 1, 2, \ldots, n, \quad x \le M, \quad n - x \le N - M,$$

$$n, M, N, \in N, \quad 1 \le n \le N, \quad 1 \le M \le N, \quad N = 1, 2, \ldots,$$

$$\mu = n\frac{M}{N}, \quad \sigma^2 = \left(\frac{N-n}{N-1}\right)n\frac{M}{N}\left(1 - \frac{M}{N}\right), \quad \beta_1 = \frac{(N-2M)(N-2n)\sqrt{N-1}}{(N-2)\sqrt{nM(N-M)(N-n)}},$$

$$\beta_2 = \frac{N^2(N-1)}{(N-2)(N-3)nM(N-M)(N-n)}$$

$$\times \left\{N(N+1) - 6n(N-n) + 3\frac{M}{N^2}(N-M)[N^2(n-2) - Nn^2 + 6n(N-n)]\right\},$$

$$M(t) = \frac{(N-M)!(N-n)!}{N!}F(., e^t), \quad \phi(t) = \frac{(N-M)!(N-n)!}{N!}F(., e^{it}),$$

$$P(t) = \left(\frac{N-M}{N}\right)^n F(., t),$$

where $F(\alpha, \beta, \gamma, x)$ is the hypergeometric function. $\alpha = -n, \beta = -M, \gamma = N - M - n + 1$.

Negative Binomial Distribution

$$p(x) = \binom{x+r-1}{r-1} p^r q^x, \quad x = 0, 1, 2, \ldots, \quad r = 1, 2, \ldots, \quad 0 \le p \le 1, \quad q = 1 - p,$$

$$\mu = \frac{rq}{p}, \quad \sigma^2 = \frac{rq}{p^2}, \quad \beta_1 = \frac{2-p}{\sqrt{rq}}, \quad \beta_2 = 3 + \frac{p^2 + 6q}{rq},$$

$$M(t) = \left(\frac{p}{1 - qe^t}\right)^r, \quad \phi(t) = \left(\frac{p}{1 - qe^{it}}\right)^r, \quad P(t) = \left(\frac{p}{1 - qt}\right)^r.$$

Poisson Distribution

$$p(x) = \frac{e^{-\mu}\mu^x}{x!}, \quad x = 0, 1, 2, \ldots, \quad \mu > 0,$$

$$\mu = \mu, \quad \sigma^2 = \mu, \quad \beta_1 = \frac{1}{\sqrt{\mu}}, \quad \beta_2 = 3 + \frac{1}{\mu},$$

$$M(t) = \exp[\mu(e^t - 1)], \quad \sigma(t) = \exp[\mu(e^{it} - 1)], \quad P(t) = \exp[\mu(t - 1)].$$

Rectangular (Discrete Uniform) Distribution

$$p(x) = \frac{1}{n}, \quad x = 1, 2, \ldots, n, \quad n \in N,$$

$$\mu = \frac{n+1}{2}, \quad \sigma^2 = \frac{n^2 - 1}{12}, \quad \beta_1 = 0, \quad \beta_2 = \frac{3}{5}\left(3 - \frac{4}{n^2 - 1}\right),$$

$$M(t) = \frac{e^t(1 - e^{nt})}{n(1 - e^t)}, \quad \phi(t) = \frac{e^{it}(1 - e^{nit})}{n(1 - e^{it})}, \quad P(t) = \frac{t(1 - t^n)}{n(1 - t)}.$$

Continuous Distributions

Probability density function, $f(x)$

Mean, μ

Variance, σ^2

Coefficient of skewness, β_1

Coefficient of kurtosis, β_2

Moment-generating function, $M(t)$

Characteristic function, $\phi(t)$

Arcsin Distribution

$$f(x) = \frac{1}{\pi\sqrt{x(1-x)}}, \quad 0 < x < 1,$$

$$\mu = \frac{1}{2}, \quad \sigma^2 = \frac{1}{8}, \quad \beta_1 = 0, \quad \beta_2 = \frac{3}{2}.$$

Beta Distribution

$$f(x) = \frac{\Gamma(\alpha + \beta)}{\Gamma(\alpha)\Gamma(\beta)} x^{\alpha-1}(1-x)^{\beta-1}, \quad 0 < x < 1, \ \alpha, \beta > 0,$$

$$\mu = \frac{\alpha}{\alpha+\beta}, \quad \sigma^2 = \frac{\alpha\beta}{(\alpha+\beta)^2(\alpha+\beta+1)}, \quad \beta_1 = \frac{2(\beta-\alpha)\sqrt{\alpha+\beta+1}}{\sqrt{\alpha\beta}(\alpha+\beta+2)},$$

$$\beta_2 = \frac{3(\alpha+\beta+1)[2(\alpha+\beta)^2+\alpha\beta(\alpha+\beta-6)]}{\alpha\beta(\alpha+\beta+2)(\alpha+\beta+3)}.$$

Cauchy Distribution

$$f(x) = \frac{1}{b\pi(1+((x-a)/b)^2)}, \quad -\infty < x < \infty, \ -\infty < a < \infty, \ b > 0,$$

$\mu, \sigma^2, \beta_1, \beta_2, M(t)$ do not exist.

$$\phi(t) = \exp[ait - b|t|].$$

Chi Distribution

$$f(x) = \frac{x^{n-1}e^{-x^2/2}}{2^{(n/2)-1}\Gamma(n/2)}, \quad x \geq 0, \ n \in N,$$

$$\mu = \frac{\Gamma((n+1)/2)}{\Gamma(n/2)}, \quad \sigma^2 = \frac{\Gamma((n+2)/2)}{\Gamma(n/2)} - \left[\frac{\Gamma((n+1)/2)}{\Gamma(n/2)}\right]^2.$$

Chi-Square Distribution

$$f(x) = \frac{e^{-x/2}x^{(\nu/2)-1}}{2^{\nu/2}\Gamma(\nu/2)}, \quad x \geq 0, \ \nu \in N,$$

$$\mu = \nu, \quad \sigma^2 = 2\nu, \quad \beta_1 = 2\sqrt{\frac{2}{\nu}}, \quad \beta_2 = 3 + \frac{12}{\nu}, \quad M(t) = (1-2t)^{-\nu/2}, \quad t < \frac{1}{2},$$

$$\phi(t) = (1-2it)^{-\nu/2}.$$

Erlang Distribution

$$f(x) = \frac{1}{\beta^n(n-1)!} x^{n-1}e^{-x/\beta}, \quad x \geq 0, \ \beta > 0, \ n \in N,$$

$$\mu = n\beta, \quad \sigma^2 = n\beta^2, \quad \beta_1 = \frac{2}{\sqrt{n}}, \quad \beta_2 = 3 + \frac{6}{n},$$

$$M(t) = (1-\beta t)^{-n}, \quad \phi(t) = (1-\beta it)^{-n}.$$

Exponential Distribution

$$f(x) = \lambda e^{-\lambda x}, \quad x \geq 0, \quad \lambda > 0,$$

$$\mu = \frac{1}{\lambda}, \quad \sigma^2 = \frac{1}{\lambda^2}, \quad \beta_1 = 2, \quad \beta_2 = 9, \quad M(t) = \frac{\lambda}{\lambda - t},$$

$$\phi(t) = \frac{\lambda}{\lambda - it}.$$

Extreme-Value Distribution

$$f(x) = \exp\left[-e^{-(x-\alpha)/\beta}\right], \quad -\infty < x < \infty, -\infty < \alpha < \infty, \beta > 0,$$

Here $\mu = \alpha + \gamma\beta$, $\gamma \doteq 0.5772\ldots$ is Euler's constant $\sigma^2 = \pi^2\beta^2/6$, then

$$\beta_1 = 1.29857, \quad \beta_2 = 5.4,$$

$$M(t) = e^{\alpha t}\Gamma(1 - \beta t), \quad t < \frac{1}{\beta}, \quad \phi(t) = e^{\alpha it}\Gamma(1 - \beta it).$$

F Distribution

$$f(x)\frac{\Gamma[(\nu_1 + \nu_2)/2]\nu_1^{\nu_1/2}\nu_2^{\nu_2/2}}{\Gamma(\nu_1/2)\Gamma(\nu_2/2)}x^{(\nu_1/2)-1}(\nu_2 + \nu_1 x)^{-(\nu_1+\nu_2)/2},$$

$x > 0$ and $\nu_1, \nu_2 \in N$:

$$\mu = \frac{\nu_2}{\nu_2 - 2}, \quad \nu_2 \geq 3, \quad \sigma^2 = \frac{2\nu_2^2(\nu_1 + \nu_2 - 2)}{\nu_1(\nu_2 - 2)^2(\nu_2 - 4)}, \quad \nu_2 \geq 5,$$

$$\beta_1 = \frac{(2\nu_1 + \nu_2 - 2)\sqrt{8(\nu_2 - 4)}}{\sqrt{\nu_1}(\nu_2 - 6)\sqrt{\nu_1 + \nu_2 - 2}}, \quad \nu_2 \geq 7,$$

$$\beta_2 = 3 + \frac{12[(\nu_2 - 2)^2(\nu_2 - 4) + \nu_1(\nu_1 + \nu_2 - 2)(5\nu_2 - 22)]}{\nu_1(\nu_2 - 6)(\nu_2 - 8)(\nu_1 + \nu_2 - 2)}, \quad \nu_2 \geq 9.$$

$M(t)$ does not exist.

$$\phi\left(\frac{\nu_1}{\nu_2}t\right) = \frac{G(\nu_1, \nu_2, t)}{B(\nu_1/2, \nu_2/2)}.$$

$B(a, b)$ is the Beta function. G is defined by

$$(m + n - 2)G(m, n, t) = (m - 2)G(m - 2, n, t) + 2itG(m, n - 2, t), \quad m, n > 2,$$
$$mG(m, n, t) = (n - 2)G(m + 2, n - 2, t) - 2itG(m + 2, n - 4, t), \quad n > 4,$$
$$nG(2, n, t) = 2 + 2itG(2, n - 2, t), \quad n > 2.$$

Gamma Distribution

$$f(x) = \frac{1}{\beta^\alpha \Gamma(\alpha)} x^{\alpha-1} e^{-x/\beta}, \quad x \geq 0, \ \alpha, \beta > 0,$$

$$\mu = \alpha\beta, \quad \sigma^2 = \alpha\beta^2, \quad \beta_1 = \frac{2}{\sqrt{\alpha}}, \quad \beta_2 = 3\left(1 + \frac{2}{\alpha}\right),$$

$$M(t) = (1 - \beta t)^{-\alpha}, \quad \phi(t) = (1 - \beta i t)^{-\alpha}.$$

Half-Normal Distribution

$$f(x) = \frac{2\theta}{\pi} \exp\left[-\left(\frac{\theta^2 x^2}{\pi}\right)\right], \quad x \geq 0, \ \theta > 0,$$

$$\mu = \frac{1}{\theta}, \quad \sigma^2 = \left(\frac{\pi - 2}{2}\right)\frac{1}{\theta^2}, \quad \beta_1 = \frac{4 - \pi}{\theta^3}, \quad \beta_2 = \frac{3\pi^2 - 4\pi - 12}{4\theta^4}.$$

Laplace (Double Exponential) Distribution

$$f(x) = \frac{1}{2\beta} \exp\left[-\frac{|x - \alpha|}{\beta}\right], \quad -\infty < x < \infty, \ -\infty < \alpha < \infty, \ \beta > 0,$$

$$\mu = \alpha, \quad \sigma^2 = 2\beta^2, \quad \beta_1 = 0, \quad \beta_2 = 6,$$

$$M(t) = \frac{e^{\alpha t}}{1 - \beta^2 t^2}, \quad \phi(t) = \frac{e^{\alpha i t}}{1 + \beta^2 t^2}.$$

Logistic Distribution

$$f(x) = \frac{\exp[(x - \alpha)/\beta]}{\beta(1 + \exp[(x - \alpha)/\beta])^2},$$

$$-\infty < x < \infty, \quad -\infty < \alpha < \infty, \quad -\infty < \beta < \infty,$$

$$\mu = \alpha, \quad \sigma^2 = \frac{\beta^2 \pi^2}{3}, \quad \beta_1 = 0, \quad \beta_2 = 4.2,$$

$$M(t) = e^{\alpha t} \pi\beta t \csc(\pi\beta t), \quad \phi(t) = e^{\alpha i t} \pi\beta i t \csc(\pi\beta i t).$$

Lognormal Distribution

$$f(x) = \frac{1}{\sqrt{2\pi}\sigma x} \exp\left[-\frac{1}{2\sigma^2}(\ln x - \mu)^2\right],$$

$$x > 0, \quad -\infty < \mu < \infty, \quad \sigma > 0,$$

$$\mu = e^{\mu + \sigma^2/2}, \quad \sigma^2 = e^{2\mu + \sigma^2}(e^{\sigma^2} - 1),$$

$$\beta_1 = (e^{\sigma^2} + 2)(e^{\sigma^2} - 1)^{1/2}, \quad \beta_2 = (e^{\sigma^2})^4 + 2(e^{\sigma^2})^3 + 3(e^{\sigma^2})^2 - 3.$$

Noncentral Chi-Square Distribution

$$f(x) = \frac{\exp[-\frac{1}{2}(x+\lambda)]}{2^{\nu/2}} \sum_{j=0}^{\infty} \frac{x^{(\nu/2)+j-1}\lambda^j}{\Gamma(\nu/2+j)2^{2j}j!},$$

$$x > 0, \quad \lambda > 0, \quad \nu \in N,$$

$$\mu = \nu + \lambda, \quad \sigma^2 = 2(\nu + 2\lambda), \quad \beta_1 = \frac{\sqrt{8}(\nu+3\lambda)}{(\nu+2\lambda)^{3/2}}, \quad \beta_2 = 3 + \frac{12(\nu+4\lambda)}{(\nu+2\lambda)^2},$$

$$M(t) = (1-2t)^{-\nu/2} \exp\left[\frac{\lambda t}{1-2t}\right], \quad \phi(t) = (1-2it)^{-\nu/2} \exp\left[\frac{\lambda it}{1-2it}\right].$$

Noncentral F Distribution

$$f(x) = \sum_{i=0}^{\infty} \frac{\Gamma(\frac{1}{2}(2i+\nu_1+\nu_2))(\nu_1/\nu_2)^{(2i+\nu_1)/2} x^{(2i+\nu_1-2)/2} e^{-\lambda/2}(\frac{1}{2}\lambda)}{\Gamma(\frac{1}{2}\nu_2)\Gamma(\frac{1}{2}(2i+\nu_1))\nu_1!(1+(\nu_1/\nu_2)x)^{(2i+\nu_1+\nu_2)/2}},$$

$$x > 0, \quad \nu_1, \nu_2 \in N, \quad \lambda > 0,$$

$$\mu = \frac{(\nu_1+\lambda)\nu_2}{(\nu_2-2)\nu_1}, \quad \nu_2 > 2,$$

$$\sigma^2 = \frac{(\nu_1+\lambda)^2 + 2(\nu_1+\lambda)\nu_2^2}{(\nu_2-2)(\nu_2-4)\nu_1^2} - \frac{(\nu_1+\lambda)^2\nu_2^2}{(\nu_2-2)^2\nu_1^2}, \quad \nu_2 > 4.$$

Noncentral t-Distribution

$$f(x) = \frac{\nu^{\nu/2}}{\Gamma(\frac{1}{2}\nu)} \frac{e^{-\delta^2/2}}{\sqrt{\pi}(\nu+x^2)^{(\nu+1)/2}} \sum_{i=0}^{\infty} \Gamma\left(\frac{\nu+i+1}{2}\right)\left(\frac{\delta^i}{i!}\right)\left(\frac{2x^2}{\nu+x^2}\right)^{i/2},$$

$$-\infty < x < \infty, \quad -\infty < \delta < \infty, \quad \nu \in N,$$

$$\mu_r' = c_r \frac{\Gamma((\nu-r)/2)\nu^{r/2}}{2^{r/2}\Gamma(\nu/2)}, \quad \nu > r, \quad c_{2r-1} = \sum_{i=1}^{r} \frac{(2r-1)!\delta^{2r-1}}{(2i-1)!(r-i)!2^{r-i}},$$

$$c_{2r} = \sum_{i=0}^{r} \frac{(2r)!\delta^{2i}}{(2i)!(r-i)!2^{r-i}}, \quad r = 1, 2, 3, \ldots.$$

Normal Distribution

$$f(x) = \frac{1}{\sigma\sqrt{2\pi}} \exp\left[-\frac{(x-\mu)^2}{2\sigma^2}\right],$$

$$-\infty < x < \infty, \quad -\infty < \mu < \infty, \quad \sigma > 0,$$

$$\text{mean} = \mu, \quad \text{variance} = \sigma^2, \quad \beta_1 = 0, \quad \beta_2 = 3, \quad M(t) = \exp\left[\mu t + \frac{t^2\sigma^2}{2}\right],$$

$$\phi(t) = \exp\left[\mu it - \frac{t^2\sigma^2}{2}\right].$$

Pareto Distribution

$$f(x) = \frac{\theta a^\theta}{x^{\theta+1}}, \quad x \geq a, \quad \theta > 0, \quad a > 0,$$

$$\mu = \frac{\theta a}{\theta - 1}, \quad \theta > 1, \quad \sigma^2 = \frac{\theta a^2}{(\theta-1)^2(\theta-2)}, \quad \theta > 2.$$

In this case, $M(t)$ does not exist.

Rayleigh Distribution

$$f(x) = \frac{x}{\sigma^2} \exp\left[-\frac{x^2}{2\sigma^2}\right], \quad x \geq 0, \quad \sigma = 0,$$

$$\mu = \sigma\sqrt{\frac{\pi}{2}}, \quad \sigma^2 = 2\sigma^2\left(1 - \frac{\pi}{4}\right), \quad \beta_1 = \frac{\sqrt{\pi}}{4} \frac{(\pi-3)}{(1-(\pi/4))^{3/2}},$$

$$\beta_2 = \frac{2 - (3/16)\pi^2}{(1-(\pi/4))^2}.$$

t-Distribution

$$f(x) = \frac{1}{\sqrt{\pi v}} \frac{\Gamma((v+1)/2)}{\Gamma(v/2)} \left(1 + \frac{x^2}{v}\right)^{-(v+1)/2}, \quad -\infty < x < \infty, \quad v \in N,$$

$$\mu = 0, \quad v \geq 2, \quad \sigma^2 = \frac{v}{v-2}, \quad v \geq 3, \quad \beta_1 = 0, \quad v \geq 4,$$

$$\beta_2 = 3 + \frac{6}{v-4}, \quad v \geq 5,$$

$M(t)$ does not exist.

$$\phi(t) = \frac{\sqrt{\pi}\Gamma(v/2)}{\Gamma((v+1)/2)} \int_{-\infty}^{\infty} \frac{e^{itz\sqrt{v}}}{(1+z^2)^{(v+1)/2}} dz.$$

Triangular Distribution

$$f(x) = \begin{cases} 0, & x \leq a, \\ \dfrac{4(x-a)}{(b-a)^2}, & a < x \leq \dfrac{1}{2}(a+b), \\ \dfrac{4(b-x)}{(b-a)^2}, & \dfrac{1}{2}(a+b) < x < b, \\ 0, & x \geq b, \end{cases}$$

$-\infty < a < b < \infty:$

$$\mu = \frac{a+b}{2}, \quad \sigma^2 = \frac{(b-a)^2}{24}, \quad \beta_1 = 0, \quad \beta_2 = \frac{12}{5},$$

$$M(t) = -\frac{4(e^{at/2} - e^{bt/2})^2}{t^2(b-a)^2}, \quad \phi(t) = \frac{4(e^{ait/2} - e^{bit/2})^2}{t^2(b-a)^2}.$$

Uniform Distribution

$$f(x) = \frac{1}{b-a}, \quad a \leq x \leq b, \quad -\infty < a < b < \infty,$$

$$\mu = \frac{a+b}{2}, \quad \sigma^2 = \frac{(b-a)^2}{12}, \quad \beta_1 = 0, \quad \beta_2 = \frac{9}{5},$$

$$M(t) = \frac{e^{bt} - e^{at}}{(b-a)t}, \quad \phi(t) = \frac{e^{bit} - e^{ait}}{(b-a)it}.$$

Weibull Distribution

$$f(x) = \frac{\alpha}{\beta^\alpha} x^{\alpha-1} e^{-(x/\beta)^\alpha}, \quad x \geq 0, \quad \alpha, \beta > 0,$$

$$\mu = \beta\Gamma\left(1 + \frac{1}{\alpha}\right), \quad \sigma^2 = \beta^2\left[\left(1 + \frac{2}{\alpha}\right) - \Gamma^2\left(1 + \frac{1}{\alpha}\right)\right],$$

$$\beta_1 = \frac{\Gamma(1+3/\alpha) - 3\Gamma(1+1/\alpha)\Gamma(1+2/\alpha) + 2\Gamma^3(1+1/\alpha)}{[\Gamma(1+2/\alpha) - \Gamma^2(1+1/\alpha)]^{3/2}},$$

$$\beta_2 = \frac{\Gamma(1+4/\alpha) - 4\Gamma(1+1/\alpha)\Gamma(1+3/\alpha) + 6\Gamma^2(1+1/\alpha)\Gamma(1+2/\alpha) - 3\Gamma^4(1+1/\alpha)}{[\Gamma(1+2/\alpha) - \Gamma^2(1+1/\alpha)]^2}.$$

Distribution Parameters

Average

$$\bar{x} = \frac{1}{n}\sum_{i=1}^{n} x_i.$$

Variance

$$s^2 = \frac{1}{n-1} \sum_{i=1}^{n} (x_i - \bar{x})^2.$$

Standard Deviation

$$s = \sqrt{s^2}.$$

Standard Error

$$\frac{s}{\sqrt{n}}.$$

Skewness

$$\frac{n \sum_{i=1}^{n} (x_i - \bar{x})^3}{(n-1)(n-2)s^3}$$

(missing if $s = 0$ or $n < 3$).

Standardized Skewness

$$\frac{\text{skewness}}{\sqrt{6/n}}.$$

Kurtosis

$$\frac{n(n+1) \sum_{i=1}^{n} (x_i - \bar{x})^4}{(n-1)(n-2)(n-3)s^4} - \frac{3(n-1)^2}{(n-2)(n-3)}$$

(missing if $s = 0$ or $n < 4$).

Standardized Kurtosis

$$\frac{\text{Kurtosis}}{\sqrt{24/n}}.$$

Weighted Average

$$\frac{\sum_{i=1}^{n} x_i w_i}{\sum_{i=1}^{n} w_i}.$$

Estimation and Testing

100(1−α)% Confidence Interval for Mean

$$\bar{x} \pm t_{n-1;\alpha/2} \frac{s}{\sqrt{n}}.$$

100(1 − α)% Confidence Interval for Variance

$$\left[\frac{(n-1)s^2}{\chi^2_{n-1;\alpha/2}}, \frac{(n-1)s^2}{\chi^2_{n-1;1-\alpha/2}} \right].$$

100(1 − α)% Confidence Interval for Difference in Means

Equal Variance

$$(\bar{x}_1 - \bar{x}_2) \pm t_{n_1+n_2-2;\alpha/2}s_p\sqrt{\frac{1}{n_1} + \frac{1}{n_2}},$$

where

$$s_p = \sqrt{\frac{(n_1-1)s_1^2 + (n_2-1)s_2^2}{n_1+n_2-2}}.$$

Unequal Variance

$$\left[(\bar{x}_1 - \bar{x}_2) \pm t_{m;\alpha/2}\sqrt{\frac{s_1^2}{n_1} + \frac{s_2^2}{n_2}}\right],$$

where

$$\frac{1}{m} = \frac{c^2}{n_1-1} + \frac{(1-c)^2}{n_2-1},$$

and

$$c = \frac{(s_1^2/n_1)}{(s_1^2/n_1) + (s_2^2/n_2)}.$$

100(1 − α)% Confidence Interval for Ratio of Variances

$$\left(\frac{s_1^2}{s_2^2}\right)\left(\frac{1}{F_{n_1-1,n_2-1;\alpha/2}}\right), \quad \left(\frac{s_1^2}{s_2^2}\right)\left(\frac{1}{F_{n_1-1,n_2-1;\alpha/2}}\right).$$

Normal Probability Plot

The data are sorted from the smallest to the largest value to compute order statistics. A scatter plot is then generated where

$$\text{horizontal position} = x_{(i)},$$

$$\text{vertical position} = \Phi\left(\frac{i-3/8}{n+1/4}\right).$$

The labels for the vertical axis are based upon the probability scale using:

$$100\left(\frac{i-3/8}{n+1/4}\right).$$

Comparison of Poisson Rates

Let n_j be the number of events in sample j and t_j be the length of sample j, then

$$\text{Rate estimates:} \quad r_j = \frac{n_j}{t_j},$$

$$\text{Rate ratio:} \quad \frac{r_1}{r_2},$$

$$\text{Test statistic:} \quad z = \max\left(0, \frac{|n_1 - ((n_1 + n_2)/2)| - 1/2}{\sqrt{(n_1 + n_2)/4}}\right),$$

where z follows the standard normal distribution.

Distribution Functions—Parameter Estimation

Bernoulli

$$\hat{p} = \bar{x}.$$

Binomial

$$\hat{p} = \frac{\bar{x}}{n},$$

where n is the number of trials.

Discrete Uniform

$$\hat{a} = \min x_i$$

and

$$\hat{b} = \max x_i.$$

Geometric

$$\hat{p} = \frac{1}{1 + \bar{x}}.$$

Negative Binomial

$$\hat{p} = \frac{k}{\bar{x}},$$

where k is the number of successes.

Poisson

$$\hat{\beta} = \bar{x}.$$

Beta

$$\hat{\alpha} = \overline{x} \left[\frac{\overline{x}(1 - \overline{x})}{s^2} - 1 \right]$$

and

$$\hat{\beta} = (1 - \overline{x}) \left(\frac{\overline{x}(1 - \overline{x})}{s^2} - 1 \right).$$

Chi-Square

$$\text{d.f.}\,\overline{v} = \overline{x}.$$

Erlang

$$\hat{\alpha} = \text{round}(\hat{\alpha} \text{ from Gamma}),$$

$$\hat{\beta} = \frac{\hat{\alpha}}{x}.$$

Exponential

$$\hat{\beta} = \frac{1}{\overline{x}}.$$

Note: system displays $1/\hat{\beta}$.

F Distribution

$$\text{number of d.f.:} \quad \hat{v} = \frac{2\hat{w}^3 - 4\hat{w}^2}{(s^2(\hat{w} - 2)^2(\hat{w} - 4)) - 2\hat{w}^2},$$

$$\text{den.d.f.:} \quad \hat{w} = \frac{\max(1, 2\overline{x})}{-1 + \overline{x}}.$$

Gamma

$$R = \log \left(\frac{\text{arithmetic mean}}{\text{geometric mean}} \right).$$

If $0 < R \leq 0.5772$, then

$$\hat{\alpha} = R^{-1}(0.5000876 + 0.1648852R - 0.0544274R)^2$$

or if $R > 0.5772$, then

$$\hat{\alpha} = R^{-1}(17.79728 + 11.968477R + R^2)^{-1}(8.898919 + 9.059950R + 0.9775373R^2)$$

and

$$\hat{\beta} = \frac{\hat{\alpha}}{x}.$$

This is an approximation of the method of maximum likelihood solution from Johnson and Kotz (1970).

Log–Normal

$$\widehat{\mu} = \frac{1}{n} \sum_{i=1}^{n} \log x_i$$

and

$$\widehat{\alpha} = \sqrt{\frac{1}{n-1} \sum_{i=1}^{n} (\log x_i - \widehat{\mu})^2}.$$

System displays:

means: $\exp\left(\widehat{\mu} + \dfrac{\widehat{\alpha}^2}{2}\right)$

and

standard deviation: $\sqrt{\exp(2\widehat{\mu} + \widehat{\alpha}^2)[\exp(\widehat{\alpha}^2) - 1]}.$

Normal

$$\widehat{\mu} = \bar{x}$$

and

$$\widehat{\sigma} = s.$$

Student's *t*

If $s^2 \le 1$ or if $\widehat{v} \le 2$, then the system indicates that data are inappropriate.

$$s^2 = \frac{\sum_{i=1}^{n} x_i^2}{n}$$

and

$$\widehat{v} = \frac{2s^2}{-1 + s^2}.$$

Triangular

$$\widehat{a} = \min x_i,$$

$$\widehat{c} = \max x_i,$$

and

$$\widehat{b} = 3\bar{x} - \widehat{a} - \bar{x}.$$

Uniform

$$\widehat{a} = \min x_i$$

and

$$\widehat{b} = \max x_i.$$

Weibull

By solving the simultaneous equations, we obtain

$$\hat{\alpha} = \frac{n}{\left[\frac{1}{3}\sum_{i=1}^{n} x_i^{\hat{\alpha}} \log x_i - \sum_{i=1}^{n} \log x_i\right]}$$

and

$$\hat{\beta} = \left(\frac{\sum_{i=1}^{n} x_i^{\hat{\alpha}}}{n}\right)^{1/\hat{\alpha}}.$$

Chi-Square Test for Distribution Fitting

Divide the range of data into nonoverlapping classes. The classes are aggregated at each end to ensure that classes have an expected frequency of at least 5.

Let O_i is the observed frequency in class i, E_i the expected frequency in class i from fitted distribution, and k the number of classes after aggregation.

Test statistic

$$\chi^2 = \sum_{i=1}^{k} \frac{(O_i - E_i)^2}{E_i}$$

follow a chi-square distribution with the degrees of freedom equal to $k - 1$ number of estimated parameters.

Kolmogorov–Smirnov Test

$$D_n^+ = \max_{1 \le i \le n} \left\{\frac{i}{n} - \widehat{F}(x_i)\right\},$$

$$D_n^- = \max_{1 \le i \le n} \left\{\widehat{F}(x_i) - \frac{i-1}{n}\right\},$$

$$D_n = \max\{D_n^+, D_n^-\},$$

where $\widehat{F}(x_i)$ is the estimated cumulative distribution at x_i.

ANOVA

Notation

d.f.   degrees of freedom for the error term $= \left(\sum_{t=1}^{k} n_t\right) - k$

k number of treatments

MSE   mean square error $\left(= \sum_{t=1}^{k} (n_t - 1)s_t^2 \Big/ \left(\left(\sum_{t=1}^{k} n_t\right) - k\right)\right)$

$\bar{n}$ average treatment size $\left(= n/k, \text{ where } n = \sum_{t=1}^{k} n_t\right)$

n_t number of observations for treatment t

x_{it} ith observation in treatment i

$\bar{x}_t$ treatment mean $\left(= \sum_{i=1}^{n_t} x_{it}/n_t\right)$

s_t^2 treatment variance $\left(= \sum_{i=1}^{n_t} (x_{it} - \bar{x}_t^2)/(n_t - 1)\right)$

Standard Error (Internal)

$$\sqrt{\frac{s_t^2}{n_t}}.$$

Standard Error (Pooled)

$$\sqrt{\frac{MSE}{n_t}}.$$

Interval Estimates

$$\bar{x}_t \pm M \sqrt{\frac{MSE}{n_t}}$$

where the confidence interval is

$$M = t_{n-k;\alpha/2}$$

and the LSD interval is

$$M = \frac{1}{\sqrt{2}} t_{n-k;\alpha/2}.$$

Tukey Interval

$$M = \frac{1}{2} q_{n-k,k;\alpha},$$

where $q_{n-k,k;\alpha}$ is the value of the studentized range distribution with $n-k$ degrees of freedom and k samples such that the cumulative probability equals $1 - \alpha$.

Scheffe Interval

$$M = \frac{\sqrt{k-1}}{\sqrt{2}} \sqrt{F_{k-1,n-k;\alpha}}.$$

Cochran C-Test

Follow F distribution with $\bar{n} - 1$ and $(\bar{n} - 1)(k - 1)$ degrees of freedom.

$$\text{Test statistic:} \quad F = \frac{(k-1)C}{1-C},$$

where

$$C = \frac{\max s_t^2}{\sum_{t=1}^{k} s_t^2}.$$

Bartlett Test

Test statistic:

$$B = 10^{M/(n-k)},$$

$$M = (n-k)\log_{10}\text{MSE} - \sum_{t=1}^{k}(n_t - 1)\log_{10}s_t^2.$$

The significance test is based on

$$\frac{M(\ln 10)}{1 + (1/3(k-1))\left[\sum_{t=1}^{k}(1/(n_t - 1)) - (1/N - k)\right]^{X_{k-1}^2}},$$

which follows a chi-square distribution with $k-1$ degrees of freedom.

Hartley's Test

$$H = \frac{\max(s_t^2)}{\min(s_t^2)}.$$

Kruskal–Wallis Test

Average rank of treatment:

$$\bar{R}_t = \frac{\sum_{i=1}^{n_t} R_{it}}{n_t}.$$

If there are no ties:

$$\text{test statistic:} \quad w = \left(\frac{12}{n}\sum_{i=1}^{k} n_t \bar{R}_t^{\,2}\right) - 3(n+1).$$

Adjustment for Ties

Let u_j be the number of observations tied at any rank for $j = 1, 2, 3, \ldots, m$, where m is the number of unique values in the sample.

$$W = \frac{w}{1 - (\sum_{j=1}^{m} u_j^3 - \sum_{j=1}^{m} u_j)/n(n^2 - 1)}.$$

Significance level: W follows a chi-square distribution with $k-1$ degrees of freedom.

Freidman Test

Let X_{it} be the observation in the ith row and tth column, where $i = 1, 2, \ldots, n, t = 1, 2, \ldots, k$, R_{it} the rank of X_{it} within its row, and n the common treatment size (all treatment sizes must be the same for this test), then

$$R_t = \sum_{i=1}^{n} R_{it},$$

and the average rank is

$$\bar{R}_t = \frac{\sum_{i=1}^{n_t} R_{it}}{n_t},$$

where data are ranked within each row separately.

test statistic: $\quad Q = \dfrac{12S(k-1)}{nk(k^2-1) - (\sum u^3 - \sum u)},$

where

$$S = \left(\sum_{t=1}^{k} R_i^2\right) - \frac{n^2 k(k+1)^2}{4}.$$

Here Q follows a chi-square distribution with k degrees of freedom.

Regression

Notation

$\underset{\sim}{Y}$ vector of n observations for the dependent variable $\sim$

$\underset{\sim}{X}$ n by p matrix of observations for p independent variables, including constant term, if any

$\sim$ a vector or matrix

$$\bar{Y} = \frac{\sum_{i=1}^{n} Y_i}{n}.$$

Regression Statistics

1. Estimated coefficients
   Note: estimated by a modified Gram–Schmidt orthogonal decomposition with tolerance $= 1.0\text{E}-08$.

$$\underset{\sim}{b} = (\underset{\sim}{X}'\underset{\sim}{X})^{-1}\underset{\sim}{X}\underset{\sim}{Y}.$$

2. Standard errors

$$S(\underset{\sim}{b}) = \sqrt{\text{diagonal elements of } (\underset{\sim}{X}'\underset{\sim}{X})^{-1}\text{MSE}},$$

where $\text{SSE} = \underset{\sim}{Y}'\underset{\sim}{Y} - \underset{\sim}{b}'\underset{\sim}{X}'\underset{\sim}{Y}$.

$$\text{MSE} = \frac{\text{SSE}}{n-p}.$$

3. t-Values

$$t = \frac{b}{S(b)}.$$

4. Significance level

t-Values follow the Student's t-distribution with $n - p$ degrees of freedom.

5. R^2

$$R^2 = \frac{\text{SSTO} - \text{SSE}}{\text{SSTO}},$$

where

$$\text{SSTO} = \begin{cases} Y' - n\overline{Y}^2 & \text{if constant,} \\ Y'Y & \text{if no constant.} \end{cases}$$

Note: When the no-constant option is selected, the total sum of square is uncorrected for the mean. Thus, the R^2 value is of little use, because the sum of the residuals is not zero.

6. Adjusted R^2

$$1 - \left(\frac{n-1}{n-p}\right)(1 - R^2).$$

7. Standard error of estimate

$$\text{SE} = \sqrt{\text{MSE}}.$$

8. Predicted values

$$\hat{Y} = Xb.$$

9. Residuals

$$e = Y - \hat{Y}.$$

10. Durbin–Watson statistic

$$D = \frac{\sum_{i=1}^{n-1}(e_{i+1} - e_i)^2}{\sum_{i=1}^{n} e_i^2}.$$

11. Mean absolute error

$$\frac{\sum_{i=1}^{n} e_i}{n}.$$

Predictions

Let X_h be the m by p matrix of independent variables for m predictions.

1. Predicted value

$$\hat{Y}_h = X_h b.$$

2. Standard error of predictions

$$S(\hat{Y}_{h(\text{new})}) = \sqrt{\text{diagonal elements of MSE}(1 + X_h(X'X)^{-1}X')}.$$

3. Standard error of mean response

$$S(\hat{\underset{\sim}{Y}}_h) = \sqrt{\text{diagonal elements of MSE}(\underset{\sim}{X}_h(\underset{\sim}{X}'\underset{\sim}{X})^{-1}\underset{\sim}{X}_h)}.$$

4. Prediction matrix results
   Column 1 = index numbers of forecasts
   $$2 = \hat{\underset{\sim}{Y}}_h.$$
   $$3 = (\hat{\underset{\sim}{Y}}_{h(\text{new})}).$$
   $$4 = (\hat{\underset{\sim}{Y}}_h - t_{n-p,\alpha/2}S(\hat{\underset{\sim}{Y}}_{h(\text{new})})).$$
   $$5 = (\hat{\underset{\sim}{Y}}_h + t_{n-p,\alpha/2}S(\hat{\underset{\sim}{Y}}_{h(\text{new})})).$$
   $$6 = \hat{\underset{\sim}{Y}}_h - t_{n-p,\alpha/2}S(\hat{\underset{\sim}{Y}}_h).$$
   $$7 = \hat{\underset{\sim}{Y}}_h + t_{n-p,\alpha/2}S(\hat{\underset{\sim}{Y}}_h).$$

Nonlinear Regression

$F(X, \hat{\beta})$ are values of nonlinear function using parameter estimates $\hat{\beta}$.

1. Estimated coefficients
   Obtained by minimizing the residual sum of squares using a search procedure suggested by Marquardt. This is a compromise between Gauss–Newton and steepest descent methods. The user specifies:
   a. initial estimates $\underset{\sim}{\beta}_0$
   b. initial value of Marquardt parameter λ, which is modified at each iteration. As $\lambda \to 0$, procedure approaches Gauss–Newton $\lambda \to \infty$, procedure approaches steepest descent
   c. scaling factor used to multiply Marquardt parameter after each iteration
   d. maximum value of Marquardt parameter
   Partial derivatives of F with respect to each parameter are estimated numerically.
2. Standard errors
   Estimated from residual sum of squares and partial derivatives.
3. Ratio
   $$\text{ratio} = \frac{\text{coefficient}}{\text{standard error}}.$$
4. R^2
   $$R^2 = \frac{\text{SSTO} - \text{SSE}}{\text{SSTO}},$$
   where $\text{SSTO} = \underset{\sim}{Y}'\underset{\sim}{Y} - n\overline{Y}^2$ and SSE is the residual sum of squares.

Ridge Regression

Additional notation:

$\underset{\sim}{Z}$ matrix of independent variables standardized so that $\underset{\sim}{Z}'\underset{\sim}{Z}$ equals the correlation matrix

θ value of the ridge parameter

Parameter estimates

$$\underset{\sim}{b}(\theta) = (\underset{\sim}{Z}'\underset{\sim}{Z} + \theta I_p)^{-1} \underset{\sim}{Z}' \underset{\sim}{Y},$$

where I_p is a p by p identity matrix.

Quality Control

For all quality control formulas:

Let k is the number of subgroups, n_j the number of observations in subgroup j, $j = 1, 2, \ldots, k$, and x_{ij} the ith observation in subgroup j.

All formulas below for quality control assume 3-sigma limits. If other limits are specified, the formulas are adjusted proportionally based on sigma for the selected limits. Also, average sample size is used unless otherwise specified.

Subgroup statistics

Subgroup means

$$\bar{x}_j = \frac{\sum_{i=1}^{n_j} x_{ij}}{n_j}.$$

Subgroup standard deviations

$$s_j = \sqrt{\frac{\sum_{i=1}^{n_j} (x_{ij} - \bar{x}_j)^2}{(n_j - 1)}}.$$

Subgroup range

$$R_j = \max\{x_{ij} | 1 \le i \le n_j\} - \min\{x_{ij} | 1 \le i \le n_j\}.$$

X Bar Charts

Compute

$$\bar{\bar{x}} \sin^{-1} \theta = \frac{\sum_{j=1}^{k} n_i \bar{x}_j}{\sum_{j=1}^{k} n_i},$$

$$\bar{R} = \frac{\sum_{j=1}^{k} n_i R_j}{\sum_{j=1}^{k} n_i},$$

$$s_p = \sqrt{\frac{\sum_{j=1}^{k} (n_j - 1) s_j^2}{\sum_{j=1}^{k} (n_j - 1)}},$$

$$\bar{n} = \frac{1}{k} \sum_{j=1}^{k} n_i.$$

For a chart based on range:

$$\text{UCL} = \bar{\bar{x}} + A_2 \bar{R}$$

and

$$LCL = \bar{\bar{x}} - A_2\bar{R}.$$

For a chart based on sigma:

$$UCL = \bar{\bar{x}} + \frac{3s_p}{\sqrt{n}}$$

and

$$LCL = \bar{\bar{x}} - \frac{3s_p}{\sqrt{n}}.$$

For a chart based on known sigma:

$$UCL = \bar{\bar{x}} + 3\frac{\sigma}{\sqrt{n}}$$

and

$$LCL = \bar{\bar{x}} - 3\frac{\sigma}{\sqrt{n}}.$$

If other than 3-sigma limits are used, such as 2-sigma limits, all bounds are adjusted proportionately. If average sample size is not used, then uneven bounds are displayed based on

$$\frac{1}{\sqrt{n_j}}$$

rather than $1/\sqrt{n}$.

If the data are normalized, each observation is transformed according to

$$z_{ij} = \frac{x_{ij} - \bar{\bar{x}}}{\hat{\alpha}},$$

where $\hat{\alpha}$ is the estimated standard deviation.

Capability Ratios

Note: The following indices are useful only when the control limits are placed at the specification limits. To override the normal calculations, specify a subgroup size of 1 and select the "known standard deviation" option. Then enter the standard deviation as half of the distance between the USL and LSL. Change the position of the centerline to be the midpoint of the USL and LSL and specify the upper and lower control line at one sigma.

$$C_p = \frac{USL - LSL}{6\hat{\alpha}},$$

$$C_R = \frac{1}{C_P},$$

$$C_{pk} = \min\left(\frac{USL - \bar{\bar{x}}}{3\hat{\alpha}}, \frac{\bar{\bar{x}} - LSL}{3\hat{\alpha}}\right).$$

R Charts

$$\mathrm{CL} = \bar{R},$$

$$\mathrm{UCL} = D_4\bar{R},$$

$$\mathrm{LCL} = \max(0, D_3\bar{R}).$$

S Charts

$$\mathrm{CL} = s_\mathrm{p},$$

$$\mathrm{UCL} = s_\mathrm{P}\sqrt{\frac{\chi^2_{\bar{n}-1;\alpha}}{\bar{n}-1}},$$

$$\mathrm{LCL} = s_\mathrm{P}\sqrt{\frac{\chi^2_{\bar{n}-1;\alpha}}{\bar{n}-1}}.$$

C Charts

$$\bar{c} = \sum u_j, \quad \mathrm{UCL} = \bar{c} + 3\sqrt{\bar{c}},$$

$$\sum n_j, \quad \mathrm{LCL} = \bar{c} - 3\sqrt{\bar{c}},$$

where u_i is the number of defects in the jth sample.

U Charts

$$\bar{u} = \frac{\text{number of defects in all samples}}{\text{number of units in all samples}} = \frac{\sum u_j}{\sum n_j},$$

$$\mathrm{UCL} = \bar{u} + \frac{3\sqrt{\bar{u}}}{\sqrt{n}},$$

$$\mathrm{LCL} = \bar{u} - \frac{3\sqrt{\bar{u}}}{\sqrt{n}}.$$

P Charts

$$p = \frac{\text{number of defective units}}{\text{number of units inspected}},$$

$$\bar{p} = \frac{\text{number of defectives in all samples}}{\text{number of units in all samples}} = \frac{\sum p_j n_j}{\sum n_j},$$

$$\mathrm{UCL} = \bar{p} + \frac{3\sqrt{\bar{p}(1-\bar{p})}}{\sqrt{n}},$$

$$\mathrm{LCL} = \bar{p} - \frac{3\sqrt{\bar{p}(1-\bar{p})}}{\sqrt{n}}.$$

NP Charts

$$\bar{p} = \frac{\sum d_j}{\sum n_j},$$

where d_j is the number of defectives in the jth sample.

$$\text{UCL} = \overline{np} + 3\sqrt{\overline{np}(1-\bar{p})}$$

and

$$\text{LCL} = \overline{np} - 3\sqrt{\overline{np}(1-\bar{p})}.$$

CuSum Chart for the Mean

Control mean $= \mu$

Standard deviation $= \alpha$

Difference to detect $= \Delta$

Plot cumulative sums C_t versus t where

$$C_t = \sum_{i=1}^{t} (\bar{x}_i - \mu) \quad \text{for} \quad t = 1, 2, \ldots, n.$$

The V-mask is located at distance

$$d = \frac{2}{\Delta} \left[\frac{\alpha^2/\bar{n}}{\Delta} \ln \frac{1-\beta}{\alpha/2} \right]$$

in front of the last data point.

$$\text{Angle of mast} = 2\tan^{-1}\frac{\Delta}{2}.$$

$$\text{Slope of the lines} = \pm\frac{\Delta}{2}.$$

Multivariate Control Charts

Let $\underset{\sim}{X}$ be the matrix of n rows and k columns containing n observations for each of k variable, S the sample covariance matrix, $\underset{\sim}{X}_t$ the observation vector at time t, and $\overline{\underset{\sim}{X}}$ the vector of column average, then

$$T_t^2 = (\underset{\sim}{X}_t - \overline{\underset{\sim}{X}})S^{-1}(\underset{\sim}{X}_t - \overline{\underset{\sim}{X}}),$$

$$\text{UCL} = \left(\frac{k(n-1)}{n-k} \right) F_{k,n-k;\alpha}.$$

Time-Series Analysis

Notation

x_t or y_t observation at time t, $t = 1, 2, \ldots, n$

n number of observations

Autocorrelation at Lag k

$$r_k = \frac{c_k}{c_0},$$

where

$$c_k = \frac{1}{n} \sum_{t=1}^{n-k} (y_t - \bar{y})(y_{t+k} - \bar{y})$$

and

$$\bar{y} = \frac{\sum_{t=1}^{n} y_t}{n}.$$

$$\text{Standard error} = \sqrt{\frac{1}{n} \left\{ 1 + 2 \sum_{v=1}^{k-1} r_v^2 \right\}}.$$

Partial autocorrelation at Lag k

$\hat{\theta}_{kk}$ is obtained by solving the Yule–Walker equations:

$$r_j = \hat{\theta}_{k1} r_{j-1} + \hat{\theta}_{k2} r_{j-2} + \cdots + \hat{\theta}_{k(k-1)} r_{j-k+1} + \hat{\theta}_{kk} r_{j-k},$$
$$j = 1, 2, \ldots, k.$$

$$\text{Standard error} = \sqrt{\frac{1}{n}}.$$

Cross-Correlation at Lag k

Let x be the input time series and y the output time series, then

$$r_{xy}(k) = \frac{c_{xy}(k)}{s_x s_y}, \quad k = 0, \pm 1, \pm 2, \ldots,$$

where

$$c_{xy}(k) = \begin{cases} \frac{1}{n} \sum_{t=1}^{n-k} (x_t - \bar{x})(y_{t+k} - \bar{y}), & k = 0, 1, 2, \ldots, \\ \frac{1}{n} \sum_{t=1}^{n+k} (x_t - \bar{x})(y_{t-k} - \bar{y}), & k = 0, -1, -2, \ldots \end{cases}$$

and

$$S_x = \sqrt{c_{xx}(0)},$$
$$S_y = \sqrt{c_{yy}(0)}.$$

Box-Cox

$$yt = \frac{(y + \lambda_2)^{\lambda_1} - 1}{\lambda_1 g^{(\lambda_1 - 1)}} \quad \text{if} \quad \lambda_1 > 0,$$

$$yt = g \ln(y + \lambda_2) \quad \text{if} \quad \lambda_1 = 0,$$

where g is the sample geometric mean $(y + \lambda_2)$.

Periodogram (Computed using Fast Fourier Transform)

If n is odd:

$$I(f_i) = \frac{n}{2}(a_i^2 + b_i^2), \quad i = 1, 2, \ldots, \left[\frac{n-1}{2}\right],$$

where

$$a_i = \frac{2}{n}\sum_{t=1}^{n} t_t \cos 2\pi f_i t,$$

$$b_i = \frac{2}{n}\sum_{t=1}^{n} y_t \sin 2\pi f_i t,$$

$$f_i = \frac{i}{n}.$$

If n is even, an additional term is added:

$$I(0.5) = n\left(\frac{1}{n}\sum_{t=1}^{n}(-1)^t Y_t\right)^2.$$

Categorical Analysis

Notation

c number of columns in table
f_{ij} frequency in position (row i, column j)
r number of rows in table
x_i distinct values of row variable arranged in ascending order; $i = 1, \ldots, r$
y_j distinct values of column variable arranged in ascending order, $j = 1, \ldots, c$

Totals

$$R_j = \sum_{j=1}^{c} f_{ij}, \quad C_j = \sum_{i=1}^{r} f_{ij},$$

$$N = \sum_{i=1}^{r}\sum_{j=1}^{c} f_{ij}.$$

Note: any row or column which totals zero is eliminated from the table before calculations are performed.

Chi-Square

$$\chi^2 = \sum_{i=1}^{r}\sum_{j=1}^{c}\frac{(f_{ij} - E_{ij})^2}{E_{ij}},$$

where

$$E_{ij} = \frac{R_i C_j}{N} \sim \chi^2_{(r-1)(c-1)}$$

A warning is issued if any $E_{ij} < 2$ or if 20% or more of all $E_{ij} < 5$. For 2×2 tables, a second statistic is printed using Yates' continuity correction.

Fisher's Exact Test

Run for a 2×2 table, when N is less than or equal to 100.

Lambda

$$\lambda = \frac{\sum_{j=1}^{c} f_{\max,j} - R_{\max}}{N - R_{\max}} \quad \text{(with row-dependent)},$$

$$\lambda = \frac{\sum_{i=1}^{r} f_{i,\max} - C_{\max}}{N - C_{\max}} \quad \text{(with column - dependent)},$$

$$\lambda = \frac{\sum_{i=1}^{r} f_{i,\max} + \sum_{j=1}^{c} f_{\max,j} - C_{\max} - R_{\max}}{2N - R_{\max} - C_{\max}},$$

when symmetric, where $f_{i\,\max}$ is the largest value in row i, $f_{\max\,j}$ the largest value in column j, $R_{\max}$ the largest row total, and $C_{\max}$ the largest column total.

Uncertainty Coefficient

$$U_R = \frac{U(R) + U(C) - U(RC)}{U(R)} \quad \text{(with rows dependent)},$$

$$U_C = \frac{U(R) + U(C) - U(RC)}{U(C)} \quad \text{(with columns dependent)},$$

$$U = 2 \left(\frac{U(R) + U(C) - U(RC)}{U(R) + U(C)} \right) \quad \text{(when symmetric)},$$

where

$$U(R) = - \sum_{i=1}^{r} \frac{R_i}{N} \log \frac{R_i}{N},$$

$$U(C) = - \sum_{j=1}^{c} \frac{C_j}{N} \log \frac{C_j}{N},$$

$$U(RC) = - \sum_{i=1}^{r} \sum_{j=1}^{c} \frac{f_{ij}}{N} \log \frac{f_{ij}}{N} \quad \text{for} \quad f_{ij} > 0.$$

Somer's D

$$D_R = \frac{2(P_C - P_D)}{N^2 - \sum_{j=1}^{c} C_j^2} \quad \text{(with row-dependent),}$$

$$D_C = \frac{2(P_C - P_D)}{N^2 - \sum_{i=1}^{r} R_i^2} \quad \text{(with column-dependent),}$$

$$D = \frac{4(P_C - P_D)}{(N^2 - \sum_{i=1}^{r} R_i^2) + (N^2 - \sum_{j=1}^{c} C_j^2)} \quad \text{(when symmetric),}$$

where the number of concordant pairs is

$$P_C = \sum_{i=1}^{r} \sum_{j=1}^{c} f_{ij} \sum_{h<i} \sum_{k<j} f_{hk}$$

and the number of discordant pairs is

$$P_D = \sum_{i=1}^{r} \sum_{j=1}^{c} f_{ij} \sum_{h<i} \sum_{k>j} f_{hk}.$$

Eta

$$E_R = \sqrt{1 - \frac{SS_{RN}}{SS_R}} \quad \text{(with row-dependent),}$$

where the total corrected sum of squares for the rows is

$$SS_R = \sum_{i=1}^{r} \sum_{j-1}^{c} x_i^2 f_{ij} - \frac{\left(\sum_{i=1}^{r} \sum_{j-1}^{c} x_i f_{ij}\right)^2}{N}$$

and the sum of squares of rows within categories of columns is

$$SS_{RN} = \sum_{j=1}^{c} \left(\sum_{i=1}^{r} x_i^2 f_{ij} - \frac{\left(\sum_{i=1}^{r} x_i^2 f_{ij}\right)^2}{C_j} \right),$$

$$E_C = \sqrt{1 - \frac{SS_{CN}}{SS_C}} \quad \text{(with column-dependent),}$$

where the total corrected sum of squares for the columns is

$$SS_C = \sum_{i=1}^{r} \sum_{j=1}^{c} y_i^2 f_{ij} - \frac{\left(\sum_{i=1}^{r} \sum_{j=1}^{c} y_i f_{ij}\right)^2}{N}$$

and the sum of squares of columns within categories of rows is

$$SS_{CN} = \sum_{i=1}^{r} \left(\sum_{j=1}^{c} y_i^2 f_{ij} - \frac{\left(\sum_{j=1}^{c} y_j^2 f_{ij}\right)^2}{R_i} \right) j.$$

Contingency Coefficient

$$C = \sqrt{\frac{\chi^2}{(\chi^2 + N)}}.$$

Cramer's V

$$V = \sqrt{\frac{\chi^2}{N}} \quad \text{for} \quad 2 \times 2 \text{ table},$$

$$V = \sqrt{\frac{\chi^2}{N(m-1)}} \quad \text{for all others},$$

where $m = \min(r, c)$.

Conditional Gamma

$$G = \frac{P_C - P_D}{P_C + P_D}.$$

Pearson's R

$$R = \frac{\sum_{j=1}^{c} \sum_{i=1}^{r} x_i y_j f_{ij} - (\sum_{j=1}^{c} \sum_{i=1}^{r} x_i f_{ij})(\sum_{j=1}^{c} \sum_{i=1}^{r} y_j f_{ij})/N}{\sqrt{SS_R SS_C}}.$$

If $R = 1$, no significance is printed. Otherwise, the one-sided significance is base on

$$\tau = R\sqrt{\frac{N-2}{1-R^2}}.$$

Kendall's Tau b

$$\tau = \frac{2(P_C - P_D)}{\sqrt{(N^2 - \sum_{i=1}^{r} R_i^2)(N^2 - \sum_{j=1}^{c} C_j^2)}}.$$

Tau C

$$\tau_C = \frac{2m(P_C - P_D)}{(m-1)N^2}.$$

Probability Terminology

Experiment: An experiment is an activity or occurrence with an observable result.
Outcome: The result of the experiment.
Sample point: An outcome of an experiment.
Event: An event is a set of outcomes (a subset of the sample space) to which a probability is assigned.

Basic Probability Principles

Consider a random sampling process in which all the outcomes solely depend on chance, that is, each outcome is equally likely to happen. If S is a uniform sample space and the collection of desired outcomes is E, the probability of the desired outcomes is

$$P(E) = \frac{n(E)}{n(S)},$$

where $n(E)$ is the number of favorable outcomes in E and $n(S)$ the number of possible outcomes in S.

Since E is a subset of S,

$$0 \le n(E) \le n(S),$$

the probability of the desired outcome is

$$0 \le P(E) \le 1.$$

Random Variable

A random variable is a rule that assigns a number to each outcome of a chance experiment.

Example

1. A coin is tossed six times. The random variable X is the number of tails that are noted. X can only take the values $1, 2, \ldots, 6$, so X is a discrete random variable.
2. A light bulb is burned until it burns out. The random variable Y is its lifetime in hours. Y can take any positive real value, so Y is a continuous random variable.

Mean Value $\hat{x}$ or Expected Value μ

The mean value or expected value of a random variable indicates its average or central value. It is a useful summary value of the variable's distribution.

1. If random variable X is a discrete mean value,

$$\hat{x} = x_1 p_1 + x_2 p_2 + \cdots + x_n p_n = \sum_{i=1}^{n} x_1 p_1,$$

where p_i is the probability densities.
2. If X is a continuous random variable with probability density function $f(x)$, then the expected value of X is

$$\mu = E(X) = \int_{-\infty}^{+\infty} x f(x) dx,$$

where f(x) is the probability densities.

Discrete Distribution Formulas

Probability mass function, $p(x)$
Mean, μ
Variance, σ^2
Coefficient of skewness, β_1
Coefficient of kurtosis, β_2
Moment-generating function, $M(t)$
Characteristic function, $\phi(t)$
Probability-generating function, $P(t)$

Bernoulli Distribution

$$p(x) = p^x q^{x-1} x = 0, 1, \quad 0 \leq p \leq 1, \quad q = 1 - p,$$

$$\mu = p, \quad \sigma^2 = pq, \quad \beta_1 = \frac{1 - 2p}{\sqrt{pq}}, \quad \beta_2 = 3 + \frac{1 - 6pq}{pq},$$

$$M(t) = q + pe^t, \quad \phi(t) = q + pe^{it}, \quad P(t) = q + pt.$$

Beta Binomial Distribution

$$p(x) = \frac{1}{n+1} \frac{B(a+x, b+n-x)}{B(x+1, n-x+1)B(a,b)}, \quad x = 0, 1, 2, \ldots, n, \quad a > 0, \quad b > 0,$$

$$\mu = \frac{na}{a+b}, \quad \sigma^2 = \frac{nab(a+b+n)}{(a+b)^2(a+b+1)},$$

where $B(a, b)$ is the Beta function.

Beta Pascal Distribution

$$p(x) = \frac{\Gamma(x)\Gamma(v)\Gamma(\rho + v)\Gamma(v + x - (\rho + r))}{\Gamma(r)\Gamma(x - r + 1)\Gamma(\rho)\Gamma(v - \rho)\Gamma(v + x)}, \quad x = r, r+1, \ldots, \quad v > p > 0,$$

$$\mu = r\frac{v - 1}{\rho - 1}, \quad \rho > 1, \quad \sigma^2 = r(r + \rho - 1)\frac{(v - 1)(v - \rho)}{(\rho - 1)^2(\rho - 2)}, \quad \rho > 2.$$

Binomial Distribution

$$p(x) = \binom{n}{x} p^x q^{n-x}, \quad x = 0, 1, 2, \ldots, n, \quad 0 \leq p \leq 1, \quad q = 1 - p,$$

$$\mu = np, \quad \sigma^2 = npq, \quad \beta_1 = \frac{1 - 2p}{\sqrt{npq}}, \quad \beta_2 = 3 + \frac{1 - 6pq}{npq},$$

$$M(t) = (q + pe^t)^n, \quad \phi(t) = (q + pe^{it})^n, \quad P(t) = (q + pt)^n.$$

Discrete Weibull Distribution

$$p(x) = (1 - p)^{x^\beta} - (1 - p)^{(x+1)^\beta}, \quad x = 0, 1, \ldots, \quad 0 \leq p \leq 1, \quad \beta > 0.$$

Geometric Distribution

$$p(x) = pq^{1-x}, \quad x = 0, 1, 2, \ldots, \quad 0 \le p \le 1, \quad q = 1 - p,$$

$$\mu = \frac{1}{p}, \quad \sigma^2 = \frac{q}{p^2}, \quad \beta_1 = \frac{2-p}{\sqrt{q}}, \quad \beta_2 = \frac{p^2 + 6q}{q},$$

$$M(t) = \frac{p}{1 - qe^t}, \quad \phi(t) = \frac{p}{1 - qe^{it}}, \quad P(t) = \frac{p}{1 - qt}.$$

Hypergeometric Distribution

$$p(x) = \frac{\binom{M}{x}\binom{N-M}{n-x}}{\binom{N}{n}}, \quad x = 0, 1, 2, \ldots, n, \quad x \le M, \quad n - x \le N - M,$$

$$n, M, N, \in N, \quad 1 \le n \le N, \quad 1 \le M \le N, \quad N = 1, 2, \ldots,$$

$$\mu = n\frac{M}{N}, \quad \sigma^2 = \left(\frac{N-n}{N-1}\right)n\frac{M}{N}\left(1 - \frac{M}{N}\right), \quad \beta_1 = \frac{(N - 2M)(N - 2n)\sqrt{N-1}}{(N-2)\sqrt{nM(N-M)(N-n)}},$$

$$\beta_2 = \frac{N^2(N-1)}{(N-2)(N-3)nM(N-M)(N-n)},$$

$$\left\{ N(N+1) - 6n(N-n) + 3\frac{M}{N^2}(N-M)[N^2(n-2) - Nn^2 + 6n(N-n)] \right\},$$

$$M(t) = \frac{(N-M)!(N-n)!}{N!}F(., e^t),$$

$$\phi(t) = \frac{(N-M)!(N-n)!}{N!}F(., e^{it}),$$

$$P(t) = \left(\frac{N-M}{N}\right)^n F(., t),$$

where $F(\alpha, \beta, \gamma, x)$ is the hypergeometric function.

$$\alpha = -n; \quad \beta = -M; \quad \gamma = N - M - n + 1.$$

Negative Binomial Distribution

$$p(x) = \binom{x + r - 1}{r - 1} p^r q^x, \quad x = 0, 1, 2, \ldots, \quad r = 1, 2, \ldots, \quad 0 \le p \le 1, \quad q = 1 - p,$$

$$\mu = \frac{rq}{p}, \quad \sigma^2 = \frac{rq}{p^2}, \quad \beta_1 = \frac{2-p}{\sqrt{rq}}, \quad \beta_2 = 3 + \frac{p^2 + 6q}{rq},$$

$$M(t) = \left(\frac{p}{1 - qe^t}\right)^r, \quad \phi(t) = \left(\frac{p}{1 - qe^{it}}\right)^r, \quad P(t) = \left(\frac{p}{1 - qt}\right)^r.$$

Poisson Distribution

$$p(x) = \frac{e^{-\mu}\mu^x}{x!}, \quad x = 0, 1, 2, \ldots, \quad \mu > 0,$$

$$\mu = \mu, \quad \sigma^2 = \mu, \quad \beta_1 = \frac{1}{\sqrt{\mu}}, \quad \beta_2 = 3 + \frac{1}{\mu},$$

$$M(t) = \exp[\mu(e^t - 1)],$$
$$\sigma(t) = \exp[\mu(e^{it} - 1)],$$
$$P(t) = \exp[\mu(t - 1)].$$

Rectangular (Discrete Uniform) Distribution

$$p(x) = \frac{1}{n}, \quad x = 1, 2, \ldots, n, \quad n \in N,$$

$$\mu = \frac{n+1}{2}, \quad \sigma^2 = \frac{n^2 - 1}{12}, \quad \beta_1 = 0, \quad \beta_2 = \frac{3}{5}\left(3 - \frac{4}{n^2 - 1}\right),$$

$$M(t) = \frac{e^t(1 - e^{nt})}{n(1 - e^t)},$$

$$\phi(t) = \frac{e^{it}(1 - e^{nit})}{n(1 - e^{it})},$$

$$P(t) = \frac{t(1 - t^n)}{n(1 - t)}.$$

Continuous Distribution Formulas

Probability density function, $f(x)$
Mean, μ
Variance, σ^2
Coefficient of skewness, β_1
Coefficient of kurtosis, β_2
Moment-generating function, $M(t)$
Characteristic function, $\phi(t)$

Arcsin Distribution

$$f(x) = \frac{1}{\pi\sqrt{x(1-x)}}, \quad 0 < x < 1,$$

$$\mu = \frac{1}{2}, \quad \sigma^2 = \frac{1}{8}, \quad \beta_1 = 0, \quad \beta_2 \frac{3}{2}.$$

Beta Distribution

$$f(x) = \frac{\Gamma(\alpha+\beta)}{\Gamma(\alpha)\Gamma(\beta)} x^{\alpha-1}(1-x)^{\beta-1}, \quad 0 < x < 1, \ \alpha, \beta > 0,$$

$$\mu = \frac{\alpha}{\alpha+\beta}, \quad \sigma^2 = \frac{\alpha\beta}{(\alpha+\beta)^2(\alpha+\beta+1)}, \quad \beta_1 = \frac{2(\beta-\alpha)\sqrt{\alpha+\beta+1}}{\sqrt{\alpha\beta}(\alpha+\beta+2)},$$

$$\beta_2 = \frac{3(\alpha+\beta+1)\left[2(\alpha+\beta)^2 + \alpha\beta(\alpha+\beta-6)\right]}{\alpha\beta(\alpha+\beta+2)(\alpha+\beta+3)}.$$

Cauchy Distribution

$$f(x) = \frac{1}{b\pi(1+((x-a)/b)^2)}, \quad -\infty < x < \infty, \ -\infty < a < \infty, \ b > 0.$$

$\mu, \sigma^2, \beta_1, \beta_2, M(t)$ do not exist and $\phi(t) = \exp[ait - b|t|]$.

Chi Distribution

$$f(x) = \frac{x^{n-1}e^{-x^2/2}}{2^{(n/2)-1}\Gamma(n/2)}, \quad x \geq 0, \ n \in N,$$

$$\mu = \frac{\Gamma((n+1)/2)}{\Gamma(n/2)}, \quad \sigma^2 = \frac{\Gamma((n+2)/2)}{\Gamma(n/2)} - \left[\frac{\Gamma((n+1)/2)}{\Gamma(n/2)}\right]^2.$$

Chi-Square Distribution

$$f(x) = \frac{e^{-x/2}x^{(v/2)-1}}{2^{v/2}\Gamma(v/2)}, \quad x \geq 0, \ v \in N.$$

$$\mu = v, \quad \sigma^2 = 2v, \quad \beta_1 = 2\sqrt{\frac{2}{v}}, \quad \beta_2 = 3 + \frac{12}{v}, \quad M(t) = (1-2t)^{-v/2}, \quad t < \frac{1}{2},$$

$$\phi(t) = (1-2it)^{-v/2}.$$

Erlang Distribution

$$f(x) = \frac{1}{\beta^n(n-1)!} x^{n-1}e^{-x/\beta}, \quad x \geq 0, \ \beta > 0, \ n \in N.$$

$$\mu = n\beta, \quad \sigma^2 = n\beta^2, \quad \beta_1 = \frac{2}{\sqrt{n}}, \quad \beta_2 = 3 + \frac{6}{n}.$$

$$M(t) = (1-\beta t)^{-n}, \quad \phi(t) = (1-\beta it)^{-n}.$$

Exponential Distribution

$$f(x) = \lambda e^{-\lambda x}, \quad x \geq 0, \ \lambda > 0,$$

$$\mu = \frac{1}{\lambda}, \quad \sigma^2 = \frac{1}{\lambda^2}, \quad \beta_1 = 2, \quad \beta_2 = 9, \quad M(t) = \frac{\lambda}{\lambda - t},$$

$$\phi(t) = \frac{\lambda}{\lambda - it}.$$

Extreme-Value Distribution

$$f(x) = \exp[-e^{-(x-\alpha)/\beta}], \quad -\infty < x < \infty, \quad -\infty < \alpha < \infty, \quad \beta > 0,$$

$\mu = \alpha + \gamma\beta, \gamma \doteq 0.5772\ldots$ is Euler's constant $\sigma^2 = \pi^2\beta^2/6.$

$$\beta_1 = 1.29857, \quad \beta_2 = 5.4,$$

$$M(t) = e^{\alpha t}\Gamma(1 - \beta t), \quad t < \frac{1}{\beta}, \quad \phi(t) = e^{\alpha it}\Gamma(1 - \beta it).$$

F Distribution

$$f(x)\frac{\Gamma[(\nu_1 + \nu_2)/2]\nu_1^{\nu_1/2}\nu_2^{\nu_2/2}}{\Gamma(\nu_1/2)\Gamma(\nu_2/2)}x^{(\nu_1/2)-1}(\nu_2 + \nu_1 x)^{-(\nu_1+\nu_2)/2},$$

$x > 0, \quad \nu_1, \nu_2 \in N:$

$$\mu = \frac{\nu_2}{\nu_2 - 2}, \quad \nu_2 \geq 3, \quad \sigma^2 = \frac{2\nu_2^2(\nu_1 + \nu_2 - 2)}{\nu_1(\nu_2 - 2)^2(\nu_2 - 4)}, \quad \nu_2 \geq 5,$$

$$\beta_1 = \frac{(2\nu_1 + \nu_2 - 2)\sqrt{8(\nu_2 - 4)}}{\sqrt{\nu_1}(\nu_2 - 6)\sqrt{\nu_1 + \nu_2 - 2}}, \quad \nu_2 \geq 7,$$

$$\beta_2 = 3 + \frac{12[(\nu_2 - 2)^2(\nu_2 - 4) + \nu_1(\nu_1 + \nu_2 - 2)(5\nu_2 - 22)]}{\nu_1(\nu_2 - 6)(\nu_2 - 8)(\nu_1 + \nu_2 - 2)}, \quad \nu_2 \geq 9.$$

$M(t)$ does not exist.

$$\phi\left(\frac{\nu_1}{\nu_2}t\right) = \frac{G(\nu_1, \nu_2, t)}{B(\nu_1/2, \nu_2/2)}.$$

$B(a, b)$ is the Beta function. G is defined by

$$(m + n - 2)G(m, n, t) = (m - 2)G(m - 2, n, t) + 2itG(m, n - 2, t), \quad m, n > 2,$$

$$mG(m, n, t) = (n - 2)G(m + 2, n - 2, t) - 2itG(m + 2, n - 4, t), \quad n > 4,$$

$$nG(2, n, t) = 2 + 2itG(2, n - 2, t), \quad n > 2.$$

Gamma Distribution

$$f(x) = \frac{1}{\beta^\alpha\Gamma(\alpha)}x^{\alpha-1}e^{-x/\beta}, \quad x \geq 0, \quad \alpha, \beta > 0,$$

$$\mu = \alpha\beta, \quad \sigma^2 = \alpha\beta^2, \quad \beta_1 = \frac{2}{\sqrt{\alpha}}, \quad \beta_2 = 3\left(1 + \frac{2}{\alpha}\right),$$

$$M(t) = (1 - \beta t)^{-\alpha}, \quad \phi(t) = (1 - \beta it)^{-\alpha}.$$

Half-Normal Distribution

$$f(x) = \frac{2\theta}{\pi}\exp\left[-\left(\frac{\theta^2 x^2}{\pi}\right)\right], \quad x \geq 0, \quad \theta > 0,$$

$$\mu = \frac{1}{\theta}, \quad \sigma^2 = \left(\frac{\pi - 2}{2}\right)\frac{1}{\theta^2}, \quad \beta_1 = \frac{4 - \pi}{\theta^3}, \quad \beta_2 = \frac{3\pi^2 - 4\pi - 12}{4\theta^4}.$$

Laplace (Double Exponential) Distribution

$$f(x) = \frac{1}{2\beta} \exp\left[-\frac{|x - \alpha|}{\beta}\right], \quad -\infty < x < \infty, \ -\infty < \alpha < \infty, \ \beta > 0,$$

$$\mu = \alpha, \quad \sigma^2 = 2\beta^2, \quad \beta_1 = 0, \quad \beta_2 = 6,$$

$$M(t) = \frac{e^{\alpha t}}{1 - \beta^2 t^2}, \quad \phi(t) = \frac{e^{\alpha i t}}{1 + \beta^2 t^2}.$$

Logistic Distribution

$$f(x) = \frac{\exp[(x - \alpha)/\beta]}{\beta(1 + \exp[(x - \alpha)/\beta])^2},$$

$$-\infty < x < \infty, \quad -\infty < \alpha < \infty, \quad -\infty < \beta < \infty,$$

$$\mu = \alpha, \quad \sigma^2 = \frac{\beta^2 \pi^2}{3}, \quad \beta_1 = 0, \quad \beta_2 = 4.2,$$

$$M(t) = e^{\alpha t} \pi \beta t \csc(\pi \beta t), \quad \phi(t) = e^{\alpha i t} \pi \beta i t \csc(\pi \beta i t).$$

Lognormal Distribution

$$f(x) = \frac{1}{\sqrt{2\pi}\sigma x} \exp\left[-\frac{1}{2\sigma^2}(\ln x - \mu)^2\right],$$

$$x > 0, \quad -\infty < \mu < \infty, \quad \sigma > 0,$$

$$\mu = e^{\mu + \sigma^2/2}, \quad \sigma^2 = e^{2\mu + \sigma^2}(e^{\sigma^2} - 1),$$

$$\beta_1 = (e^{\sigma^2} + 2)(e^{\sigma^2} - 1)^{1/2}, \quad \beta_2 = (e^{\sigma^2})^4 + 2(e^{\sigma^2})^3 + 3(e^{\sigma^2})^2 - 3.$$

Noncentral Chi-Square Distribution

$$f(x) = \frac{\exp[-(1/2)(x + \lambda)]}{2^{\nu/2}} \sum_{j=0}^{\infty} \frac{x^{(\nu/2)+j-1}\lambda^j}{\Gamma((\nu/2) + j)2^{2j}j!},$$

$$x > 0, \quad \lambda > 0, \quad \nu \in N,$$

$$\mu = \nu + \lambda, \quad \sigma^2 = 2(\nu + 2\lambda), \quad \beta_1 = \frac{\sqrt{8}(\nu + 3\lambda)}{(\nu + 2\lambda)^{3/2}}, \quad \beta_2 = 3 + \frac{12(\nu + 4\lambda)}{(\nu + 2\lambda)^2},$$

$$M(t) = (1 - 2t)^{-\nu/2} \exp\left[\frac{\lambda t}{1 - 2t}\right], \quad \phi(t) = (1 - 2it)^{-\nu/2} \exp\left[\frac{\lambda i t}{1 - 2it}\right].$$

Noncentral F Distribution

$$f(x) = \sum_{i=0}^{\infty} \frac{\Gamma((2i + v_1 + v_2)/2)(v_1/v_2)^{(2i+v_1)/2} x^{(2i+v_1-2)/2} e^{-\lambda/2}(\lambda/2)}{\Gamma(v_2/2)\Gamma((2i+v_1)/2) v_1! (1 + (v_1/v_2)x)^{(2i+v_1+v_2)/2}},$$

$$x > 0, \quad v_1, v_2 \in N, \quad \lambda > 0,$$

$$\mu = \frac{(v_1 + \lambda)v_2}{(v_2 - 2)v_1}, \quad v_2 > 2,$$

$$\sigma^2 = \frac{(v_1 + \lambda)^2 + 2(v_1 + \lambda)v_2^2}{(v_2 - 2)(v_2 - 4)v_1^2} - \frac{(v_1 + \lambda)^2 v_2^2}{(v_2 - 2)^2 v_1^2}, \quad v_2 > 4.$$

Noncentral t-Distribution

$$f(x) = \frac{v^{v/2}}{\Gamma(v/2)} \frac{e^{-\delta^2/2}}{\sqrt{\pi}(v + x^2)^{(v+1)/2}} \sum_{i=0}^{\infty} \Gamma\left(\frac{v + i + 1}{2}\right) \left(\frac{\delta^i}{i!}\right) \left(\frac{2x^2}{v + x^2}\right)^{i/2},$$

$$-\infty < x < \infty, \quad -\infty < \delta < \infty, \quad v \in N,$$

$$\mu'_r = c_r \frac{\Gamma((v - r)/2)v^{r/2}}{2^{r/2}\Gamma(v/2)}, \quad v > r, \quad c_{2r-1} = \sum_{i=1}^{r} \frac{(2r - 1)!\delta^{2r-1}}{(2i - 1)!(r - i)!2^{r-i}},$$

$$c_{2r} = \sum_{i=0}^{r} \frac{(2r)!\delta^{2i}}{(2i)!(r - i)!2^{r-i}}, \quad r = 1, 2, 3, \dots.$$

Normal Distribution

$$f(x) = \frac{1}{\sigma\sqrt{2\pi}} \exp\left[-\frac{(x - \mu)^2}{2\sigma^2}\right],$$

$$-\infty < x < \infty, \quad -\infty < \mu < \infty, \quad \sigma > 0,$$

$$\mu = \mu, \quad \sigma^2 = \sigma^2, \quad \beta_1 = 0, \quad \beta_2 = 3, \quad M(t) = \exp\left[\mu t + \frac{t^2\sigma^2}{2}\right],$$

$$\phi(t) = \exp\left[\mu it - \frac{t^2\sigma^2}{2}\right].$$

Pareto Distribution

$$f(x) = \frac{\theta a^{\theta}}{x^{\theta+1}}, \quad x \geq a, \quad \theta > 0, \quad a > 0,$$

$$\mu = \frac{\theta a}{\theta - 1}, \quad \theta > 1, \quad \sigma^2 = \frac{\theta a^2}{(\theta - 1)^2(\theta - 2)}, \quad \theta > 2.$$

$M(t)$ does not exist.

Rayleigh Distribution

$$f(x) = \frac{x}{\sigma^2} \exp\left[-\frac{x^2}{2\sigma^2}\right], \quad x \geq 0, \ \sigma = 0,$$

$$\mu = \sigma\sqrt{\pi/2}, \quad \sigma^2 = 2\sigma^2\left(1 - \frac{\pi}{4}\right), \quad \beta_1 = \frac{\sqrt{\pi}}{4} \frac{(\pi - 3)}{(1 - \pi/4)^{3/2}},$$

$$\beta_2 = \frac{2 - (3/16)\pi^2}{(1 - \pi/4)^2}.$$

t-Distribution

$$f(x) = \frac{1}{\sqrt{\pi\nu}} \frac{\Gamma((\nu+1)/2)}{\Gamma(\nu/2)} \left(1 + \frac{x^2}{\nu}\right)^{-(\nu+1)/2}, \quad -\infty < x < \infty, \ \nu \in N,$$

$$\mu = 0, \quad \nu \geq 2, \quad \sigma^2 = \frac{\nu}{\nu - 2}, \quad \nu \geq 3, \quad \beta_1 = 0, \quad \nu \geq 4,$$

$$\beta_2 = 3 + \frac{6}{\nu - 4}, \quad \nu \geq 5,$$

$M(t)$ does not exist.

$$\phi(t) = \frac{\sqrt{\pi}\Gamma(\nu/2)}{\Gamma((\nu+1)/2)} \int_{-\infty}^{\infty} \frac{e^{itz\sqrt{\nu}}}{(1 + z^2)^{(\nu+1)/2}} dz.$$

Triangular Distribution

$$f(x) = \begin{cases} 0, & x \leq a, \\ \dfrac{4(x-a)}{(b-a)^2}, & a < x \leq \dfrac{a+b}{2}, \\ \dfrac{4(b-x)}{(b-a)^2}, & \dfrac{a+b}{2} < x < b, \\ 0, & x \geq b, \end{cases}$$

$-\infty < a < b < \infty:$

$$\mu = \frac{a+b}{2}, \quad \sigma^2 = \frac{(b-a)^2}{24}, \quad \beta_1 = 0, \quad \beta_2 = \frac{12}{5},$$

$$M(t) = -\frac{4(e^{at/2} - e^{bt/2})^2}{t^2(b-a)^2}, \quad \phi(t) = \frac{4(e^{ait/2} - e^{bit/2})^2}{t^2(b-a)^2}.$$

Uniform Distribution

$$f(x) = \frac{1}{b-a}, \quad a \leq x \leq b, \ -\infty < a < b < \infty,$$

$$\mu = \frac{a+b}{2}, \quad \sigma^2 = \frac{(b-a)^2}{12}, \quad \beta_1 = 0, \quad \beta_2 = \frac{9}{5},$$

$$M(t) = \frac{e^{bt} - e^{at}}{(b-a)t}, \quad \phi(t) = \frac{e^{bit} - e^{ait}}{(b-a)it}.$$

Weibull Distribution

$$f(x) = \frac{\alpha}{\beta^\alpha} x^{\alpha-1} e^{-(x/\beta)^\alpha}, \quad x \geq 0, \alpha, \beta > 0,$$

$$\mu = \beta \Gamma\left(1 + \frac{1}{\alpha}\right), \quad \sigma^2 = \beta^2 \left[\Gamma\left(1 + \frac{2}{\alpha}\right) - \Gamma^2\left(1 + \frac{1}{\alpha}\right)\right],$$

$$\beta_1 = \frac{\Gamma(1 + 3/\alpha) - 3\Gamma(1 + 1/\alpha)\Gamma(1 + 2/\alpha) + 2\Gamma^3(1 + 1/\alpha)}{[\Gamma(1 + 2/\alpha) - \Gamma^2(1 + 1/\alpha)]^{3/2}},$$

$$\beta_2 = \frac{\Gamma(1 + 4/\alpha) - 4\Gamma(1 + 1/\alpha)\Gamma(1 + 3/\alpha) + 6\Gamma^2(1 + 1/\alpha)\Gamma(1 + 2/\alpha) - 3\Gamma^4(1 + 1/\alpha)}{[\Gamma(1 + 2/\alpha) - \Gamma^2(1 + 1/\alpha)]^2}.$$

Variate Generation Techniques*

Notation

Let $h(t)$ and $H(t) = \int_0^t h(\tau)d\tau$ be the hazard and cumulative hazard functions, respectively, for a continuous nonnegative random variable T, the lifetime of the item under study. The $q \times 1$ vector z contains covariates associated with a particular item or individual. The covariates are linked to the lifetime by the function $\Psi(z)$, which satisfies $\Psi(0 = 1)$ and $\Psi(z) \geq 0$ for all z. A popular choice is $\Psi(z) = e^{\beta'z}$, where β is a $q \times 1$ vector of regression coefficients.

The cumulative hazard function for T in the *accelerated life* model (Leemis, 1987) is

$$H(t) = H_0(t\Psi(z)),$$

where H_0 is a baseline cumulative hazard function. Note that when $z = 0, H_0 \equiv H$. In this model, the covariates accelerate ($\Psi(z) > 1$) or decelerate ($\Psi(z) < 1$), the rate at which the item moves through time. The *proportional* hazards model

$$H(t) = \Psi(z)H_0(t)$$

increases ($\Psi(z) > 1$) or decreases ($\Psi(z) < 1$) the failure rate of the item by the factor $\Psi(z)$ for all values of t.

Variate Generation Algorithms

The literature shows that the cumulative hazard function, $H(T)$, has a unit exponential distribution. Therefore, a random variate t corresponding to a cumulative hazard function

---

* From Leemis, L. M. 1987. Variate generation for accelerated life and proportional hazards models. *Oper Res* 35 (6).

$H(t)$ can be generated by

$$t = H^{-1}(-\log(u)),$$

where u is uniformly distributed between 0 and 1. In the accelerated life model, since time is being expanded or contracted by a factor $\Psi(z)$, variates are generated by

$$t = \frac{H_0^{-1}(-\log(u))}{\Psi(z)}.$$

In the proportional hazards model, equating $-\log(u)$ to $H(t)$ yields the variate generation formula

$$t = H_0^{-1}\left(\frac{-\log(u)}{\Psi(z)}\right).$$

Table 3.2 shows formulas for generating event times from a renewal or nonhomogeneous Poisson process (NHPP). In addition to generating individual lifetimes, these variate generation techniques may also be applied to point processes. A renewal process, for example, with time between events having a cumulative hazard function $H(t)$, can be simulated by using the appropriate generation formula for the two cases just shown. These variate generation formulas must be modified, however, to generate variates from a NHPP.

In an NHPP, the hazard function, $h(t)$, is equivalent to the intensity function, which governs the rate at which events occur. To determine the appropriate method for generating values from an NHPP, assume that the last even in a point process has occurred at time a. The cumulative hazard function for the time of the next event conditioned on survival to time a is

$$H_{T|T>a}(t) = H(t) - H(a), \quad t > a.$$

In the accelerate life model, where $H(t) = H_0(t\Psi(z))$, the time of the next event is generated by

$$t = \frac{H_0^{-1}(H_0(a\Psi(z)) - \log(u))}{\Psi(z)}.$$

If we equate the conditional cumulative hazard function to $-\log(u)$, the time of the next event in the proportional hazards case is generated by

$$t = H_0^{-1}\left(H_0(a) - \frac{\log(u)}{\Psi(z)}\right).$$

TABLE 3.2   Formulas for generating event times from a renewal or NHPP

	Renewal	NHPP
Accelerated life	$t = a + \dfrac{H_0^{-1}(-\log(u))}{\Psi(z)}$	$t = \dfrac{H_0^{-1}(H_0(a\Psi(z)) - \log(u))}{\Psi(z)}$
Proportional hazards	$t = a + H_0^{-1}\left(\dfrac{-\log(u)}{\Psi(z)}\right)$	$t = H_0^{-1}\left(H_0(a) - \dfrac{\log(u)}{\Psi(z)}\right)$

Example

The exponential power distribution (Leemis, 1987) is a flexible two-parameter distribution with cumulative hazard function

$$H(t) = e^{(t/\alpha)^\gamma} - 1, \quad \alpha > 0, \ \gamma > 0, \ t > 0$$

and inverse cumulate hazard function

$$H^{-1}(y) = \alpha[\log(y+1)]^{1/\gamma}.$$

Assume that the covariates are linked to survival by the function $\Psi(z) = e^{\beta'z}$ in the accelerated life model. If an NHPP is to be simulated, the baseline hazard function has the exponential power distribution with parameters α and γ, and the previous event has occurred at time a, then the next event is generated at time

$$t = \alpha e^{-\beta'z}[\log(e^{(ae^{\beta'z}/\alpha)^\gamma} - \log(u))]^{1/\gamma},$$

where u is uniformly distributed between 0 and 1.

References

Johnson, N. L. and Kotz, S. 1970. *Distributions in statistics: Continuous univariate distributions,* John Wiley & Sons, New York, NY.

Leemis, L. M. 1987. Variate generation for accelerated life and proportional hazards models. *Oper Res* 35 (6), 892–894.

4

Computations with Descriptive Statistics

Sample Average

$$\overline{x} = \frac{1}{n} \sum_{i=1}^{n} x_i.$$

Application Areas

Quality control, simulation, facility design, productivity measurement.

Sample calculations

Given:

$x_i : 25, 22, 32, 18, 21, 27, 22, 30, 26, 20$

$n = 10$

$$\sum_{i=1}^{10} x_i = 25 + 22 + 32 + 18 + 21 + 27 + 22 + 30 + 26 + 20 = 243,$$

$$\overline{x} = \frac{243}{10} = 24.30.$$

Sample Variance

$$s^2 = \frac{1}{n-1} \sum_{i=1}^{n} (x_i - \overline{x})^2.$$

Application Areas

1. Quality control
2. Simulation
3. Facility design
4. Productivity measurement

The variance and the closely related standard deviation are measures of the extent of the spread of elements in a data distribution. In other words, they are measures of variability in the data set.

Sample Calculations

Given:

$x_i : 25, 22, 32, 18, 21, 27, 22, 30, 26, 20$

$$n = 10$$

$$n - 1 = 9$$

$$\sum_{i=1}^{10} x_i = 25 + 22 + 32 + 18 + 21 + 27 + 22 + 30 + 26 + 20 = 243,$$

$$\bar{x} = \frac{243}{10} = 24.30,$$

$$s^2 = \frac{1}{9\{(25 - 24.3)^2 + (22 - 24.3)^2 + \cdots + (32 - 24.3)^2\}}$$

$$= \frac{1}{9\{182.10\}} = 20.2333.$$

Alternate Formulas:

$$S^2 = \left[\frac{\sum x_i^2 - \left(\sum x_i\right)^2 / n}{n - 1} \right],$$

$$S^2 = \left[\frac{n \left(\sum x_i^2\right) - \left(\sum x_i\right)^2}{n(n - 1)} \right].$$

Sample Standard Deviation

$$s = \sqrt{s^2}.$$

Application Areas

The standard deviation formula is simply the square root of the variance. It is the most commonly used measure of spread. An important attribute of the standard deviation as a measure of spread is that if the mean and standard deviation of a normal distribution are known, it is possible to compute the percentile rank associated with any given score. In a normal distribution:

- 68.27% of the data is within one standard deviation of the mean.
- 95.46% of the data is within two standard deviations of the mean.
- 99.73% of the data is within 3 standard deviations.
- 99.99% of the data is within 4 standard deviations.
- 99.99985% of the data is within 6 standard deviations (i.e., within six sigma).

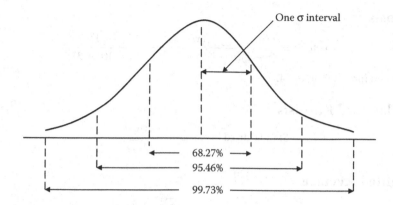

FIGURE 4.1 Deviation spread of normal distribution.

Figure 4.1 illustrates the deviation spread of a normal distribution. The standard deviation is used extensively as a measure of spread because it is computationally simple to understand and use. Many formulas in inferential statistics use the standard deviation.

Sample Standard Error of the Mean

$$s_\mathrm{m} = \frac{s}{\sqrt{n}}.$$

Application Areas

The standard error of the mean is the standard deviation of the sampling distribution of the mean, where s is the standard deviation of the original distribution and n the sample size (the number of data points that each mean is based upon). This formula does not assume a normal distribution. However, many of the uses of the formula do assume a normal distribution. The formula shows that the larger the sample size, the smaller the standard error of the mean. In other words, the size of the standard error of the mean is inversely proportional to the square root of the sample size.

Skewness

$$\text{Skewness} = \frac{n\sum_{i=1}^{n}(x_i - \bar{x})^3}{(n-1)(n-2)s^3}.$$

Undefined for $s = 0$ or $n < 3$.

Standardized Skewness

$$\text{Standardized skewness} = \frac{\text{Skewness}}{\sqrt{6/n}}.$$

Kurtosis

$$\text{Kurtosis} = \frac{n(n+1)\sum_{i=1}^{n}(x_i - \bar{x})^4}{(n-1)(n-2)(n-3)s^4} - \frac{3(n-1)^2}{(n-2)(n-3)}.$$

Undefined for $s = 0$ or $n < 4$.

Standardized Kurtosis

$$\text{Standardized kurtosis} = \frac{\text{Kurtosis}}{\sqrt{24/n}}.$$

Weighted Average

$$\text{Weighted average} = \frac{\sum_{i=2}^{n} x_i w_i}{\sum_{i=1}^{n} w_i}.$$

Estimation and Testing

100(1 − α)% Confidence Interval for Mean

$$CI = \bar{x} \pm t_{n-1;\alpha/2} \frac{s}{\sqrt{n}}.$$

100(1 − α)% Confidence Interval for Variance

$$CI = \left[\frac{(n-1)s^2}{\chi^2_{n-1;\alpha/2}}, \frac{(n-1)s^2}{\chi^2_{n-1;1-\alpha/2}} \right].$$

100(1 − α)% Confidence Interval for Difference in Means

For Equal Variance:

$$CI = (\bar{x}_1 - \bar{x}_2) \pm t_{n_1+n_2-2;\alpha/2}\, s_p \sqrt{\frac{1}{n_1} + \frac{1}{n_2}},$$

where

$$s_p = \sqrt{\frac{(n_1 - 1)s_1^2 + (n_2 - 1)s_2^2}{n_1 + n_2 - 2}}.$$

For Unequal Variance:

$$CI = \left[(\bar{x}_1 - \bar{x}_2) \pm t_{m;\alpha/2} \sqrt{\frac{s_1^2}{n_1} + \frac{s_2^2}{n_2}} \right],$$

where

$$\frac{1}{m} = \frac{c^2}{n_1 - 1} + \frac{(1-c)^2}{n_2 - 1}$$

and

$$c = \frac{s_1^2/n_1}{s_1^2/n_1 + s_2^2/n_1}.$$

100(1 − α)% Confidence Interval for Ratio of Variances

$$CI = \left(\frac{s_1^2}{s_2^2}\right)\left(\frac{1}{F_{n_1-1,n_2-1;\alpha/2}}\right), \left(\frac{s_1^2}{s_2^2}\right)\left(\frac{1}{F_{n_1-1,n_2-1;\alpha/2}}\right).$$

Normal Probability Plot

The input data are first sorted from the smallest to the largest value to compute order statistics. A scatterplot is then generated where the axis positions are computed as follow:

$$\text{Horizontal position} = x_{(i)},$$

$$\text{Vertical position} = \Phi\left(\frac{i-3/8}{n+1/4}\right).$$

The labels for the vertical axis are based upon the probability scale using the following expression:

$$100\left(\frac{i-3/8}{n+1/4}\right).$$

Comparison of Poisson Rates

Let n_j be the number of events in sample j and t_j the length of sample j, then

$$\text{Rate estimates:} \quad r_j = \frac{n_j}{t_j},$$

$$\text{Rate ratio:} \quad \frac{r_1}{r_2},$$

$$\text{Test statistic:} \quad z = \max\left(0, \frac{\left|n_1 - (n_1 + n_2)/2\right| - 1/2}{\sqrt{(n_1 + n_2)/4}}\right),$$

where z follows the standard normal distribution.

Distribution Functions and Parameter Estimation

Bernoulli Distribution

$$\hat{p} = \bar{x}.$$

Binomial Distribution

$$\hat{p} = \frac{\bar{x}}{n},$$

where n is the number of trials.

Discrete Uniform Distribution

$$\hat{a} = \min\, x_i$$

and

$$\hat{b} = \max\, x_i.$$

Geometric Distribution

$$\hat{p} = \frac{1}{1 + \bar{x}}.$$

Negative Binomial Distribution

$$\hat{p} = \frac{k}{\bar{x}},$$

where k is the number of successes.

Poisson Distribution

$$\hat{\beta} = \bar{x}.$$

Beta Distribution

$$\hat{\alpha} = \bar{x}\left[\frac{\bar{x}(1 - \bar{x})}{s^2} - 1\right],$$

$$\hat{\beta} = (1 - \bar{x})\left(\frac{\bar{x}(1 - \bar{x})}{s^2} - 1\right).$$

Chi-Square Distribution

If $X_1, \ldots, X_k$ are k independent, normally distributed random variables with mean 0 and variance 1, then the random variable, Q, defined as follows, is distributed according to the chi-square distribution with k degrees of freedom:

$$Q = \sum_{i=1}^{k} X_i^2.$$

The chi-square distribution is a special case of the gamma distribution and it is represented as

$$Q \sim \chi_k^2.$$

The distribution has one parameter, k, which is a positive integer that specifies the number of degrees of freedom (i.e., the number of X_i's).

Erlang Distribution

$$\hat{\alpha} = \text{round}\,(\hat{\alpha}\ \text{from Gamma}),$$

$$\hat{\beta} = \frac{\hat{\alpha}}{\bar{x}}.$$

Exponential Distribution

$$\hat{\beta} = \frac{1}{\bar{x}} \quad \text{and} \quad \bar{x} = \frac{1}{\hat{\beta}}.$$

Application Areas

Common applications of the exponential distribution are for description of the times between events in a Poisson process, in which events occur continuously and independently at a constant average rate, such as queuing analysis and forecasting.

F Distribution

$$\text{number of d.f.:} \quad \hat{v} = \frac{2\hat{w}^3 - 4\hat{w}^2}{(s^2(\hat{w}-2)^2(\hat{w}-4)) - 2\hat{w}^2},$$

$$\text{den. d.f.:} \quad \hat{w} = \frac{\max(1, 2\bar{x})}{-1 + \bar{x}}.$$

Gamma Distribution

$$R = \log\left(\frac{\text{arithmetic mean}}{\text{geometric mean}}\right).$$

If $0 < R \le 0.5772$, then

$$\hat{\alpha} = R^{-1}(0.5000876 + 0.1648852R - 0.0544274R)^2$$

or if $R > 0.5772$, then

$$\hat{\alpha} = R^{-1}(17.79728 + 11.968477R + R^2)^{-1}(8.898919 + 9.059950R + 0.9775373R^2),$$

$$\hat{\beta} = \frac{\hat{\alpha}}{\bar{x}}.$$

Log–Normal Distribution

$$\hat{\mu} = \frac{1}{n}\sum_{i=1}^{n} \log x_i,$$

$$\hat{\alpha} = \sqrt{\frac{1}{n-1}\sum_{i=1}^{n}(\log x_i - \hat{\mu})^2}.$$

$$\text{Mean:} \quad \exp\left(\hat{\mu} + \frac{\hat{\alpha}^2}{2}\right).$$

$$\text{Standard deviation:} \quad \sqrt{\exp(2\hat{\mu} + \hat{\alpha}^2)[\exp(\hat{\alpha}^2) - 1]}.$$

Normal Distribution

$$\hat{\mu} = \bar{x}$$

and

$$\hat{\sigma} = s.$$

Description of Equation: **Student's *t***

$$s^2 = \frac{\sum_{i=1}^{n} x_i^2}{n},$$

$$\hat{v} = \frac{2s^2}{-1 + s^2}.$$

Triangular Distribution

$$\hat{a} = \min x_i,$$
$$\hat{c} = \max x_i,$$
$$\hat{b} = 3\bar{x} - \hat{a} - \hat{c}.$$

Uniform Distribution

$$\hat{a} = \min x_i,$$

and

$$\hat{b} = \max x_i.$$

Weibull Distribution

$$\hat{\alpha} = \frac{n}{(1/\hat{\beta})^{\hat{\alpha}} \sum_{i=1}^{n} x_i^{\hat{\alpha}} \log x_i - \sum_{i=1}^{n} \log x_i},$$

$$\hat{\beta} = \left(\frac{\sum_{i=1}^{n} x_i^{\hat{\alpha}}}{n} \right)^{1/\hat{\alpha}}.$$

Chi-Square Test for Distribution Fitting

Divide the range of data into nonoverlapping classes. The classes are aggregated at each end to ensure that classes have an expected frequency of at least 5.

Let O_i is the observed frequency in class i, E_i the expected frequency in class i from fitted distribution, and k the number of classes after aggregation.

$$\text{Test statistic:} \quad x^2 = \sum_{i=1}^{k} \frac{(O_i - E_i)^2}{E_i}$$

follow a chi-square distribution with the degrees of freedom equal to $k - 1$ number of estimated parameters.

Kolmogorov–Smirnov Test

$$D_n^+ = \max \left\{ \frac{i}{n} - \hat{F}(x_i) \right\}, \quad 1 \leqslant i \leqslant n,$$

$$D_n^- = \max \left\{ \hat{F}(x_i) - \frac{i-1}{n} \right\}, \quad 1 \leqslant i \leqslant n,$$

$$D_n = \max \left\{ D_n^+, D_n^- \right\},$$

where $\hat{F}(x_i)$ is the estimated cumulative distribution at x_i.

ANOVA

Notation

d.f.   degrees of freedom for the error term $\left(= \left(\sum_{t=1}^{k} n_t \right) - k \right)$

k number of treatments

MSE   mean square error $\left(= \sum_{t=1}^{k} (n_t - 1)s_t^2 / \left(\left(\sum_{t=1}^{k} n_t \right) - k \right) \right)$

$\bar{n}$ average treatment size $\left(= n/k, \text{ where } n = \sum_{t=1}^{k} n_t \right)$

n_t number of observations for treatment t

x_{it} ith observation in treatment i

$\bar{x}_t$ treatment mean $\left(= \sum_{i=1}^{n_t} x_{it} / n_t \right)$

s_t^2 treatment variance $\left(= \sum_{i=1}^{n_t} (x_{it} - \bar{x}_t^2) / (n_t - 1) \right)$

Standard Error

$$\sqrt{\frac{s_t^2}{n_t}}.$$

Description of Equation: **Standard error (pooled)**

Formula:

$$\sqrt{\frac{\text{MSE}}{n_t}}.$$

Interval Estimates

$$\bar{x}_t \pm M \sqrt{\frac{\text{MSE}}{n_t}},$$

where the confidence interval

$$M = t_{n-k;\alpha/2}.$$

Least significant difference interval is

$$M = \frac{1}{\sqrt{2}} t_{n-k;\alpha/2}.$$

Tukey Interval

$$M = \frac{1}{2}q_{n-k,k;\alpha},$$

where $q_{n-k,k;\alpha}$ is the value of the studentized range distribution with $n - k$ degrees of freedom and k samples such that the cumulative probability equals $1 - \alpha$.

Scheffe Interval

$$M = \frac{\sqrt{k-1}}{\sqrt{2}}\sqrt{F_{k-1,n-k;\alpha}}.$$

Cochran C-test

This follows F distribution with $\check{n} - 1$ and $(\check{n} - 1)(k - 1)$ degrees of freedom.

$$\text{Test statistic:} \quad F = \frac{(k-1)C}{1-C},$$

where

$$C = \frac{\max s_t^2}{\sum_{t=1}^{k} s_t^2}.$$

Bartlett Test

$$\text{Test statistic:} \quad B = 10^{M/(n-k)},$$

where

$$M = (n-k)\log_{10} \text{MSE} - \sum_{t=1}^{k}(n_t - 1)\log_{10} s_t^2.$$

The significance test is based on

$$\frac{M(\ln 10)}{1 + \frac{1}{3(k-1)}\left[\sum_{t=1}^{k} 1/(n_t - 1) - 1/N - k\right]^{X_{k-1}^2}},$$

which follows a chi-square distribution with $k - 1$ degrees of freedom.

Hartley's Test

$$H = \frac{\max(s_t^2)}{\min(s_t^2)}.$$

Kruskal–Wallis Test

Average rank of treatment:

$$\overline{R}_t = \frac{\sum_{i=1}^{n_t} R_{it}}{n_t}.$$

If there are no ties, test statistic is

$$w = \left(\frac{12}{n}\sum_{i=1}^{k} n_t \overline{R}_t{}^2\right) - 3(n+1).$$

Adjustment for Ties

Let u_j is the number of observations tied at any rank for $j = 1, 2, 3, \ldots, m$, where m is the number of unique values in the sample.

$$W = \frac{w}{1 - \sum_{j=1}^{m} u_j^3 - \sum_{j=1}^{m} u_j/n(n^2 - 1)}.$$

Significance level: W follows a chi-square distribution with $k - 1$ degrees of freedom.

Freidman Test

Let X_{it} be the observation in the ith row and ith column, where $i = 1, 2, \ldots, n$, $t = 1, 2, \ldots, k$, $R_i t$ the rank of X_{it} within its row, and n the common treatment size (all treatment sizes must be the same for this test), then

$$R_t = \sum_{i=1}^{n} R_{it},$$

and the average rank is

$$\overline{R}_t = \frac{\sum_{i=1}^{n_t} R_{it}}{n_t},$$

where data are ranked within each row separately.

Test statistic: $$Q = \frac{12S(k-1)}{nk(k^2 - 1) - \left(\sum u^3 - \sum u\right)},$$

where

$$S = \left(\sum_{t=1}^{k} R_i^2\right) - \frac{n^2 k(k+1)^2}{4}.$$

Here Q follows a chi-square distribution with k degrees of freedom.

Regression

Notation

Y vector of n observations for the dependent variable

X n by p matrix of observations for p independent variables, including constant term, if any

Then

$$\overline{Y} = \frac{\sum_{i=1}^{n} Y_i}{n}.$$

Description of Equation: **Regression Statistics**

Estimated coefficients

$$b = (X'X)^{-1}XY.$$

Standard errors

$$S(b) = \sqrt{\text{diagonal elements of } (X'X)^{-1} \text{ MSE}},$$

where

$$SSE = Y'Y - b'X'Y \quad \text{and} \quad MSE = \frac{SSE}{n-p}.$$

t-values

$$t = \frac{b}{S(b)}.$$

Significance level: t-Values follow the Student's t distribution with $n - p$ degrees of freedom.

R^2

$$R^2 = \frac{SSTO - SSE}{SSTO},$$

where

$$SSTO = \begin{cases} Y'Y - n\overline{Y}^2 & \text{if constant,} \\ Y'Y & \text{if no constant.} \end{cases}$$

In the case of no constant, the total sum of square is uncorrected for the mean and the R^2 value is of little use, since the sum of the residuals is not zero.

Adjusted R^2

$$1 - \left(\frac{n-1}{n-p} \right)(1 - R^2).$$

Standard error of estimate

$$SE = \sqrt{MSE}.$$

Predicted values

$$\hat{Y} = Xb.$$

Residuals

$$e = Y - \hat{Y}.$$

Durbin–Watson statistic

$$D = \frac{\sum_{i=1}^{n-1} (e_{i+1} - e_i)^2}{\sum_{i=1}^{n} e_i^2}.$$

Mean absolute error

$$\frac{\sum_{i=1}^{n} e_i}{n}.$$

Predictions

Let X be the m by p matrix of independent variables for m predictions.

Predicted value

$$\hat{Y} = Xb.$$

Standard error of predictions

$$S(\hat{Y}) = \sqrt{\text{diagonal elements of MSE } (1 + X(X'X)^{-1}X')}.$$

Standard error of mean response

$$S(Y') = \sqrt{\text{diagonal elements of MSE}(X(X'X)^{-1}X')}.$$

Statistical Quality Control

Let k is the number of subgroups, n_j the number of observations in subgroup j, $j = 1, 2, \ldots, k$, and x_{ij} the ith observation in subgroup j.

Subgroup Statistics

Subgroup means

$$\overline{x}_j = \frac{\sum_{i=1}^{n_j} x_{ij}}{n_j}.$$

Subgroup standard deviations

$$s_j = \sqrt{\frac{\sum_{i=1}^{n_j} (x_{ij} - \overline{x}_j)^2}{(n_j - 1)}}.$$

Subgroup range

$$R_j = \max\{x_{ij} | 1 \leqslant i \leqslant n_j\} - \min\{x_{ij} | 1 \leqslant i \leqslant n_j\}.$$

X-Bar Charts

$$\bar{\bar{x}} = \frac{\sum_{j=1}^{k} n_i \bar{x}_j}{\sum_{j=1}^{k} n_i},$$

$$\bar{R} = \frac{\sum_{j=1}^{k} n_i R_j}{\sum_{j=1}^{k} n_i},$$

$$s_p = \sqrt{\frac{\sum_{j=1}^{k} (n_j - 1)s_j^2}{\sum_{j=1}^{k} (n_j - 1)}},$$

$$\bar{n} = \frac{1}{k} \sum_{j=1}^{k} n_i.$$

Chart based on range

$$\text{UCL} = \bar{\bar{x}} + A_2 \bar{R}$$

and

$$\text{LCL} = \bar{\bar{x}} - A_2 \bar{R}.$$

Chart based on sigma

$$\text{UCL} = \bar{\bar{x}} + \frac{3s_p}{\sqrt{\bar{n}}}$$

$$\text{LCL} = \bar{\bar{x}} - \frac{3s_p}{\sqrt{\bar{n}}}.$$

Chart based on known sigma:

$$\text{UCL} = \bar{\bar{x}} + 3\frac{\sigma}{\sqrt{\bar{n}}}$$

and

$$\text{LCL} = \bar{\bar{x}} - 3\frac{\sigma}{\sqrt{\bar{n}}}.$$

Capability Ratios

$$C_p = \frac{\text{USL} - \text{LSL}}{6\hat{a}},$$

$$C_R = \frac{1}{C_P},$$

$$C_{pk} = \min\left(\frac{\text{USL} - \bar{\bar{x}}}{3\hat{a}}, \frac{\bar{\bar{x}} - \text{LSL}}{3\hat{a}}\right).$$

R Charts

$$CL = \overline{R},$$
$$UCL = D_4\overline{R},$$
$$LCL = \max(0, D_3\overline{R}).$$

S Charts

$$CL = s_p,$$

$$UCL = s_p\sqrt{\frac{\chi^2_{\overline{n}-1;\alpha}}{\overline{n}-1}},$$

$$LCL = s_p\sqrt{\frac{\chi^2_{\overline{n}-1;\alpha}}{\overline{n}-1}}.$$

C Charts

$$\overline{c} = \sum u_j, \quad UCL \qquad\qquad = \overline{c} + 3\sqrt{\overline{c}},$$
$$\sum n_j, \quad LCL \qquad\qquad = \overline{c} - 3\sqrt{\overline{c}},$$

where u_j is the number of defects in the jth sample.

U Charts

$$\overline{u} = \frac{\text{number of defects in all samples}}{\text{number of units in all samples}} = \frac{\sum u_j}{\sum n_j},$$

$$UCL = \overline{u} + \frac{3\sqrt{\overline{u}}}{\sqrt{\overline{n}}},$$

$$LCL = \overline{u} - \frac{3\sqrt{\overline{u}}}{\sqrt{\overline{n}}}.$$

P Charts

$$p = \frac{\text{number of defective units}}{\text{number of units inspected}},$$

$$\overline{p} = \frac{\text{number of defectives in all samples}}{\text{number of units in all samples}} = \frac{\sum p_j n_j}{\sum n_j},$$

$$UCL = \overline{p} + \frac{3\sqrt{\overline{p}(1-\overline{p})}}{\sqrt{\overline{n}}},$$

$$LCL = \overline{p} - \frac{3\sqrt{\overline{p}(1-\overline{p})}}{\sqrt{\overline{n}}}.$$

NP Charts

$$\bar{p} = \frac{\sum d_j}{\sum n_j},$$

where d_j is the number of defectives in the jth sample.

$$\text{UCL} = \overline{np} + 3\sqrt{\overline{np}(1 - \bar{p})}$$

and

$$\text{LCL} = \overline{np} - 3\sqrt{\overline{np}(1 - \bar{p})}.$$

CuSum Chart for the Mean

$$\text{Control mean} = \mu$$
$$\text{Standard deviation } = \alpha$$
$$\text{Difference to detect} = \Delta$$

Plot cumulative sums C_t versus t, where

$$C_t = \sum_{i=1}^{t} (\bar{x}_i - \mu) \quad \text{for } t = 1, 2, \dots, n.$$

The V-mask is located at distance

$$d = \frac{2}{\Delta} \left[\frac{\alpha^2/\overline{n}}{\Delta} \ln \frac{1 - \beta}{\alpha/2} \right]$$

in front of the last data point.

$$\text{Angle of mast} = 2 \tan^{-1} \frac{\Delta}{2}.$$

$$\text{Slope of the lines} = \pm \frac{\Delta}{2}.$$

Time-Series Analysis

Notation

x_t or y_t observation at time t, $t = 1, 2, \dots, n$
n number of observations

Autocorrelation at Lag k

$$r_k = \frac{c_k}{c_0},$$

where

$$c_k = \frac{1}{n} \sum_{t=1}^{n-k} (y_t - \bar{y})(y_{t+k} - \bar{y})$$

and

$$\bar{y} = \frac{\sum_{t=1}^{n} y_t}{n}.$$

$$\text{Standard error} = \sqrt{\frac{1}{n} \left\{ 1 + 2 \sum_{v=1}^{k-1} r_v^2 \right\}}.$$

Partial Autocorrelation at Lag k

$\hat{\theta}_{kk}$ is obtained by solving the following equation:

$$r_j = \hat{\theta}_{k1} r_{j-1} + \hat{\theta}_{k2} r_{j-2} + \cdots + \hat{\theta}_{k(k-1)} r_{j-k+1} + \hat{\theta}_{kk} r_{j-k},$$

where $j = 1, 2, \ldots, k$.

$$\text{Standard error} = \sqrt{1n}.$$

Cross-Correlation at Lag k

Let x be the input time series and y the output time series, then

$$r_{xy}(k) = \frac{c_{xy}(k)}{s_x s_y}, \quad k = 0, \pm 1, \pm 2, \ldots,$$

where

$$c_{xy}(k) = \begin{cases} \dfrac{1}{n} \displaystyle\sum_{t=1}^{n-k} (x_t - \bar{x})(y_{t+k} - \bar{y}), & k = 0, 1, 2, \ldots, \\ \dfrac{1}{n} \displaystyle\sum_{t=1}^{n+k} (x_t - \bar{x})(y_{t-k} - \bar{y}), & k = 0, -1, -2, \ldots \end{cases}$$

and

$$S_x = \sqrt{c_{xx}(0)},$$
$$S_y = \sqrt{c_{yy}(0)},$$

Box-Cox Computation

$$yt = \frac{(y + \lambda_2)^{\lambda_1 - 1}}{\lambda_1 g^{(\lambda_1 - 1)}} \quad \text{if } \lambda_1 > 0.$$

$$yt = g \ln(y + l_2) \quad \text{if } \lambda_1 = 0,$$

where g is the sample geometric mean $(y + \lambda_2)$.

Periodogram (Computed using Fast Fourier Transform)

If n is odd:

$$I(f_1) = \frac{n}{2}(a_i^2 + b_i^2), \quad i = 1, 2, \ldots, \left[\frac{n-1}{2}\right],$$

where

$$a_i = \frac{2}{n}\sum_{t=1}^{n} t_t \cos 2\pi f_i t,$$

$$b_i = \frac{2}{n}\sum_{t=1}^{n} y_t \sin 2\pi f_i t.$$

$$f_i = \frac{i}{n}.$$

If n is even, an additional term is added:

$$I(0.5) = n\left(\frac{1}{n}\sum_{t=1}^{n}(-1)^t Y_t\right)^2.$$

Categorical Analysis

Notation

c number of columns in table
f_{ij} frequency in position (row i, column j)
r number of rows in table
x_i distinct values of row variable arranged in ascending order; $i = 1, \ldots, r$
y_j distinct values of column variable arranged in ascending order, $j = 1, \ldots, c$

Totals

$$R_j = \sum_{j=1}^{c} f_{ij}, \qquad C_j = \sum_{i=1}^{\gamma} f_{ij},$$

$$N = \sum_{i=1}^{r}\sum_{j=1}^{c} f_{ij}.$$

Chi-Square

$$\chi^2 = \sum_{i=1}^{r}\sum_{j=1}^{c} \frac{(f_{ij} - E_{ij})^2}{E_{ij}},$$

where

$$E_{ij} = \frac{R_i C_j}{N} \sim \chi^2_{(r-1)(c-1)}$$

Lambda

$$\lambda = \frac{\sum_{j=1}^{c} f_{\max,j} - R_{\max}}{N - R_{\max}}$$

with rows dependent.

$$\lambda = \frac{\sum_{i=1}^{\gamma} f_{i,\max} - C_{\max}}{N - C_{\max}}$$

with columns dependent.

$$\lambda = \frac{\sum_{i=1}^{\gamma} f_{i,\max} + \sum_{j=1}^{c} f_{\max,j} - C_{\max} - R_{\max}}{2N - R_{\max} - C_{\max}}$$

when symmetric, where $f_{i\,\max}$ is the largest value in row i, $f_{\max j}$ the largest value in column j, $R_{\max}$ the largest row total, and $C_{\max}$ the largest column total.

Uncertainty Coefficient

$$U_R = \frac{U(R) + U(C) - U(RC)}{U(R)}$$

with rows dependent,

$$U_C = \frac{U(R) + U(C) - U(RC)}{U(C)}$$

with columns dependent, and

$$U = 2\left(\frac{U(R) + U(C) - U(RC)}{U(R) + U(C)}\right)$$

when symmetric, where

$$U(R) = -\sum_{i=1}^{r} \frac{R_i}{N} \log \frac{R_i}{N},$$

$$U(C) = -\sum_{j=1}^{c} \frac{C_j}{N} \log \frac{C_j}{N},$$

$$U(RC) = -\sum_{i=1}^{r} \sum_{j=1}^{c} \log \frac{f_{ij}}{N} \quad \text{for } f_{ij} > 0,$$

Somer's D Measure

$$D_R = \frac{2(P_c - P_D)}{N^2 - \sum_{j=1}^{c} C_j^2}$$

with rows dependent,

$$D_C = \frac{2(P_c - P_D)}{N^2 - \sum_{i=1}^{r} R_i^2}$$

with columns dependent, and

$$D = \frac{4(P_C - P_D)}{\left(N^2 - \sum_{i=1}^{r} R_i^2\right) + \left(N^2 - \sum_{j=1}^{c} C_j^2\right)}$$

when symmetric.

Where the number of concordant pairs is

$$P_C = \sum_{i=1}^{r} \sum_{j=1}^{c} f_{ij} \sum \sum_{h<i\ k<j} f_{hk}.$$

and the number of discordant pairs is

$$P_D = \sum_{i=1}^{r} \sum_{j=1}^{c} f_{ij} \sum \sum_{h<i\ k>j} f_{hk}.$$

Eta

$$E_R = \sqrt{1 - \frac{SS_{RN}}{SS_R}}$$

with rows dependent, where the total corrected sum of squares for the rows is

$$SS_R = \sum_{i=1}^{r} \sum_{j-1}^{c} x_i^2 f_{ij} - \frac{\left(\sum_{i=1}^{r} \sum_{j-1}^{c} x_i f_{ij}\right)^2}{N}$$

and the sum of squares of rows within categories of columns is

$$SS_{RN} = \sum_{j=1}^{c} \left(\sum_{i=1}^{r} x_i^2 f_{ij} - \frac{\left(\sum_{i=1}^{r} x_i^2 f_{ij}\right)^2}{C_j}\right),$$

$$E_C = \sqrt{1 - \frac{SS_{CN}}{SS_C}}$$

with columns dependent.

Where the total corrected sum of squares for the columns is

$$SS_C = \sum_{i=1}^{r}\sum_{j=1}^{c} x_i^2 f_{ij} - \frac{\left(\sum_{i=1}^{r}\sum_{j=1}^{c} y_i f_{ij}\right)^2}{N}$$

and the sum of squares of columns within categories of rows is

$$SS_{CN} = \sum_{i=1}^{r}\left(\sum_{j=1}^{c} y_i^2 f_{ij} - \frac{\left(\sum_{j=1}^{c} y_j^2 f_{ij}\right)^2}{R_i}\right)j.$$

Contingency Coefficient

$$C = \sqrt{\frac{\chi^2}{\chi^2 + N}}$$

Cramer's *V* Measure

$$V = \sqrt{\frac{\chi^2}{N}}$$

for 2 × 2 table.

$$V = \sqrt{\frac{\chi^2}{N(m-1)}}$$

for all others where $m = \min(r, c)$.

Conditional Gamma

$$G = \frac{P_C - P_D}{P_C + P_D}$$

Pearson's *R* Measure

$$R = \frac{\sum_{j=1}^{c}\sum_{i=1}^{r} x_i y_i f_{ij} - \left(\sum_{j=1}^{c}\sum_{i=1}^{r} x_i f_{ij}\right)\left(\sum_{j=1}^{c}\sum_{i=1}^{r} y_i f_{ij}\right)/N}{\sqrt{SS_R SS_C}}$$

If $R = 1$, no significance is printed. Otherwise, the one-sided significance is based on

$$t = R\sqrt{\frac{N-2}{1-R^2}}.$$

Kendall's Tau *b* Measure

$$\tau = \frac{2(P_C - P_D)}{\sqrt{(N^2 - \sum_{i=1}^{r} R_i^2)(N^2 - \sum_{j=1}^{c} C_j^2)}}$$

Tau C Measure

$$\tau_C = \frac{2m(P_C - P_D)}{(m-1)N^2}.$$

Overall Mean

$$\bar{x} = \frac{n_1\bar{x}_1 + n_2\bar{x}_2 + n_3\bar{x}_3 + \cdots + n_k\bar{x}_k}{n_1 + n_2 + n_3 + \cdots + n_k} = \frac{\sum n\bar{x}}{\sum n}.$$

Chebyshev's Theorem

$$1 - \frac{1}{k^2}.$$

Permutation

$$P_\gamma^n = \frac{n!}{(n-r)!}.$$

!! = factorial operation:

$$n! = n(n-1)(n-2)\cdots(3)(2)(1).$$

Combination

$$C_r^n = \frac{n!}{r!(n-r)!}.$$

Failure

$$q = 1 - p = \frac{n-s}{n}.$$

5

Computations for Economic Analysis

Fundamentals of Economic Analysis

Simple Interest

Interest is the fee paid for the use of someone else's money. Simple interest is interest paid only on the amount deposited and not on past interest.

The formula for simple interest is

$$I = Prt, \tag{5.1}$$

where I is the interest, P the principal amount, r the interest rate in percent/year, and t the time in years.

Example

Find the simple interest for $1500 at 8% for 2 years.

SOLUTION

$$P = \$1500, \quad r = 8\% = 0.08, \quad \text{and} \quad t = 2 \text{ years,}$$
$$I = I = Prt = 1500 \times 0.08 \times 2 = 240 \text{ or } \$240.$$

Future Value

If P dollars are deposited at interest rate r for t years, the money earns interest. When this interest is added to the initial deposit, the total amount in the account is

$$A = P + I = P + Ptr = P(I + rt). \tag{5.2}$$

This amount is called the future value or maturity value.

Example

Find the maturity value of $10,000 at 8% for 6 months.

SOLUTION

$$P = \$10,000, \quad r = 8\% = 0.08, \quad t = 6/12 = 0.5 \text{ year.}$$

The maturity value is

$$A = P(1 + rt) = 10,000[1 + 0.08(0.5)] = 10,400 \text{ or } \$10,400.$$

Compound Interest

Simple interest is normally used for loans or investment of a year or less. For longer period, compound interest is used. The compound amount at the end of t years is given by the compound interest formula

$$A = P(1 + i)^n, \tag{5.3}$$

where i is the interest rate per compounding period $(= r/m)$, n the number of conversion periods for t years $(= mt)$, A the compound amount at the end of n conversion period, P the principal amount, r the nominal interest per year, m the number of conversion periods per year, and t the term (number of years).

Example

Suppose $15,000 is deposited at 8% and compounded annually for 5 years. Find the compound amount.

SOLUTION

$$P = \$15,000, \quad r = 8\% = 0.08, \quad m = 1, \quad n = 5,$$

$$A = P(1 + i)^n = 1500\left[1 + \left(\frac{0.08}{1}\right)\right]^5 = 15,000[1.08]^5$$

$$= 22039.92 \quad \text{or} \quad \$22,039.92.$$

Continuous Compound Interest

The compound amount A for a deposit of P at an interest rate r per year compounded continuously for t years is given by

$$A = Pe^{rt}, \tag{5.4}$$

where P is the principal amount, r the annual interest rate compounded continuously, t the time in years, A the compound amount at the end of t years, and $e = 2.7182818$.

Effective Rate

The effective rate is the simple interest rate that would produce the same accumulated amount in 1 year as the nominal rate compounded m times a year.

The formula for effective rate of interest is

$$r_{\text{eff}} = \left(1 + \frac{r}{m}\right)^m - 1, \tag{5.5}$$

where r_{eff} is the effective rate of interest, r the nominal interest rate per year, and m the number of conversion periods per year.

Sample

Find the effective rate of interest corresponding to a nominal rate of 8% compounded quarterly.

SOLUTION

$$r = 8\% = 0.08, \quad m = 4,$$

then

$$r_{\text{eff}} = \left(1 + \frac{r}{m}\right)^m - 1 = \left(1 + \frac{0.08}{4}\right)^4 - 1 = 0.082432.$$

So the corresponding effective rate in this case is 8.243% per year.

Present Value with Compound Interest

The principal amount P is often referred to as the present value and the accumulated value A is called the future value because it is realized at a future date. The present value is given by

$$P = \frac{A}{(1+i)^n} = A(1+i)^{-n}. \tag{5.6}$$

Example

How much money should be deposited in a bank paying interest at the rate of 3% per year compounding monthly so that at the end of 5 years the accumulated amount will be $15,000?

SOLUTION

Here:

- nominal interest per year $r = 3\% = 0.03$
- number of conversion per year $m = 12$
- interest rate per compounding period $i = 0.03/12 = 0.0025$

- number of conversion periods for t years $n = (5)(12) = 60$
- accumulated amount $A = 15{,}000$

$$P = A(1 + i)^{-n} = 15{,}000(1 + 0.0025)^{-60},$$

$$P = 12{,}913.03 \quad \text{or} \quad \$12{,}913.$$

Annuities

An annuity is a sequence of payments made at regular time intervals. This is the typical situation in finding the relationship between the amount of money loaned and the size of the payments.

Present Value of Annuity

The present value P of an annuity of n payments of R dollars each, paid at the end of each investment period into an account that earns interest at the rate of i per period, is

$$P = R\left[\frac{1 - (1 + i)^{-n}}{i}\right], \qquad (5.7)$$

where P is the present value of annuity, R the regular payment per month, n the number of conversion periods for t years, and i the annual interest rate.

Sample

What size loan could Bob get if he can afford to pay $1000 per month for 30 years at 5% annual interest?
Here:

$$R = 1000, \quad i = 0.05/12 = 0.00416, \quad n = (12)(30) = 360,$$

$$P = R\left[\frac{1 - (1 + i)^{-n}}{i}\right] = 1000\left[\frac{1 - (1 + 0.00416)^{-360}}{0.00416}\right],$$

$$P = 186579.61 \quad \text{or} \quad \$186{,}576.61.$$

Under these terms, Bob would end up paying a total of $360,000, so the total interest paid would be $360,000 to $186,579.61 = $173,420.39.

Future Value of an Annuity

The future value S of an annuity of n payments of R dollars each, paid at the end of each investment period into an account that earns interest at the rate of i per period is

$$S = R\left[\frac{(1 + i)^n - 1}{i}\right]. \qquad (5.8)$$

Example

Let us consider the future value of $1000 paid at the end of each month into an account paying 8% annual interest for 30 years. How much will accumulate?

SOLUTION

This is a future value calculation with $R = 1000$, $n = 360$, and $i = 0.05/12 = 0.00416$. This account will accumulate as follows:

$$S = R\left[\frac{(1+i)^n - 1}{i}\right] = 1000\left[\frac{(1+0.00416)^{360} - 1}{0.00416}\right],$$

where

$$S = 831028.59 \quad \text{or} \quad \$831,028.59.$$

Note: This is much larger than the sum of the payments, because many of those payments are earning interest for many years.

Amortization of Loans

The periodic payment R_a on a loan of P dollars to be amortized over n periods with interest charge at the rate of i per period is

$$R_a = \frac{Pi}{1 - (1+i)^{-n}}. \tag{5.9}$$

Example

Bob borrowed $120,000 from a bank to buy the house. The bank charges interest at a rate of 5% per year. Bob has agreed to repay the loan in equal monthly installments over 30 years. How much should each payment be if the loan is to be amortized at the end of the time?

SOLUTION

This is a periodic payment calculation with

$$P = 120,000, \quad i = 0.05/12 = 0.00416, \quad \text{and} \quad n = (30)(12) = 360.$$

$$R_a = \frac{Pi}{1 - (1+i)^{-n}} = \frac{(120,000)(0.00416)}{1 - (1.00416)^{-360}} = 643.88 \quad \text{or} \quad \$643.88.$$

Interest and Equity Computations

It is always of interest to calculate the portion or percentage of annuity payment going into equity accrual and interest charge, respectively. The computational procedure follows the steps outlined below:

1. Given a principal amount, P, a periodic interest rate, i (in decimals), and a discrete time span of n periods, the uniform series of equal end-of-period *annuity* payments needed to amortize P is computed as

$$A = \frac{P[i(1+i)^n]}{(1+i)^n - 1}.$$

   It is assumed that the loan is to be repaid in equal monthly payments. Thus, $A(t) = A$, for each period t throughout the life of the loan.

2. The *unpaid balance* after making t installment payments is given by

$$U(t) = \frac{A[1 - (1+i)^{(t-n)}]}{i}.$$

3. The amount of *equity* or principal amount paid with installment payment number t is given by

$$E(t) = A(1+i)^{t-n-1}.$$

4. The amount of *interest charge* contained in installment payment number t is derived to be

$$I(t) = A[1 - (1+i)^{t-n-1}],$$

   where $A = E(t) + I(t)$.

5. The *cumulative total payment* made after t periods is denoted by

$$C(t) = \sum_{k=1}^{t} A(k)$$

$$= \sum_{k=1}^{t} A = (A)(t).$$

6. The *cumulative interest payment* after t periods is given by

$$Q(t) = \sum_{x=1}^{t} I(x).$$

7. The *cumulative principal payment* after t periods is computed as

$$S(t) = \sum_{k=1}^{t} E(k)$$

$$= A \sum_{k=1}^{t} (1+i)^{-(n-k+1)}$$

$$= A \left[\frac{(1+i)^t - 1}{i(1+i)^n} \right],$$

where

$$\sum_{n=1}^{t} x^n = \frac{x^{x+1} - x}{x - 1}.$$

8. The *percentage of interest charge* contained in installment payment number t is

$$f(t) = \frac{I(t)}{A}(100\%).$$

9. The *percentage of cumulative interest charge* contained in the cumulative total payment up to and including payment number t is

$$F(t) = \frac{Q(t)}{C(t)} \times 100.$$

10. The *percentage of cumulative principal payment* contained in the cumulative total payment up to and including payment number t is

$$\begin{aligned}
H(t) &= \frac{S(t)}{C(t)} \\
&= \frac{C(t) - Q(t)}{C(t)} \\
&= 1 - \frac{Q(t)}{C(t)} \\
&= 1 - F(t).
\end{aligned}$$

Example

Suppose a manufacturing productivity improvement project is to be financed by borrowing $500,000 from an industrial development bank. The annual nominal interest rate for the loan is 10%. The loan is to be repaid in equal monthly installments over a period of 15 years. The first payment on the loan is to be made exactly 1 month after financing is approved.

The tabulated result shows a monthly payment of $5373.03 on the loan. If time $t = 10$ months, one can see the following results:

$U(10) = \$487,473.83$ (unpaid balance)
$A(10) = \$5373.03$ (monthly payment)
$E(10) = \$1299.91$ (equity portion of the tenth payment)
$I(10) = \$4073.11$ (interest charge contained in the tenth payment)
$C(10) = \$53,730.26$ (total payment to date)
$S(10) = \$12,526.17$ (total equity to date)
$f(10) = 75.81\%$ (percentage of the tenth payment going into interest charge)
$F(10) = 76.69\%$ (percentage of the total payment going into interest charge)

Thus, over 76% of the sum of the first 10 installment payments goes into interest charges. The analysis shows that by time $t = 180$, the unpaid balance has been reduced to zero. That is, $U(180) = 0.0$. The total payment made on the loan is \$967,144.61 and the total interest charge is \$967,144.61 − \$500,000 = \$467,144.61. So, 48.30% of the total payment goes into interest charges. The information about interest charges might be very useful for tax purposes. The tabulated output shows that equity builds up slowly, whereas unpaid balance decreases slowly. Note that very little equity is accumulated during the first 3 years of the loan schedule. The effects of inflation, depreciation, property appreciation, and other economic factors are not included in the analysis presented above, but the decision analysis should include such factors whenever they are relevant to the loan situation.

Equity Break-Even Formula

The point at which the curves intersect is referred to as the *equity break-even point*. It indicates when the unpaid balance is exactly equal to the accumulated equity or the cumulative principal payment. For the example, the equity break-even point is approximately 120 months (10 years). The importance of the equity break-even point is that any equity accumulated after that point represents the amount of ownership or equity that the debtor is entitled to after the unpaid balance on the loan is settled with project collateral. The implication of this is very important, particularly in the case of mortgage loans. "Mortgage" is a word of French origin, meaning "death pledge," which, perhaps, is an ironic reference to the burden of mortgage loans. The equity break-even point can be calculated directly from the formula derived below.

Let the equity break-even point, x, be defined as the point where $U(x) = S(x)$. That is,

$$A\left[\frac{1-(1+i)^{-(n-x)}}{i}\right] = A\left[\frac{(1+i)^x - 1}{i(1+i)^n}\right].$$

Multiplying both the numerator and denominator of the left-hand side of the above expression by $(1+i)^n$ and then simplifying yields

$$\frac{(1+i)^n - (1+i)^x}{i(1+i)^n}$$

on the left-hand side. Consequently, we have

$$(1+i)^n - (1+i)^x = (1+i)^x - 1,$$

$$(1+i)^x = \frac{(1+i)^n + 1}{2},$$

which yields the equity break-even formula:

$$x = \frac{\ln[0.5(1+i)^n + 0.5]}{\ln(1+i)},$$

where ln is the natural log function, n is the number of periods in the life of the loan, and i is the interest rate per period.

Sinking Fund Payment

The sinking fund calculation is used to compute the periodic payments that will accumulate to yield a specific value at a future point in time, so that investors can be certain that the funds will be available at maturity.

The periodic payment R required to accumulate a sum of S dollars over n periods, with interest charged at the rate of i per period, is

$$R = \frac{iS}{(1+i)^n - 1}, \tag{5.10}$$

where S is the future value, i the annual interest rate, and n the number of conversion periods for t years.

Internal Rate of Return

If we let i^* denote the internal rate of return, then we have

$$FW_{t=n} = \sum_{t=0}^{n} (\pm A_t)(1+i^*)^{n-t} = 0,$$

$$PW_{t=0} = \sum_{t=0}^{n} (\pm A_t)(1+i^*)^{-t} = 0,$$

where the "+" sign is used in the summation for positive cash flow amounts or receipts and the "−" sign is used for negative cash flow amounts or disbursements. A_t denotes the cash flow amount at time t, which may be a receipt ($+$) or a disbursement ($-$). The value of i^* is referred to as *internal rate of return*.

Benefit–Cost Ratio

$$B/C = \frac{\sum_{t=0}^{n} B_t(1+i)^{-t}}{\sum_{t=0}^{n} C_t(1+i)^{-t}}$$

$$= \frac{PW_{\text{benefits}}}{PW_{\text{costs}}} = \frac{AW_{\text{benefits}}}{AW_{\text{costs}}},$$

where B_t is the benefit (receipt) at time t and C_t is the cost (disbursement) at time t.

Simple Payback Period

The payback period is defined as the smallest value of n ($n_{\min}$) that satisfies the following expression:

$$\sum_{t=1}^{n_{\min}} R_t \geq C,$$

where R_t is the revenue at time t and C_0 the initial investment.

Discounted Payback Period

Discounted payback period is

$$\sum_{t=1}^{n_{\min}} R_t(1+i)^{n_{\min}-1} \geq \sum_{t=0}^{n_{\min}} C_t.$$

Economic Methods of Comparing Investment Alternatives

Present Value Analysis

The guidelines for using the present value analysis for investment alternatives are:

- *For one alternative:* Calculate Net Present Value (NPV) at the minimum attractive rate of return (MARR). If NPV $\geq$ 0, the requested MARR is met or exceeded and the alternative is economically viable.
- *For two or more alternatives:* Calculate the NPV of each alternative at the MARR. Select the alternative with the *numerically largest* NPV value. Numerically largest indicates a lower NPV of cost cash flows (less negative) or larger NPV of net cash flows (more positive).
- *For independent projects:* Calculate the NPV of each alternative. Select all projects with NPV $\geq$ 0 at the given MARR.

Let

- P cash flow value at the present time period; this usually occurs at time 0
- F cash flow value at some time in the future
- A series of equal, consecutive, and end-of-period cash flow; this is also called annuity
- G a uniform arithmetic gradient increase in period-by-period cash flow
- n the total number of time periods, which can be in days, weeks, months, or years
- i interest rate per time period expressed as a percentage
- Z face, or par, value of a bond
- C redemption or disposal price (usually equal to Z)
- r bond rate (nominal interest rate) per interest period
- NCF_t estimated net cash flow for each year t
- NCF_A estimated equal amount net cash flow for each year
- n_p discounted payback period

The general equation for the present value analysis is

$$NPV = A_0 + A(P/A, i, n) + F(P/F, i, n) + \cdots + [A_1(P/A, i, n)$$
$$+ G(P/G, i, n)] + A_1(P/A, g, i, n). \tag{5.11}$$

Annual Value Analysis

The net annual value (NAV) method selects the viable alternative(s) based on the following guidelines:

- *One alternative:* Select alternative with NAV $\geq$ 0 since MARR is met or exceeded.

- *Two or more alternatives:* Choose alternative with the lowest cost or the highest revenue NAV value.

Let

CR   capital recovery component.
 A   annual amount component of other cash flows
 P   initial investment (first cost) of all assets
 S   estimated salvage value of the assets at the end of their useful life
 i   investment interest rate

Therefore, the general equation for the annual value analysis is

$$NAV = -CR - A$$
$$= -[P(A/P, i, n) - S(A/F, i, n)] - A. \qquad (5.12)$$

Internal Rate of Return Analysis

Let

 NPV   present value
EUAB   equivalent uniform annual benefits
EUAC   equivalent uniform annual costs

If i^* denote the internal rate of return, then the unknown interest rate can be solved for using any of the following expressions:

$$PW(benefits) - PW(costs) = 0,$$
$$EUAB - EUAC = 0. \qquad (5.13)$$

The procedure for selecting the viable alternative(s) is:

- If $i^* \geq$ MARR, accept the alternative as economically viable project
- If $i^* <$ MARR, the alternative is not economically viable

External Rate of Return Analysis

The expression for calculating ERR is given by

$$F = P(1 + i')^n. \qquad (5.14)$$

Incremental Analysis

The steps involve in using incremental analysis are:

1. If IRR (B/C ratio) for each alternative is given, reject all alternatives with IRR < MARR (B/C < 1.0)
2. Arrange other alternatives in increasing order of initial cost (total costs)
3. Compute incremental cash flow pairwise starting with the first two alternatives

4. Compute incremental measures of value based on the appropriate equations
5. Use the following criteria for selecting the alternatives that will advance to the next stage of comparisons:
    a. If ΔIRR $\geq$ MARR, select higher cost alternative
    b. If ΔB/C $\geq$ 1.0, select higher cost alternative
6. Eliminate the defeated alternative and repeat Steps 3 through 5 for the remaining alternatives
7. Continue until only one alternative remains. This last alternative is the most economically viable alternative

Guidelines for Comparison of Alternatives

- Total cash-flow approach (ranking approach):
    a. Use the individual cash flow for each alternative to calculate the following measures of worth: PW, AW, IRR, ERR, payback period, or B/C ratio
    b. Rank the alternatives on the basis of the measures of worth
    c. Select the highest-ranking alternative as the *preferred* alternative
- Incremental cash-flow approach:
    Find the incremental cash flow needed to go from a "lower cost" alternative to a "higher cost" alternative. Then calculate the measures of worth for the incremental cash flow. The incremental measures of worth are denoted as:

$$\Delta PW, \Delta AW, \Delta IRR, \Delta ERR, \Delta PB, \Delta B/C$$

Step 1. Arrange alternatives in increasing order of initial cost
Step 2. Compute incremental cash flow pairwise starting with the first two alternatives
Step 3. Compute incremental measures of worth as explained above
Step 4. Use the following criteria for selecting the alternatives that will advance to the next stage of comparisons.
- If ΔPW $\geq$ 0, select higher cost alternative
- If ΔAW $\geq$ 0, select higher cost alternative
- If ΔFW $\geq$ 0, select higher cost alternative
- If ΔIRR $\geq$ MARR, select higher cost alternative
- If ΔERR $\geq$ MARR, select higher cost alternative
- If ΔB/C ratio $\geq$ 1.0, select higher cost alternative

Step 5. Eliminate the defeated alternative and repeat Steps 2 through 4 for the remaining alternatives. Continue until only one alternative remains. This last alternative is the *preferred* alternative.

Asset Replacement and Retention Analysis

Several factors are responsible for evaluating the replacement of an asset including:

- Deteriorating factor: Changes that occur in the physical condition of an asset as a result of aging, unexpected accident, and other factors that affect the physical condition of the asset.

- Requirements factor: Changes in production plans that affect the economics of use of the asset. For example, the production capacity of an asset may become smaller as a result of plant expansion.
- Technological factor: The impact of changes in technology can also be a factor. For example, the introduction of flash discs affects assets that are used for producing zip discs.
- Financial factor: For example, the lease of an asset may become more attractive than ownership, or outsourcing the production of a product may become more attractive than in-house production.

Replacement analysis involves several terms, including:

- *Defender:* This is the currently installed asset being considered for replacement.
- *Challenger:* This is the potential replacement.
- *Defender first cost:* The current market value of the defender is the correct estimate for this term in the replacement study. However, if the defender must be upgraded to make it equivalent to the challenger, this cost of such upgrade is added to the market value to obtain the correct estimate for this term.
- *Challenger first cost:* This is the amount that must be recovered when replacing a defender with a challenger. This may be equal to the first cost of the challenger. However, if trade-in is involved, this will be the first cost less the difference between the trade-in value and the market value of the defender. For example, let us assume we bought an asset (the defender) 5 years ago for $60,000 and a fair market value for it today is $30,000. In addition, a new asset can be purchased for $50,000 today and the seller offers a trade-in of $40,000 on the current asset; then the true investment in the challenger will be $50,000 − ($40,000 − $30,000) = $40,000. Therefore, the challenger first cost is $40,000. Note that the initial cost of the defender ($60,000) has no relevance in this computation.
- *First cost:* This is the total cost of preparing the asset for economic use. It includes the purchase price, delivery cost, installation cost, and other costs that must be incurred before the asset can be put into production. This is also called the basis.
- *Sunk cost:* This is the difference between an asset's book value (BV) and its market value (MV) at a particular period. Sunk costs have no relevance to the replacement decisions and must be neglected.
- *Outsider viewpoint:* This is the perspective that would have been taken by an impartial third party to establish a fair MV for the installed asset. This perspective forces the analyst to focus on the present and the future cash flows in a replacement study; hence, avoiding the temptations to dwell on past (sunk) costs.
- *Asset life:* The life of an asset can be divided into three: ownership life, useful life, and economic life. The ownership life of an asset is the period between when an owner acquired it and when he disposed it. The useful life, on the other hand, is the period an asset is kept in productive service. In addition, the economic service life of an asset is the number of periods that results in the minimum EUAC of owing and operating the asset. The economic life is often shorter than the useful life and it is usually one year for the defender. Of all these, only the economic life is relevant to replacement computations.

- *Marginal cost:* This is the additional cost of increasing production output by one additional unit using the current asset. This is a useful cost parameter in replacement analysis.
- *Before-tax and after-tax analysis:* Replacement analysis can be based on before-tax or after-tax cash flows; however, it is always better to use after-tax cash flows in order to account for the effect of taxes on replacement decisions.

Replacement Analysis Computation

The process involved can be depicted as shown in Figure 5.1.

The column elements in a replacement analysis table are summarized below:

Column a: EOY, $k =$ given
Column b: MV = Given

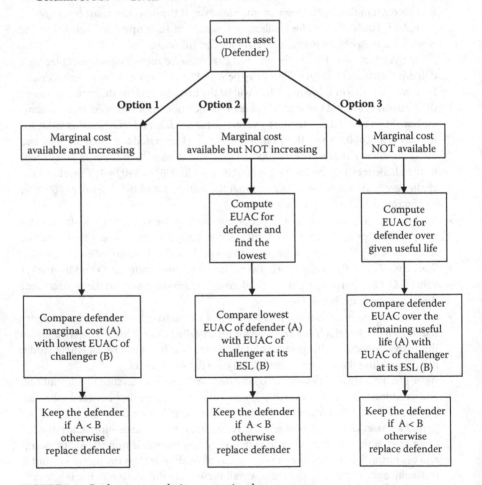

FIGURE 5.1 Replacement analysis computational process map.

Column c: Loss in MV $= MV_{(k-1)} - MV_{(k)}$
Column d: Cost of capital $= MV_{(k-1)} \times i\%$
Column e: Annual expenses $=$ given
Column f: Total marginal cost (TMC) $= [(c) + (d) + (e)]$, if not given

Column g: EUAC $= EUAC_k = \left[\sum_{j=1}^{k} TMC_j (P/F, i\%, j) \right] (A/P, i\%, k)$

Depreciation Methods

Depreciation can be defines as:

- A decline in the MV of an asset (deterioration).
- A decline in value of an asset to its owner (obsolescence).
- Allocation of the cost of an asset over its depreciable or useful life. The accountants usually uses this definition and it is adopted in economic analysis for income tax computation purposes. Therefore, depreciation is a way to claim over time an already paid expense for a depreciable asset.

For an asset to be depreciated, it must satisfy these three requirements:

- The asset must be used for business purposes to generate income
- The asset must have a useful life that can be determined and longer than 1 year
- The asset must be an asset that decays, gets used up, wears out, become obsolete, or loses value to the owner over time as a result of natural causes

Depreciation Terminology

- *Depreciation:* The annual depreciation amount, D_t, is the decreasing value of the asset to the owner. It does not represent an actual cash flow or actual usage pattern.
- *Book depreciation:* This is an internal description of depreciation. It is the reduction in the asset investment due to its usage pattern and expected useful life.
- *Tax depreciation:* This is used for after-tax economic analysis. In the United States and many other countries, the annual tax depreciation is tax deductible using the approved method of computation.
- *First cost or unadjusted basis:* This is the cost of preparing the asset for economic use. This is also called the basis. This term is used when an asset is new. Adjusted basis is used after some depreciation has been charged.
- *Book value:* This represents the difference between the basis and the accumulated depreciation charges at a particular period. It is also called the undepreciated capital investment and it is usually calculated at the end of each year.
- *Salvage value:* Estimated trade-in or MV at the end of the asset useful life. It may be positive, negative, or zero. It can be expressed as a dollar amount or as a percentage of the first cost.

- *Market value:* This is the estimated amount realizable if the asset were sold in an open market. This amount may be different from the BV.
- *Recovery period:* This is the depreciable life of an asset in years. Often there are different n values for book and tax depreciations. Both values may be different from the asset's estimated productive life.
- *Depreciation or recovery rate:* This is the fraction of the first cost removed by depreciation each year. Depending on the method of depreciation, this rate may be different for each recovery period.
- *Half-year convention:* This is used with modified accelerated cost recovery system (MACRS) depreciation method, which will be discussed later. It assumes that assets are placed in service or disposed in midyear, regardless of when these placements or disposal actually occur during the year. There are also midquarter and midmonth conventions.

Depreciation Methods

Let

B first cost, unadjusted basis, or basis
BV_t book value after period t
d depreciation rate $(= 1/n)$
D_t annual depreciable charge
MV market value
n recovery period in years
S estimated salvage value
t year $(t = 1, 2, 3, \ldots, n)$

Straight-Line (SL) Method

The annual depreciation charge is given as

$$D_t = \frac{B - S}{n} = (B - S)d. \tag{5.15}$$

The book value after t year is given as

$$BV_t = B - \frac{t}{n}(B - S) = B - tD_t. \tag{5.16}$$

Declining Balance (DB) Method

The DB annual depreciation charge is

$$D_t = \frac{1.5B}{n}\left(1 - \frac{1.5B}{n}\right)^{t-1}. \tag{5.17}$$

Total DB depreciation at the end of t years is

$$B\left[1-\left(1-\frac{1.5}{n}\right)^t\right] = B[1-(1-d)^t]. \tag{5.18}$$

Book value at the end of t years is

$$B\left(1-\frac{1.5}{n}\right)^t = B(1-d)^t. \tag{5.19}$$

For DDB (200% depreciation) method, substitute 2.0 for 1.5 in Equations 5.17 and 5.19.

Sums-of-Years' Digits (SYD) Method

The annual depreciation charge is

$$D_t = \frac{n-t+1}{\text{SUM}}(B-S) = d_t(B-S),$$
$$\text{SUM} = \frac{n(n+1)}{2}. \tag{5.20}$$

Book value at the end of t years is

$$\text{BV}_t = B - \frac{t(n-t/2+0.5)}{\text{SUM}}(B-S). \tag{5.21}$$

MACRS Method

The following information is required to depreciate an asset using the MACRS method:

- The cost basis
- The date the property was placed in service
- The property class and recovery period
- The MACRS depreciation system to be used [General Depreciation System (GDS) or Alternate Depreciation System (ADS)]
- The time convention that applies (e.g., half-year or quarter-year convention).

The steps involved in using MACRS depreciation method are:

1. Determine the property class of the asset being depreciated using published standard tables (reproduced in Tables 5.1 and 5.2). Any asset not in any of the stated class is automatically assigned a 7-year recovery period under the GDS system.
2. After the property class is known, read off the appropriate published depreciation schedule (see Table 5.3). For nonresidential real property, see Table 5.4.
3. The last step is to multiply the asset's cost basis by the depreciation schedule for each year to obtain the annual depreciation charge as stated in Equation 5.22.

TABLE 5.1 Property Classes Based on Asset Description

| | | | MACRS Property Class (Years) | |
Asset Class	Asset Description	Class Life (Years) ADR[a]	GDS	ADS
00.11	Office furniture, fixtures, and equipment	10	7	10
00.12	Information systems: computers/peripheral	6	5	6
00.22	Automobiles, taxis	3	5	6
00.241	Light general-purpose trucks	4	5	6
00.25	Railroad cars and locomotives	15	7	15
00.40	Industrial steam and electric distribution	22	15	22
01.11	Cotton gin assets	10	7	10
01.21	Cattle, breeding, or dairy	7	5	7
13.00	Offshore drilling assets	7.5	5	7.5
13.30	Petroleum refining assets	16	10	16
15.00	Construction assets	6	5	6
20.10	Manufacture of grain and grain mill products	17	10	17
20.20	Manufacture of yarn, thread, and woven fabric	11	7	11
24.10	Cutting of timber	6	5	6
32.20	Manufacture of cement	20	15	20
20.1	Manufacture of motor vehicles	12	7	12
48.10	Telephone distribution plant	24	15	24
48.2	Radio and television broadcasting equipment	6	5	6
49.12	Electric utility nuclear production plant	20	15	20
49.13	Electric utility steam production plant	28	20	28
49.23	Natural gas production plant	14	7	14
50.00	Municipal wastewater treatment plant	24	15	24
80.00	Theme and amusement assets	12.5	7	12.5

[a] ADR, asset depreciation range.

MACRS annual depreciation amount is

$$D_t = \text{(first cost)} \times \text{(tabulated depreciation schedule)}$$
$$= d_t B. \tag{5.22}$$

The annual book value is

$$BV_t = \text{first cost} - \text{sum of accumulated depreciation}$$
$$= B - \sum_{j=1}^{t} D_j. \tag{5.23}$$

Effects of Inflation and Taxes

Real interest rate (d) is defined as the desired rate of return in the absence of inflation.

TABLE 5.2 MACRS GDS Property Classes

Property Class	Personal Property (All Property Except Real Estate)
3-year property	Special handling devices for food processing and beverage manufacture
	Special tools for the manufacture of finished plastic products, fabricated metal products, and motor vehicles
	Property with ADR class life of 4 years or less
5-year property	Automobiles[a] and trucks
	Aircraft (of non-air-transport companies)
	Equipment used in research and experimentation
	Computers
	Petroleum drilling equipment
	Property with ADR class life of more than 4 years and less than 10 years
7-year property	All other property not assigned to another class
	Office furniture, fixtures, and equipment
	Property with ADR class life of 10 years or more and less than 16 years
10-year property	Assets used in petroleum refining and certain food products
	Vessels and water transportation equipment
	Property with ADR class life of 16 years or more and less than 20 years
15-year property	Telephone distribution plants
	Municipal sewage treatment plants
	Property with ADR class life of 20 years or more and less than 25 years
20-year property	Municipal sewers
	Property with ADR class life of 20 years or more and less than 25 years
Property class	Real property (real estate)
27.5 years	Residential rental property (does not include hotels and motels)
39 years	Nonresidential real property

Source: U.S. Department of the Treasury, Internal Revenue Service Publication 946, *How to Depreciate Property.* Washington, DC: U.S. Government Printing Office.

[a] The depreciation deduction for automobiles is limited to $7660 (maximum) the first tax year, $4900 the second year, $2950 the third year, and $1775 per year in subsequent years.

Combined interest rate (i) is the rate of return combining real interest rate and inflation rate.

Inflation rate j can be expressed as

$$1 + i = (1 + d)(1 + j).$$

The combined interest rate can be expressed as

$$i = d + j + dj.$$

If $j = 0$ (i.e., no inflation), then $i = d$.

Commodity escalation rate (g) is the rate at which individual commodity prices escalate. This may be greater than or less than the overall inflation rate. For the constant worth

TABLE 5.3 MACRS Depreciation for Personal Property Based on the Half-Year Convention

Recovery Year	Applicable Percentage for Property Class					
	3-Year Property	5-Year Property	7-Year Property	10-Year Property	15-Year Property	20-Year Property
1	33.33	20.00	14.29	10.00	5.00	3.750
2	44.45	32.00	24.49	18.00	9.50	7.219
3	14.81	19.20	17.49	14.40	8.55	6.677
4	7.41	11.52	12.49	11.52	7.70	6.177
5		11.52	8.93	9.22	6.93	5.713
6		5.76	8.92	7.37	6.23	5.285
7			8.93	6.55	5.90	4.888
8			4.46	6.55	5.90	4.522
9				6.56	5.91	4.462
10				6.55	5.90	4.461
11				3.28	5.91	4.462
12					5.90	4.461
13					5.91	4.462
14					5.90	4.461
15					5.91	4.462
16					2.95	4.461
17						4.462
18						4.461
19						4.462
20						4.461
21						2.231

cash flow, we have

$$C_k = T_0, \quad k = 1, 2, \ldots, n$$

and for the then-current cash flow, we have

$$T_k = T_0(1 + j)^k, \quad k = 1, 2, \ldots, n,$$

where j is the inflation rate.

TABLE 5.4 MACRS Depreciation for Real Property (Real Estate)

Recovery Year	Recovery Percentage for Nonresidential Real Property (Month Placed in Service)											
	1	2	3	4	5	6	7	8	9	10	11	12
1	2.461	2.247	2.033	1.819	1.605	1.391	1.177	0.963	0.749	0.535	0.321	0.107
2–39	2.564	2.564	2.564	2.564	2.564	2.564	2.564	2.564	2.564	2.564	2.564	2.564
40	0.107	0.321	0.535	0.749	0.963	1.177	1.391	1.605	1.819	2.033	2.247	2.461

If we are using the commodity escalation rate g, then we will have

$$T_k = T_0(1+g)^k, \quad k = 1, 2, \ldots, n.$$

Thus, a then-current cash flow may increase based on both a regular inflation rate (j) and a commodity escalation rate (g). We can convert a then-current cash flow to a constant worth cash flow by using the following relationship:

$$C_k = T_k(1+j)^{-k}, \quad k = 1, 2, \ldots, n.$$

If we substitute T_k from the commodity escalation cash flow into the expression for C_k above, we obtain

$$C_k = T_k(1+j)^{-k}$$
$$= T(1+g)^k(1+j)^{-k}$$
$$= T_0\left[\frac{1+g}{1+j}\right]^k, \quad k = 1, 2, \ldots, n.$$

Note that if $g = 0$ and $j = 0$, the $C_k = T_0$. That is, no inflationary effect. We now define effective commodity escalation rate (v) as

$$v = \left[\frac{1+g}{1+j}\right] - 1$$

and we can express the commodity escalation rate (g) as

$$g = v + j + vj.$$

Foreign Exchange Rates

Using the United States as the base country where the local business is situated, let

$i_{U.S.}$ Rate of return in terms of a market interest rate relative to a U.S. dollar

i_{fc} Rate of return in terms of a market interest rate relative to the currency of a foreign country

f_e Annual devaluation rate between the currency of a foreign country and the U.S. dollar. A positive value means that the foreign currency is being devalued relative to the U.S. dollar. A negative value means that the U.S. dollar is being devalued relative to the foreign currency

The rate of return of the local business with respect to a foreign country can be given by

$$i_{U.S.} = \frac{i_{fc} - f_e}{1 + f_e}, \quad i_{fc} = i_{U.S.} + f_e + f_e(i_{U.S.}). \tag{5.24}$$

After-Tax Economic Analysis

There are several types of taxes:

- *Income taxes:* These are taxes assessed as a function of gross revenue less allowable deductions and are levied by the federal, most state, and municipal governments.

- *Property taxes:* They are assessed as a function of the value of property owned, such as land, buildings, and equipment and are mostly levied by municipal, county, or state governments.
- *Sales taxes:* These are assessed on purchases of goods and services, hence, independent of gross income or profits. They are normally levies by state, municipal, or county governments. Sales taxes are relevant in economic analysis only to the extent that they add to the cost of items purchased.
- *Excise taxes:* These are federal taxes assessed as a function of the sale certain goods or services often considered nonnecessities. They are usually charged to the manufacturer of the goods and services, but a portion of the cost is passed on to the purchaser.

Let

TI   taxable income (amount upon which taxes are based)
T   tax rate (percentage of taxable income owned in taxes)
NPAT   net profit after taxes (taxable income less income taxes each year; this amount is returned to the company)

$$\text{TI} = \text{gross income} - \text{expenses} - \text{depreciation (depletion) deductions}, \tag{5.25}$$

$$T = \text{TI} \times \text{applicable tax rate},$$

$$\text{NPAT} = \text{TI}(1 - T),$$

$$\text{Effective tax rates (Te)} = \text{state rate} + (1 - \text{state rate})(\text{federal rate}), \tag{5.26}$$

$$T = (\text{taxable income})(\text{Te}). \tag{5.27}$$

The after-tax MARR is usually smaller than the before-tax MARR and they are related by the following equation:

$$\text{After-tax MARR} \cong (\text{before-tax MARR})(1 - \text{Te}). \tag{5.28}$$

The taxable income equation can be rewritten as

$$\text{TI} = \text{gross income} - \text{expenses} - \text{depreciation (depletion) deductions}$$
$$+ \text{depreciation recapture} + \text{capital gain} - \text{capital loss}. \tag{5.29}$$

Cost and Value Computations

Cost management in a project environment refers to the functions required to maintain effective financial control of the project throughout its life cycle. There are several cost concepts that influence the economic aspects of managing industrial projects. Within a given scope of analysis, there will be a combination of different types of cost factors as defined below.

Actual Cost of Work Performed

The cost actually incurred and recorded in accomplishing the work performed within a given time period.

Applied Direct Cost

The amounts recognized in the time period associated with the consumption of labor, material, and other direct resources, regardless of the date of commitment or the date of payment. These amounts are to be charged to work-in- process (WIP) when resources are actually consumed, material resources are withdrawn from inventory for use, or material resources are received and scheduled for use within 60 days.

Budgeted Cost for Work Performed

The sum of the budgets for completed work plus the appropriate portion of the budgets for level of effort and apportioned effort. Apportioned effort is effort that by itself is not readily divisible into short-span work packages but is related in direct proportion to measured effort.

Budgeted Cost for Work Scheduled

The sum of budgets for all work packages and planning packages scheduled to be accomplished (including WIP), plus the amount of level of effort and apportioned effort scheduled to be accomplished within a given period of time.

Direct Cost

Cost that is directly associated with actual operations of a project. Typical sources of direct costs are direct material costs and direct labor costs. Direct costs are those that can be reasonably measured and allocated to a specific component of a project.

Economies of Scale

This is a term referring to the reduction of the relative weight of the fixed cost in total cost, achieved by increasing the quantity of output. Economies of scale help to reduce the final unit cost of a product and are often simply referred to as the savings due to mass production.

Estimated Cost at Completion

This refers to the sum of actual direct costs, plus indirect costs that can be allocated to a contract, plus the estimate of costs (direct and indirect) for authorized work remaining to be done.

First Cost

The total initial investment required to initiate a project or the total initial cost of the equipment needed to start the project.

Fixed Cost

Costs incurred regardless of the level of operation of a project. Fixed costs do not vary in proportion to the quantity of output. Examples of costs that make up the fixed cost of a project are administrative expenses, certain types of taxes, insurance cost, depreciation cost, and debt servicing cost. These costs usually do not vary in proportion to quantity of output.

Incremental Cost

The additional cost of changing the production output from one level to another. Incremental costs are normally variable costs.

Indirect Cost

This is a cost that is indirectly associated with project operations. Indirect costs are those that are difficult to assign to specific components of a project. An example of an indirect cost is the cost of computer hardware and software needed to manage project operations. Indirect costs are usually calculated as a percentage of a component of direct costs. For example, the direct costs in an organization may be computed as 10% of direct labor costs.

Life-Cycle Cost

This is the sum of all costs, recurring and nonrecurring, associated with a project during its entire life cycle.

Maintenance Cost

This is a cost that occurs intermittently or periodically for the purpose of keeping project equipment in good operating condition.

Marginal Cost

Marginal cost is the additional cost of increasing production output by one additional unit. The marginal cost is equal to the slope of the total cost curve or line at the current operating level.

Operating Cost

This is a recurring cost needed to keep a project in operation during its life cycle. Operating costs may consist of items such as labor, material, and energy costs.

Opportunity Cost

This refers to the cost of forgoing the opportunity to invest in a venture that, if it had been pursued, would have produced an economic advantage. Opportunity costs are usually incurred due to limited resources that make it impossible to take advantage of all investment opportunities. It is often defined as the cost of the best-rejected opportunity. Opportunity costs can also be incurred due to a missed opportunity rather than due to an intentional rejection. In many cases, opportunity costs are hidden or implied because they typically relate to future events that cannot be accurately predicted.

Overhead Cost

These are costs incurred for activities performed in support of the operations of a project. The activities that generate overhead costs support the project efforts rather than contributing directly to the project goal. The handling of overhead costs varies widely from company to company. Typical overhead items are electric power cost, insurance premiums, cost of security, and inventory carrying cost.

Standard Cost

This is a cost that represents the normal or expected cost of a unit of the output of an operation. Standard costs are established in advance. They are developed as a composite of several component costs, such as direct labor cost per unit, material cost per unit, and allowable overhead charge per unit.

Sunk Cost

Sunk cost is a cost that occurred in the past and cannot be recovered under the present analysis. Sunk costs should have no bearing on the prevailing economic analysis and project decisions. Ignoring sunk costs can be a difficult task for analysts. For example, if $950,000 was spent 4 years ago to buy a piece of equipment for a technology-based project, a decision on whether or not to replace the equipment now should not consider that initial cost. But uncompromising analysts might find it difficult to ignore that much money. Similarly, an individual making a decision on selling a personal automobile would typically try to relate the asking price to what was paid for the automobile when it was acquired. This is wrong under the strict concept of sunk costs.

Total Cost

This is the sum of all the variable and fixed costs associated with a project.

Variable Cost

This cost varies in direct proportion to the level of operation or quantity of output. For example, the costs of material and labor required to make an item will be classified as variable costs because they vary with changes in level of output.

Cash-Flow Calculations

Economic analysis is performed when a choice must be made between mutually exclusive projects that compete for limited resources. The cost performance of each project will depend on the timing and levels of its expenditures. The techniques of computing cash-flow equivalence permit us to bring competing project cash flows to a common basis for comparison. The common basis depends on the prevailing interest rate. Two cash flows that are equivalent at a given interest rate will not be equivalent at a different interest rate. The basic techniques for converting cash flows from one point in time to another are presented in the following sections.

Cash-flow conversion involves the transfer of project funds from one point in time to another. The following notation is used for the variables involved in the conversion process:

A a uniform end-of-period cash receipt or disbursement
F a future sum of money
G a uniform arithmetic gradient increase in period-by-period payments or disbursements
i interest rate per period
n number of interest periods
P a present sum of money

In many cases, the interest rate used in performing economic analysis is set equal to the MARR of the decision-maker. The MARR is also sometimes referred to as *hurdle rate*, *required internal rate of return* (IRR), *return on investment* (ROI), or *discount rate*. The value of MARR is chosen for a project based on the objective of maximizing the economic performance of the project.

Calculations with Compound Amount Factor

The procedure for the single-payment compound amount factor finds a future amount, F, that is equivalent to a present amount, P, at a specified interest rate, i, after n periods. This is calculated by the following formula:

$$F = P(1 + i)^n.$$

A graphical representation of the relationship between P and F is shown in Figure 5.2.

Example

A sum of $5000 is deposited in a project account and left there to earn interest for 15 years. If the interest rate per year is 12%, the compound amount after 15 years can be calculated as follows:

$$F = \$5000(1 + 0.12)^{15} = \$27,367.85.$$

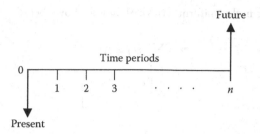

FIGURE 5.2   Single-payment compound amount cash flow.

Calculations with Present Worth Factor

The present worth factor computes P when F is given. The present worth factor is obtained by solving for P in the equation for the compound amount factor. That is,

$$P = F(1 + i)^{-n}.$$

Supposing it is estimated that $15,000 would be needed to complete the implementation of a project 5 years from now, how much should be deposited in a special project fund now so that the fund would accrue to the required $15,000 exactly 5 years from now?

If the special project fund pays interest at 9.2% per year, the required deposit would be

$$P = \$15,000(1 + 0.092)^{-5} = \$9,660.03.$$

Calculations with Uniform Series Present Worth Factor

The uniform series present worth factor is used to calculate the present worth equivalent, P, of a series of equal end-of-period amounts, A. Figure 5.3 shows the uniform series cash flow. The derivation of the formula uses the finite sum of the present worth values of the

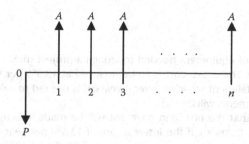

FIGURE 5.3   Uniform series cash flow.

individual amounts in the uniform series cash flow as shown below:

$$P = \sum_{t=1}^{n} A(1+i)^{-t} \tag{5.30}$$

$$= A \left[\frac{(1+i)^n - 1}{i(1+i)^n} \right]. \tag{5.31}$$

Example

Suppose the sum of $12,000 must be withdrawn from an account to meet the annual operating expenses of a multiyear project. The project account pays interest at 7.5% per year compounded on an annual basis. If the project is expected to last 10 years, how much must be deposited in the project account now so that the operating expenses of $12,000 can be withdrawn at the end of every year for 10 years? The project fund is expected to be depleted to zero by the end of the last year of the project. The first withdrawal will be made 1 year after the project account is opened, and no additional deposits will be made in the account during the project life cycle. The required deposit is calculated as

$$P = \$12,000 \left[\frac{(1+0.075)^{10} - 1}{0.075(1+0.075)^{10}} \right]$$

$$= \$82,368.92.$$

Calculations with Uniform Series Capital Recovery Factor

The capital recovery formula is used to calculate the uniform series of equal end-of-period payments, A, that are equivalent to a given present amount, P. This is the converse of the uniform series present amount factor. The equation for the uniform series capital recovery factor is obtained by solving for A in the uniform series present amount factor. That is,

$$A = P \left[\frac{i(1+i)^n}{(1+i)^n - 1} \right].$$

Example

Suppose a piece of equipment needed to launch a project must be purchased at a cost of $50,000. The entire cost is to be financed at 13.5% per year and repaid on a monthly installment schedule over 4 years. It is desired to calculate what the monthly loan payments will be.

It is assumed that the first loan payment will be made exactly 1 month after the equipment is financed. If the interest rate of 13.5% per year is compounded monthly, then the interest rate per month will be 13.5%/12 = 1.125% per month. The number of interest periods over which the loan will be repaid is 4(12) = 48 months.

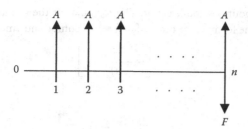

FIGURE 5.4   Uniform series compound amount cash flow.

Consequently, the monthly loan payments are calculated to be

$$A = \$50{,}000 \left[\frac{0.01125(1+0.01123)^{48}}{(1+0.01125)^{48} - 1} \right]$$

$$= \$1353.82.$$

Calculations with Uniform Series Compound Amount Factor

The series compound amount factor is used to calculate a single future amount that is equivalent to a uniform series of equal end-of-period payments. The cash flow is shown in Figure 5.4. Note that the future amount occurs at the same point in time as the last amount in the uniform series of payments. The factor is derived as shown below:

$$F = \sum_{t=1}^{n} A(1+i)^{n-t}$$

$$= A \left[\frac{(1+i)^n - 1}{i} \right].$$

Example

If equal end-of-year deposits of $5000 are made to a project fund paying 8% per year for 10 years, how much can be expected to be available for withdrawal from the account for capital expenditure immediately after the last deposit is made?

$$F = \$5000 \left[\frac{(1+0.08)^{10} - 1}{0.08} \right]$$

$$= \$72{,}432.50.$$

Calculations with Uniform Series Sinking Fund Factor

The sinking fund factor is used to calculate the uniform series of equal end-of-period amounts, A, that are equivalent to a single future amount, F. This is the reverse of the uniform series compound amount factor. The formula for the sinking fund is obtained by solving for A in the formula for the uniform series compound amount factor. That is,

$$A = F \left[\frac{i}{(1+i)^n - 1} \right].$$

Example

How large are the end-of-year equal amounts that must be deposited into a project account so that a balance of $75,000 will be available for withdrawal immediately after the 12th annual deposit is made?

The initial balance in the account is zero at the beginning of the first year. The account pays 10% interest per year. Using the formula for the sinking fund factor, the required annual deposits are

$$A = \$75,000 \left[\frac{0.10}{(1+0.10)^{12} - 1} \right]$$
$$= \$3,507.25.$$

Calculations with Capitalized Cost Formula

Capitalized cost refers to the present value of a single amount that is equivalent to a perpetual series of equal end-of-period payments. This is an extension of the series present worth factor with an infinitely large number of periods. This is shown graphically in Figure 5.5.

Using the limit theorem from calculus as n approaches infinity, the series present worth factor reduces to the following formula for the capitalized cost:

$$P = \frac{A}{i}.$$

Example

How much should be deposited in a general fund to service a recurring public service project to the tune of $6500 per year forever if the fund yields an annual interest rate of 11%?

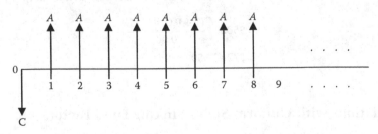

FIGURE 5.5 Capitalized cost cash flow.

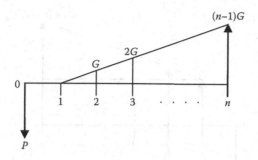

FIGURE 5.6 Arithmetic gradient cash flow with zero base amount.

Using the capitalized cost formula, the required one-time deposit to the general fund is

$$P = \frac{\$6500}{0.11}$$
$$= \$59,090.91.$$

Arithmetic Gradient Series

The gradient series cash flow involves an increase of a fixed amount in the cash flow at the end of each period. Thus, the amount at a given point in time is greater than the amount at the preceding period by a constant amount. This constant amount is denoted by G. Figure 5.6 shows the basic gradient series in which the base amount at the end of the first period is zero. The size of the cash flow in the gradient series at the end of period t is calculated as

$$A_t = (t-1)G, \quad t = 1, 2, \ldots, n.$$

The total present value of the gradient series is calculated by using the present amount factor to convert each individual amount from time t to time 0 at an interest rate of $i\%$ per period and then summing up the resulting present values. The finite summation reduces to a closed-form as shown below:

$$P = \sum_{t=1}^{n} A_t(1+i)^{-t}$$
$$= G\left[\frac{(1+i)^n - (1+ni)}{i^2(1+i)^n}\right].$$

Example

The cost of supplies for a 10-year project increases by $1500 every year starting at the end of year 2. There is no cost for supplies at the end of the first year. If interest rate is 8% per year, determine the present amount that must be set aside at time zero to take care of all the future supplies expenditures.

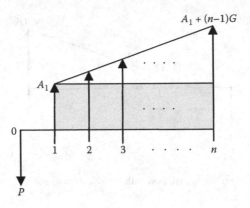

FIGURE 5.7 Arithmetic gradient cash flow with nonzero base amount.

We have $G = \$1500$, $i = 0.08$, and $n = 10$. Using the arithmetic gradient formula, we obtain

$$P = 1500 \left[\frac{1 - (1 + 10(0.08))(1 + 0.08)^{-10}}{(0.08)^2} \right]$$

$$= \$1500(25.9768)$$

$$= \$38,965.20.$$

In many cases, an arithmetic gradient starts with some base amount at the end of the first period and then increases by a constant amount thereafter. The nonzero base amount is denoted as A_1. Figure 5.7 shows this type of cash flow.

The calculation of the present amount for such cash flows requires breaking the cash flow into a uniform series cash flow of amount A_1 and an arithmetic gradient cash flow with zero base amount. The uniform series present worth formula is used to calculate the present worth of the uniform series portion, whereas the basic gradient series formula is used to calculate the gradient portion. The overall present worth is then calculated:

$$P = P_{\text{uniform series}} + P_{\text{gradient series}}$$

$$= A_1 \left[\frac{(1+i)^n - 1}{i(1+i)^n} \right] + G \left[\frac{(1+i)^n - (1+ni)}{i^2(1+i)^n} \right].$$

Internal Rate of Return

The IRR for a cash flow is defined as the interest rate that equates the future worth at time n or present worth at time 0 of the cash flow to zero. If we let i^* denote the IRR, then we have

$$\text{FW}_{t=n} = \sum_{t=0}^{n} (\pm A_t)(1 + i^*)^{n-t} = 0,$$

$$PW_{t=0} = \sum_{t=0}^{n} (\pm A_t)(1 + i^*)^{-t} = 0,$$

where the "+" sign is used in the summation for positive cash-flow amounts or receipts and the "−" sign is used for negative cash-flow amounts or disbursements. A_t denotes the cash-flow amount at time t, which may be a receipt (+) or a disbursement (−). The value of i^* is referred to as *discounted cash flow rate of return, IRR*, or *true rate of return*. The procedure above essentially calculates the net future worth or the net present worth of the cash flow. That is,

Net future worth = future worth of receipts − future worth of disbursements

$$NFW = FW_{receipts} - FW_{disbursements}$$

and

Net present worth = present worth of receipts − present worth of disbursements

$$NPW = PW_{receipts} - PW_{disbursements}.$$

Setting the NPW or NFW equal to zero and solve for the unknown variable i in the above expressions.

Benefit–Cost Ratio Analysis

The benefit–cost ratio of a cash flow is the ratio of the present worth of benefits to the present worth of costs. This is defined below:

$$B/C = \frac{\sum_{t=0}^{n} B_t(1 + i)^{-t}}{\sum_{t=0}^{n} C_t(1 + i)^{-t}}$$

$$= \frac{PW_{benefits}}{PW_{costs}},$$

where B_t is the benefit (receipt) at time t and C_t is the cost (disbursement) at time t. If the benefit–cost ratio is greater than 1, then the investment is acceptable. If the ratio is less than 1, the investment is not acceptable. A ratio of 1 indicates a break-even situation for the project.

Simple Payback Period

Payback period refers to the length of time it will take to recover an initial investment. The approach does not consider the impact of the time value of money. Consequently, it is not an accurate method of evaluating the worth of an investment. However, it is a simple technique that is used widely to perform a "quick-and-dirty" assessment of investment performance. Another limitation of the technique is that it considers only the initial cost.

Other costs that may occur after time zero are not included in the calculation. The payback period is defined as the smallest value of n (n_{min}) that satisfies the following expression:

$$\sum_{t=1}^{n_{min}} R_t \geq C,$$

where R_t is the revenue at time t and C_0 the initial investment. The procedure calls for a simple addition of the revenues period by period until enough total has been accumulated to offset the initial investment.

Example

An organization is considering installing a new computer system that will generate significant savings in material and labor requirements for order processing. The system has an initial cost of $50,000. It is expected to save the organization $20,000 a year. The system has an anticipated useful life of 5 years with a salvage value of $5000. Determine how long it would take for the system to pay for itself from the savings it is expected to generate. Since the annual savings are uniform, we can calculate the payback period by simply dividing the initial cost by the annual savings. That is,

$$n_{min} = \frac{\$50,000}{\$20,000}$$
$$= 2.5 \text{ years.}$$

Note that the salvage value of $5000 is not included in the above calculation because the amount is not realized until the end of the useful life of the asset (i.e., after 5 years). In some cases, it may be desired to consider the salvage value. In that case, the amount to be offset by the annual savings will be the net cost of the asset. In that case, we would have the following:

$$n_{min} = \frac{\$50,000 - \$5000}{\$20,000}$$
$$= 2.25 \text{ years.}$$

If there are tax liabilities associated with the annual savings, those liabilities must be deducted from the savings before the payback period is calculated.

Discounted Payback Period

In this book, we introduce the *discounted payback period* approach, in which the revenues are reinvested at a certain interest rate. The payback period is determined when enough money has been accumulated at the given interest rate to offset the initial cost as well as other interim costs. In this case, the calculation is done by the following expression:

$$\sum_{t=1}^{n_{min}} R_t(1+i)^{n_{min}-1} \geq \sum_{t=0}^{n_{min}} C_t.$$

Example

A new solar cell unit is to be installed in an office complex at an initial cost of $150,000. It is expected that the system will generate annual cost savings of $22,500 on the electricity bill. The solar cell unit will need to be overhauled every 5 years at a cost of $5000 per overhaul. If the annual interest rate is 10%, find the discounted payback period for the solar cell unit considering the time value of money. The costs of overhaul are to be considered in calculating the discounted payback period.

SOLUTION

Using the single-payment compound amount factor for one period iteratively, the following solution of cumulative savings is obtained:

Time 1: $22,500
Time 2: $22,500 + $22,500(1.10)^1 = $47,250
Time 3: $22,500 + $47,250(1.10)^1 = $74,475
Time 4: $22,500 + $74,475(1.10)^1 = $104,422.50
Time 5: $22,500 + $104,422.50(1.10)^1 − $5000 = $132,364.75
Time 6: $22,500 + $132,364.75(1.10)^1 = $168,101.23.

The initial investment is $150,000. By the end of period 6, we have accumulated $168,101.23, more than the initial cost. Interpolating between period 5 and period 6 results in n_{min} of 5.49 years. That is, it will take five-and-a-half years to recover the initial investment. The calculation is shown below:

$$n_{min} = 5 + \frac{150,000 - 132,364.75}{168,101.25 - 132,364.75}(6-5)$$

$$= 5.49.$$

Time Required to Double Investment

It is sometimes of interest to determine how long it will take a given investment to reach a certain multiple of its initial level. The *Rule of 72* is one simple approach to calculate the time required for an investment to double in value, at a given interest rate per period. The Rule of 72 gives the following formula for estimating the time required:

$$n = \frac{72}{i},$$

where i is the interest rate expressed in percentage. Referring to the single-payment compound amount factor, we can set the future amount equal to twice the present amount

TABLE 5.5 Evaluation of the Rule of 72

i%	n (Rule of 72)	n (exact value)
0.25	288.00	277.61
0.50	144.00	138.98
1.00	72.00	69.66
2.00	36.00	35.00
5.00	14.20	17.67
8.00	9.00	9.01
10.00	7.20	7.27
12.00	6.00	6.12
15.00	4.80	4.96
18.00	4.00	4.19
20.00	3.60	3.80
25.00	2.88	3.12
30.00	2.40	2.64

and then solve for n. That is, $F = 2P$. Thus,

$$2P = P(1+i)^n.$$

Solving for n in the above equation yields an expression for calculating the exact number of periods required to double P:

$$n = \frac{\ln(2)}{\ln(1+i)},$$

where i is the interest rate expressed in decimals. In general, the length of time it would take to accumulate m multiples of P is expressed as

$$n = \frac{\ln(m)}{\ln(1+i)},$$

where m is the desired multiple. For example, at an interest rate of 5% per year, the time it would take an amount, P, to double in value ($m = 2$) is 14.21 years. This, of course, assumes that the interest rate will remain constant throughout the planning horizon. Table 5.5 presents a tabulation of the values calculated from both approaches. Figure 5.8 shows a graphical comparison of the Rule of 72 to the exact calculation.

Effects of Inflation on Industrial Project Costing

Inflation can be defined as the decline in purchasing power of money and, as such, is a major player in the financial and economic analysis of projects. Multiyear projects are particularly subject to the effects of inflation. Some of the most common causes of inflation include the following:

- An increase in the amount of currency in circulation
- A shortage of consumer goods

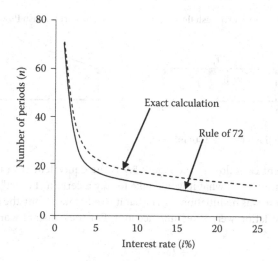

FIGURE 5.8   Evaluation of investment life for double return.

- An escalation of the cost of production
- An arbitrary increase in prices set by resellers

The general effects of inflation are felt in terms of an increase in the prices of goods and a decrease in the worth of currency. In cash-flow analysis, ROI for a project will be affected by time value of money as well as inflation. The real interest rate (d) is defined as the desired rate of return in the absence of inflation. When we talk of "today's dollars" or "constant dollars," we are referring to the use of the real interest rate. The combined interest rate (i) is the rate of return combining the real interest rate and the inflation rate. If we denote the inflation rate as j, then the relationship between the different rates can be expressed as shown below:

$$1 + i = (1 + d)(1 + j).$$

Thus, the combined interest rate can be expressed as follows:

$$i = d + j + dj.$$

Note that if $j = 0$ (i.e., no inflation), then $i = d$. We can also define commodity escalation rate (g) as the rate at which individual commodity prices escalate. This may be greater than or less than the overall inflation rate. In practice, several measures are used to convey inflationary effects. Some of these are the consumer price index, the producer price index, and the wholesale price index. A "market basket" rate is defined as the estimate of inflation based on a weighted average of the annual rates of change in the costs of a wide range of representative commodities. A "then-current" cash flow is a cash flow that explicitly incorporates the impact of inflation. A "constant worth" cash flow is a cash flow that does not incorporate the effect of inflation. The real interest rate, d, is used for analyzing constant worth cash flows. Figure 5.9 shows constant worth and then-current cash flows.

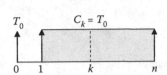

Constant-worth cash flow

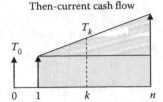

Then-current cash flow

FIGURE 5.9 Cash flows for effects of inflation.

The then-current cash flow in the Figure 5.9 is the equivalent cash flow considering the effect of inflation. C_k is what it would take to buy a certain "basket" of goods after k time periods if there was no inflation. T_k is what it would take to buy the same "basket" in k time period if inflation were taken into account. For the constant worth cash flow, we have

$$C_k = T_0, \quad k = 1, 2, \ldots, n$$

and for the then-current cash flow, we have

$$T_k = T_0(1+j)^k, \quad k = 1, 2, \ldots, n,$$

where j is the inflation rate. If $C_k = T_0 = \$100$ under the constant worth cash flow, then we have $\$100$ worth of buying power. If we are using the commodity escalation rate, g, then we will have

$$T_k = T_0(1+g)^k, \quad k = 1, 2, \ldots, n.$$

Thus, a then-current cash flow may increase based on both a regular inflation rate (j) and a commodity escalation rate (g). We can convert a then-current cash flow to a constant worth cash flow by using the following relationship:

$$C_k = T_k(1+j)^{-k}, \quad k = 1, 2, \ldots, n.$$

If we substitute T_k from the commodity escalation cash flow into the expression for C_k above, we obtain the following:

$$\begin{aligned} C_k &= T_k(1+j)^{-k} \\ &= T(1+g)^k(1+j)^{-k} \\ &= T_0\left[\frac{1+g}{1+j}\right]^k, \quad k = 1, 2, \ldots, n. \end{aligned}$$

Note that if $g = 0$ and $j = 0$, the $C_k = T_0$. That is, there is no inflationary effect. We can now define the effective commodity escalation rate (v):

$$v = \left[\frac{1+g}{1+j}\right] - 1.$$

The commodity escalation rate (g) can be expressed as follows:

$$g = v + j + vj.$$

Inflation can have a significant impact on the financial and economic aspects of an industrial project. Inflation may be defined, in economic terms, as the increase in the amount of currency in circulation. To a producer, inflation means a sudden increase in the cost of items that serve as inputs for the production process (equipment, labor, materials, etc.). To the retailer, inflation implies an imposed higher cost of finished products. To an ordinary citizen, inflation portends a noticeable escalation of prices of consumer goods. All these aspects are intertwined in a project management environment.

The amount of money supply, as a measure of a country's wealth, is controlled by the government. When circumstances dictate such action, governments often feel compelled to create more money or credit to take care of old debts and pay for social programs. When money is generated at a faster rate than the growth of goods and services, it becomes a surplus commodity, and its value (i.e., purchasing power) will fall. This means that there will be too much money available to buy only a few goods and services. When the purchasing power of a currency falls, each individual in a product's life cycle (i.e., each person or entity that spends money on a product throughout its life cycle, from production through disposal) has to use more of the currency in order to obtain the product. Some of the classic concepts of inflation are discussed below:

1. In *cost-driven* or *cost-push inflation*, increases in producer's costs are passed on to consumers. At each stage of the product's journey from producer to consumer, prices are escalated disproportionately in order to make a good profit. The overall increase in the product's price is directly proportional to the number of intermediaries it encounters on its way to the consumer.

2. In *demand-driven* or *demand-pull inflation*, excessive spending power of consumers forces an upward trend in prices. This high spending power is usually achieved at the expense of savings. The law of supply and demand dictates that the more the demand, the higher the price. This results in *demand-driven* or *demand-pull inflation.*

3. The impact of international economic forces can induce inflation on a local economy. Trade imbalances and fluctuations in currency values are notable examples of international inflationary factors.

4. In *wage-driven* or *wage-push inflation*, the increasing base wages of workers generate more disposable income and, hence, higher demands for goods and services. The high demand, consequently, creates a pull on prices. Coupled with this, employers pass the additional wage cost on to consumers through higher prices. This type of inflation is very difficult to contain because wages set by union contracts and prices set by producers almost never fall.

5. Easy availability of credit leads consumers to "buy now and pay later," thereby creating another opportunity for inflation. This is a dangerous type of inflation because the credit not only pushes prices up, but also leaves consumers with less money later to pay for the credit. Eventually, many credits become uncollectible debts, which may then drive the economy toward recession.

6. Deficit spending results in an increase in money supply and, thereby, creates less room for each dollar to get around. The popular saying which indicates that "a dollar does not go far anymore," simply refers to inflation in laymen's terms. The different levels of inflation may be categorized as discussed below.

Mild Inflation

When inflation is mild (at 2–4%), the economy actually prospers. Producers strive to produce at full capacity in order to take advantage of the high prices to the consumer. Private investments tend to be brisk, and more jobs become available. However, the good fortune may only be temporary. Prompted by the prevailing success, employers are tempted to seek larger profits and workers begin to ask for higher wages. They cite their employer's prosperous business as a reason to bargain for bigger shares of the business profit. So, we end up with a vicious cycle where the producer asks for higher prices, the unions ask for higher wages, and inflation starts an upward trend.

Moderate Inflation

Moderate inflation occurs when prices increase at 5–9%. Consumers start purchasing more as a hedge against inflation. They would rather spend their money now instead of watching it decline further in purchasing power. The increased market activity serves to fuel further inflation.

Severe Inflation

Severe inflation is indicated by price escalations of 10% or more. Double-digit inflation implies that prices rise much faster than wages do. Debtors tend to be the ones who benefit from this level of inflation because they repay debts with money that is less valuable than when they borrowed.

Hyperinflation

When each price increase signals an increase in wages and costs, which again sends prices further up, the economy has reached a stage of malignant galloping inflation or hyperinflation. Rapid and uncontrollable inflation destroys the economy. The currency becomes economically useless as the government prints it excessively to pay for obligations.

Inflation can affect any industrial project in terms of raw materials procurement, salaries and wages, and/or cost tracking dilemmas. Some effects are immediate and easily observable, whereas others are subtle and pervasive. Whatever form it takes, inflation must be taken into account in long-term project planning and control. Large projects, especially, may be adversely affected by the effects of inflation in terms of cost overruns and poor resource utilization. Managers should note that the level of inflation will determine the severity of the impact on projects.

Break-Even Analysis

Break-even analysis refers to the determination of the balanced performance level where project income is equal to project expenditure. The total cost of an operation is expressed as the sum of the fixed and variable costs with respect to output quantity. That is,

$$TC(x) = FC + VC(x),$$

where x is the number of units produced, $TC(x)$ the total cost of producing x units, FC the total fixed cost, and $VC(x)$ the total variable cost associated with producing x units. The total revenue resulting from the sale of x units is defined as

$$TR(x) = px,$$

where p is the price per unit. The profit due to the production and sale of x units of the product is calculated as

$$P(x) = TR(x) - TC(x).$$

The break-even point of an operation is defined as the value of a given parameter that will result in neither profit nor loss. The parameter of interest may be the number of units produced, the number of hours of operation, the number of units of a resource type allocated, or any other measure of interest. At the break-even point, we have the following relationship:

$$TR(x) = TC(x) \quad \text{or} \quad P(x) = 0.$$

In some cases, there may be a known mathematical relationship between cost and the parameter of interest. For example, there may be a linear cost relationship between the total cost of a project and the number of units produced. The cost expressions facilitate a straightforward break-even analysis. Figure 5.10 shows an example of a break-even point for a single project. Figure 5.11 shows examples of multiple break-even points that exist when multiple projects are compared. When two project alternatives are compared, the break-even point refers to the point of indifference between the two alternatives. In Figure 5.11, $x1$ represents the point where projects A and B are equally desirable, $x2$ represents where A and C are equally desirable, and $x3$ represents where B and C are equally desirable. The Figure 5.10 shows that if we are operating below a production level of $x2$ units, then project C is the preferred project among the three. If we are operating at a level more than $x2$ units, then project A is the best choice.

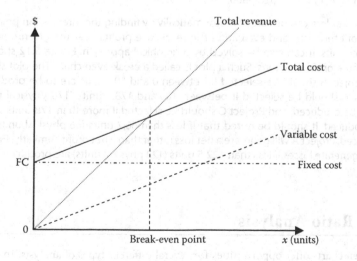

FIGURE 5.10 Break-even points for a single project.

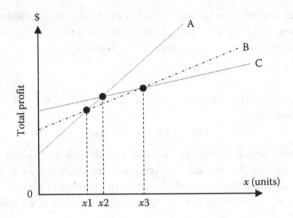

FIGURE 5.11 Break-even points for multiple projects.

Example

Three project alternatives shown in Figure 5.11 are being considered for producing a new product. The required analysis involves determining which alternative should be selected on the basis of how many units of the product are produced per year. Based on past records, there is a known relationship between the number of units produced per year, x, and the net annual profit, $P(x)$, from each alternative. The level of production is expected to be between 0 and 250 units per year. The net annual profits (in thousands of dollars) are given below for each alternative:

Project A: $P(x) = 3x - 200$
Project B: $P(x) = x$
Project C: $P(x) = (1/50)x^2 - 300$.

This problem can be solved mathematically by finding the intersection points of the profit functions and evaluating the respective profits over the given range of product units. It can also be solved by a graphical approach. Figure 5.12 shows a plot of the profit functions. Such a plot is called a break-even chart. The plot shows that Project B should be selected if between 0 and 100 units are to be produced, Project A should be selected if between 100 and 178.1 units (178 physical units) are to be produced, and Project C should be selected if more than 178 units are to be produced. It should be noted that if less than 66.7 units (66 physical units) are produced, Project A will generate a net loss rather than a net profit. Similarly, Project C will generate losses if less than 122.5 units (122 physical units) are produced.

Profit Ratio Analysis

Break-even charts offer opportunities for several different types of analysis. In addition to the break-even points, other measures of worth or criterion measures may be derived

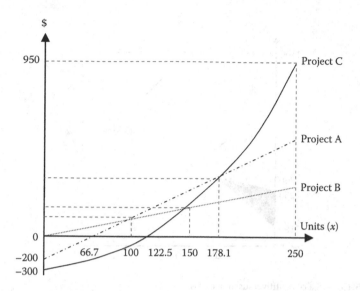

FIGURE 5.12   Plot of profit functions.

from the charts. A measure called the *profit ratio* (Badiru, 1996) is presented here for the purpose of obtaining a further comparative basis for competing projects. A profit ratio is defined as the ratio of the profit area to the sum of the profit and loss areas in a break-even chart. That is,

$$\text{Profit ratio} = \frac{\text{area of profit region}}{\text{area of profit region} + \text{area of loss region}}.$$

For example, suppose that the expected revenue and the expected total cost associated with a project are given, respectively, by the following expressions:

$$R(x) = 100 + 10x$$

and

$$TC(x) = 2.5x + 250,$$

where x is the number of units produced and sold from the project. Figures 5.13, 5.14, and 5.15 shows the break-even chart for the project. The break-even point is shown to be 20 units. Net profits are realized from the project if more than 20 units are produced and net losses are realized if less than 20 units are produced. It should be noted that the revenue function in Figure 5.16 represents an unusual case, in which a revenue of $100 is realized when zero units are produced.

Suppose it is desired to calculate the profit ratio for this project if the number of units that can be produced is limited to between 0 and 100 units. From Figure 5.16, the surface area of the profit region and the area of the loss region can be calculated by using the standard formula for finding the area of a triangle: Area = (1/2)(base)(height). Using this

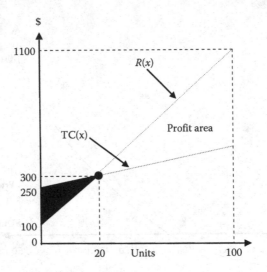

FIGURE 5.13 Area of profit versus area of loss.

formula, we have the following:

$$\text{Area of profit region} = \frac{1}{2}(\text{base})(\text{height})$$

$$= \frac{1}{2}(1100 - 500)(100 - 20)$$

$$= 24{,}000 \text{ square units,}$$

$$\text{Area of loss region} = \frac{1}{2}(\text{base})(\text{height})$$

$$= \frac{1}{2}(250 - 100)(20)$$

$$= 1500 \text{ square units.}$$

Thus, the profit ratio is computed as follows:

$$\text{Profit ratio} = \frac{24{,}000}{24{,}000 + 1500} = 0.9411 \equiv 94.11\%.$$

The profit ratio may be used as a criterion for selecting among project alternatives. If this is done, the profit ratios for all the alternatives must be calculated over the same values of the independent variable. The project with the highest profit ratio will be selected as the desired project. For example, Figure 5.14 presents the break-even chart for an alternate project, say Project II. It can be seen that both the revenue and cost functions for the project are nonlinear. The revenue and cost are defined as follows:

$$R(x) = 160x - x^2$$

and

$$TC(x) = 500 + x^2.$$

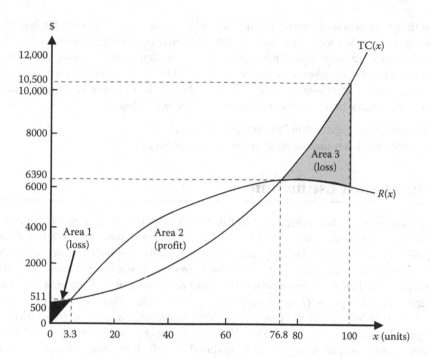

FIGURE 5.14 Break-even chart for revenue and cost functions.

If the cost and/or revenue functions for a project are not linear, the areas bounded by the functions may not be easily determined. For those cases, it may be necessary to use techniques such as definite integrals to find the areas. Figure 5.14 indicates that the project generates a loss if less than 3.3 units (3 actual units) are produced or if more than 76.8 units (76 actual units) are produced. The respective profit and loss areas on the chart are calculated as shown below:

Area 1 (loss) = 802.80 unit-dollars
Area 2 (profit) = 132,272.08 unit-dollars
Area 3 (loss) = 48,135.98 unit-dollars

Consequently, the profit ratio for Project II is computed as

$$\text{Profit ratio} = \frac{\text{total area of profit region}}{\text{total area of profit region} + \text{total area of loss region}}$$
$$= \frac{132,272.08}{802.76 + 132,272.08 + 48,135.98}$$
$$= 72.99\%.$$

The profit ratio approach evaluates the performance of each alternative over a specified range of operating levels. Most of the existing evaluation methods use single-point analysis with the assumption that the operating condition is fixed at a given production level. The profit ratio measure allows an analyst to evaluate the net yield of an alternative, given that the production level may shift from one level to another. An alternative, for example, may operate at a loss for most of its early life, but it may generate large incomes to offset those losses in its later stages. Conventional methods cannot easily capture this type of transition from one performance level to another. In addition to being used to compare alternate projects, the profit ratio may also be used for evaluating the economic feasibility of a single project. In such a case, a decision rule may be developed, such as the following:

If profit ratio is greater than 75%, accept the project
If profit ratio is less than or equal to 75%, reject the project

Project Cost Estimation

Cost estimation and budgeting help establish a strategy for allocating resources in project planning and control. Based on the desired level of accuracy, there are three major categories of cost estimation for budgeting: *order-of-magnitude estimates*, *preliminary cost estimates*, and *detailed cost estimates*. Order-of-magnitude cost estimates are usually gross estimates based on the experience and judgment of the estimator. They are sometimes called "ballpark" figures. These estimates are typically made without a formal evaluation of the details involved in the project. The level of accuracy associated with order-of-magnitude estimates can range from −50% to +50% of the actual cost. These estimates provide a quick way of getting cost information during the initial stages of a project. The estimation range is summarized as follows:

$$50\%(\text{actual cost}) \leq \text{order-of-magnitude estimate} \leq 150\%(\text{actual cost}).$$

Preliminary cost estimates are also gross estimates, but with a higher level of accuracy. In developing preliminary cost estimates, more attention is paid to some selected details of the project. An example of a preliminary cost estimate is the estimation of expected labor cost. Preliminary estimates are useful for evaluating project alternatives before final commitments are made. The level of accuracy associated with preliminary estimates can range from −20% to +20% of the actual cost, as shown below:

$$80\%(\text{actual cost}) \leq \text{preliminary estimate} \leq 120\%(\text{actual cost}).$$

Detailed cost estimates are developed after careful consideration is given to all the major details of a project. Considerable time is typically needed to obtain detailed cost estimates. Because of the amount of time and effort needed to develop detailed cost estimates, the estimates are usually developed after a firm commitment has been made that the project will take off. Detailed cost estimates are important for evaluating actual cost performance during the project. The level of accuracy associated with detailed estimates normally ranges from −5% to +5% of the actual cost.

$$95\%(\text{actual cost}) \leq \text{detailed cost} \leq 105\%(\text{actual cost}).$$

There are two basic approaches to generating cost estimates. The first one is a variant approach, in which cost estimates are based on variations of previous cost records. The other approach is the generative cost estimation, in which cost estimates are developed from scratch without taking previous cost records into consideration.

Optimistic and Pessimistic Cost Estimates

Using an adaptation of the Program Evaluation and Review Technique (PERT) formula, we can combine optimistic and pessimistic cost estimates. If O is the optimistic cost estimate, M the most likely cost estimate, and P the pessimistic cost estimate, then the estimated cost can be stated as follows:

$$E[C] = (O + 4M + P)6$$

and the cost variance can be estimated as follows:

$$V[C] = \left[\frac{P - O}{6}\right]^2.$$

Cost Performance Index

As a project progresses, costs can be monitored and evaluated to identify areas of unacceptable cost performance. Figure 5.15 shows a plot of cost versus time for projected cost and actual cost. The plot permits a quick identification of the points at which cost overruns occur in a project.

Plots similar to those presented above may be used to evaluate cost, schedule, and time performance of a project. An approach similar to the profit ratio presented earlier may be used along with the plot to evaluate the overall cost performance of a project over a specified planning horizon. Presented below is a formula for cost performance index (CPI):

$$\text{CPI} = \frac{\text{area of cost benefit}}{\text{area of cost benefit} + \text{area of cost overrun}}.$$

As in the case of the profit ratio, CPI may be used to evaluate the relative performances of several project alternatives or to evaluate the feasibility and acceptability of an individual alternative.

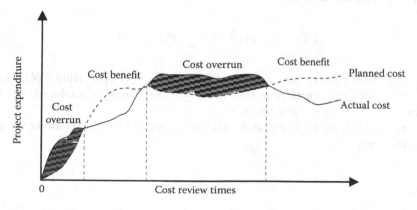

FIGURE 5.15 Evaluation of actual and projected cost.

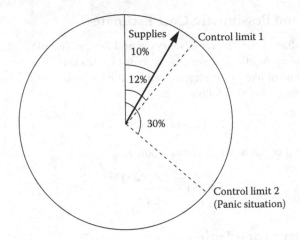

FIGURE 5.16 Cost-control pie chart.

Cost Control Limits

Figure 5.16 presents a cost-control pie chart, which can be used to track the percentage of the cost going into a specific component of a project. Control limits can be included in the pie chart to identify costs that have gone out of control. The example in Figure 5.16 shows that 10% of total cost is tied up in supplies. The control limit is located at 12% of total cost. Hence, the supplies expenditure is within control (so far, at least).

Project Balance Computation

The project balance technique helps in assessing the economic state of a project at a desired point in time in the life cycle of the project. It calculates the net cash flow of a project up to a given point in time:

$$B(i)_t = S_t - P(1+i)^t + \sum_{k=1}^{t} PW_{income}(i)_k,$$

where $B(i)_t$ is the project balance at time t at an interest rate of $i\%$ per period, PW income $(i)_k$ the present worth of net income from the project up to time k, P the initial cost of the project, and S_t the salvage value at time t.

The project balance at time t gives the net loss or net profit associated with the project up to that time.

6

Industrial Production Calculations

This chapter presents a collection of useful formulas and computational examples for industrial production planning and control. Reference sources for this chapter include Badiru (1996), Badiru and Omitaomu (2007), Badiru et al. (2008), Heragu (1997), and Moffat (1987).

Learning Curve Models and Computations

There are several models of learning curves in use in business and industry. The log–linear model is, perhaps, the most extensively used. The log–linear model states that the improvement in productivity is constant (i.e., it has a constant slope) as output increases. There are two basic forms of the log–linear model: the average cost function and the unit cost function.

The Average Cost Model

The average cost model is used more frequently than the unit cost model. It specifies the relationship between the cumulative average cost per unit and cumulative production. The relationship indicates that cumulative cost per unit will decrease by a constant percentage as the cumulative production volume doubles. The *average cost* model is expressed as

$$A_x = C_1 x^b,$$

where A_x is the cumulative average cost of producing x units, C_1 the cost of the first unit, x the cumulative production count, and b the learning curve exponent (i.e., constant slope of on log–log paper).

The relationship between the learning curve exponent, b, and the learning rate percentage, p, is given by

$$b = \frac{\log p}{\log 2} \quad \text{or} \quad p = 2^b.$$

The derivation of the above relationship can be seen by considering two production levels where one level is double the other, as shown below.

Let Level I $= x_1$ and Level II $= x_2 = 2x_1$. Then,

$$A_{x_1} = C_1(x_1)^b \quad \text{and} \quad A_{x_2} = C_1(2x_1)^b.$$

The percent productivity gain is then computed as

$$p = \frac{C_1(2x_1)^b}{C_1(x_1)^b} = 2^b.$$

On a log–log paper, the model is represented by the following straight-line equation:

$$\log A_x = \log C_1 + b \log x,$$

where b is the constant slope of the line. It is from this straight line that the name log–linear was derived.

Computational Example

Assume that 50 units of an item are produced at a cumulative average cost of $20 per unit. Suppose we want to compute the learning percentage when 100 units are produced at a cumulative average cost of $15 per unit. The learning curve analysis would proceed as follows:

Initial production level $= 50$ units; average cost $= \$20$

Double production level $= 100$ units; cumulative average cost $= \$15$.

Using the log relationship, we obtain the following equations:

$$\log 20 = \log C_1 + b \log 50,$$

$$\log 15 = \log C_1 + b \log 100.$$

Solving the equations simultaneously yields

$$b = \frac{\log 20 - \log 15}{\log 50 - \log 100} = -0.415.$$

Thus,

$$p = 2^{-0.415} = 0.75.$$

That is 75% learning rate. In general, the learning curve exponent, b, may be calculated directly from actual data or computed analytically. That is

$$b = \frac{\log A_{x_1} - \log A_{x_2}}{\log x_1 - \log x_2}$$

or

$$b = \frac{\ln p}{\ln 2},$$

where x_1 is the first production level, x_2 the second production level, A_{x_1} the cumulative average cost per unit at the first production level, A_{x_2} the cumulative average cost per unit at the second production level, and p the learning rate percentage.

Using the basic cumulative average cost function, the total cost of producing x units is computed as

$$TC_x = xA_x = xC_1x^b = C_1x^{(b+1)}.$$

The unit cost of producing the xth unit is given by

$$U_x = C_1x^{(b+1)} - C_1(x-1)^{(b+1)}$$
$$= C_1[x^{(b+1)} - (x-1)^{(b+1)}].$$

The marginal cost of producing the xth unit is given by

$$MC_x = \frac{d[TC_x]}{dx} = (b+1)C_1x^b.$$

Computational Example

Suppose in a production run of a certain product, it is observed that the cumulative hours required to produce 100 units is 100,000 h with a learning curve effect of 85%. For project planning purposes, an analyst needs to calculate the number of hours spent in producing the 50th unit. Following the notation used previously, we have the following information:

$$p = 0.85$$
$$X = 100 \text{ units}$$
$$A_x = 100,000 \text{ h}/100 \text{ units} = 1000 \text{ h/unit}$$

Now,

$$0.85 = 2^b.$$

Therefore, $b = 0.2345$. Also,

$$1000 = C_1(100)^b.$$

Therefore, $C_1 = 2944.42$ h. Thus,

$$C_{50} = C_1(50)^b$$
$$= 1176.50 \text{ h}.$$

That is, the cumulative average hours for 50 units is 1176.50 h. Therefore, cumulative total hours for 50 units = 58,824.91 h. Similarly,

$$C_{49} = C_1(49)^b = 1182.09 \text{ h}.$$

That is, the cumulative average hours for 49 units is 1182.09 h. Therefore, the cumulative total hours for 49 units = 57,922.17 h. Consequently, the number of hours for the 50th unit is given by

$$C_{50} - C_{49} = 58,824.91 \text{ h} - 57,922.17 \text{ h}$$
$$= 902.74 \text{ h}.$$

The Unit Cost Model

The unit cost model is expressed in terms of the specific cost of producing the xth unit. The unit cost formula specifies that the individual cost per unit will decrease by a constant percentage as cumulative production doubles. The formulation of the unit cost model is presented below. Define the average cost as A_x.

$$A_x = C_1 x^b.$$

The total cost is defined as

$$TC_x = xA_x = xC_1 x^b = C_1 x^{(b+1)}$$

and the marginal cost is given by

$$MC_x = \frac{d[TC_x]}{dx} = (b+1)C_1 x^b.$$

This is the cost of one specific unit. Therefore, we can define the unit cost model as

$$U_x = (1+b)C_1 x^b,$$

where U_x is the cost of producing the xth unit. We will derive the relationship between A_x and U_x using the following relationship:

$$U_x = (1+b)C_1 x^b,$$

$$\frac{U_x}{1+b} = C_1 x^b = A_x,$$

$$A_x = \frac{U_x}{1+b},$$

$$U_x = (1+b)A_x.$$

To derive an expression for finding the cost of the first unit, C_1, we will proceed as follows. Since $A_x = C_1 x^b$, we have

$$C_1 x^b = \frac{U_x}{1+b},$$

$$\therefore C_1 = \frac{U_x x^{-b}}{1+b}.$$

For the case of continuous product volume (e.g., chemical processes), we have the following corresponding expressions:

$$TC_x = \int_0^x U(z)\,dz = C_1 \int_0^x z^b\,dz = \frac{C_1 x^{(b+1)}}{b+1},$$

$$Y_x = \left(\frac{1}{x}\right)\frac{C_1 x^{(b+1)}}{b+1},$$

$$MC_x = \frac{d\,[TC_x]}{dx} = \frac{d[C_1 x^{(b+1)}/(b+1)]}{dx} = C_1 x^b.$$

Productivity Calculations Using Learning Curves

Let a_1 is the time to complete the task the first time, a_n the overall average unit time after completing the task n times, I the time improvement factor, k the learning factor, n the number of times the task has been completed, r_n the ratio of time to perform the task for the nth time divided by time to perform for the $(n-1)$th time, t_n the time to perform the task the nth time, and t_{tn} the total time to perform the task n times.

Determining Average Unit Time

$$a_n = \frac{a_1}{n^k}.$$

Calculation of the Learning Factor

$$k = \frac{\log a_1 - \log a_n}{\log n}.$$

Calculating Total Time for a Job

$$t_{tn} = t_1 n^{1-k}.$$

Time Required to Perform a Task for the nth Time

$$t_n = a_1 \left(\frac{1-k}{n^k} \right).$$

The Improvement Ratio

This is the number of times a task must be completed before improvement flattens to a given ratio:

$$n \geq \frac{1}{1 - r^{1/k}}.$$

Computational Example

Suppose a new employee can perform a task for the first time in 30 min. Assume that the learning factor for the task is 0.07. At steady-state operation, an experienced employee can perform the task such that there is less than 1% improvement in successive execution of the task. This means that the ratio of times between succeeding tasks will be 0.99 or larger. How long will it take the experienced employee to perform the task?

Solution

First, determine how many times the task must be performed before we can achieve an improvement ratio of 0.99. Then, use the resulting value of n to compute the required processing.

$$n \geq \frac{1}{1 - 0.99^{1/.07}}$$

$$\geq 7.4769 \approx 8.$$

Now, determine how long it should take to perform the task for the eighth time.

$$t_8 = 30 \left(\frac{1 - 0.07}{8^{.07}} \right)$$

$$= 24.12 \text{ min}$$

Computation for Improvement Goal in Unit Time

This is to find how many times a task should be performed before there will be a given improvement in unit time:

$$n_2 = \frac{n_1}{I^{1/k}}.$$

If this computation results in an extremely large value of n_2, it implies that the specified improvement goal cannot be achieved. This could be because the limit of improvement is reached due to leveling off of the learning curve before the goal is reached.

Computation of Machine Output Capacities

Notation

e_c	efficiency of capital equipment in percent of running time
f_a	fraction of output accepted $(= U_a/C)$
f_r	fraction of output rejected $(= U_r/U)$
n	number of machines
t	time to manufacture one unit (in min)
T_a	time per shift that machine actually produces (in min)
T_t	hours worked per day = eight times numbers of shifts (i.e., 8 h per shift)
T_y	production hours per year, usually 2080 h (i.e., 8 h/day × 260 days)
U	unit per shift $(= U_a/U_r)$
U_a	units accepted per shift
U_r	units rejected per shift
U_y	units manufactured per year

Machine Utilization Ratio: Determining How Often a Machine Is Idle

$$e_c = \frac{\text{time the machine is working}}{\text{time the machine could be working}}.$$

The above expression can also be used for calculating the portion of time a machine is idle. In that case, the numerator would be idle time instead of working time.

Calculating Number of Machines Needed to Meet Output

When total output is specified, the number of machines required can be calculated from the following formula:

$$n = \frac{1.67tU}{T_t e_c}.$$

Example

Suppose we desire to produce 1500 units per shift on machines that each complete one unit every 5 min. The production department works one 8-h shift with two 15-min breaks. Thus, the machines run 7 h and 30 min per day. How many machines will be required?

SOLUTION

Efficiency must first be calculated as a percentage of operating time in relation to shift duration. That is,

$$e_c = \left(\frac{7\,h, 30\,min}{8\,h}\right) \times 100$$
$$= 93.75\%$$

The above percent efficiency value is now substituted into the computational formula for the number of machines. That is,

$$n = \frac{1.67 \times 5 \times 1500}{8 \times 93.75}$$
$$= 16.7 \approx 17 \text{ machines.}$$

Alternate Forms for Machine Formula

The formula for calculating number of machines can be rearranged to produce the alternate forms presented below. In some cases, e_c (percentage efficiency of time usage) is replaced by actual running time in minutes (t_a):

$$t = \frac{0.6nT_t e_c}{U},$$
$$U = \frac{0.6nT_t e_c}{t},$$
$$T_t = \frac{1.67tU}{ne_c},$$
$$e_c = \frac{1.67tU}{nT_t},$$
$$t = \frac{60nt_a}{U},$$

$$U = \frac{60nt_a}{t},$$

$$T_a = \frac{0.0167tU}{n}.$$

Example

How many hours a day must you operate six machines that each produce one unit every 50 s if you need 4000 units per day?

To find hours per day (T_a), substitute directly into the last one of the alternate forms of the machine equation. Note that t is defined in minutes in the equation, but the problem gives the time in seconds. So, first, convert 50 seconds to minutes. That is, 50 seconds = 50/60 minutes. Now, hours per day is calculated as

$$T_a = \frac{0.0167(50/60)4000}{6}$$

$$= 9.28 \, \text{h}.$$

Based on this calculation, it is seen that adding more machines is not an option. We must either extend the shift to about 9.28 h or perform part of the production on another shift, or outsource.

Referring back to the first computational example, let us suppose we want to compare total production from 17 and from 18 machines. The units produced per shift can be calculated as shown below:

$$U = \frac{0.6(17)8(93.75)}{5}$$

$$= 1530 \, \text{units from 17 machines.}$$

$$U = \frac{0.6(18)8(93.75)}{5}$$

$$= 1620 \, \text{units from 18 machines.}$$

Calculating Buffer Production to Allow for Defects

If a certain fraction of output is usually rejected, a buffer can be produced to allow for rejects. The following formula shows how many to produce in order to have a given number of acceptable units:

$$U = \frac{U_a}{1 - f_r}$$

$$= \frac{U_a}{f_a}.$$

Example

A production department normally rejects 3% of a machine's output. How many pieces should be produced in order to have 1275 acceptable ones?

The reject rate of 3% translates to the fraction 0.03. So, $f_r = 0.03$. Thus, we have

$$U = \frac{1275}{1 - 0.03}$$
$$= 1315.$$

Adjusting Machine Requirements to Allow for Rejects

To determine how many machines are required for a given amount of shippable output, considering that a given portion of product is rejected, we will use the following formula:

$$n = \frac{U_y t}{0.6 e_c T_y (1 - f_r)}.$$

Example

A manufacturer of widgets must have 500,000 ready to ship each year. One out of 300 generally has a defect and must be scrapped. The production machines, each of which makes 50 widgets per hour, work 90% of the time and the plant operates on a 2080-h year. Calculate the number of machines required to meet the target production.

SOLUTION

Production time per widget is 60/50 min per widget.

$$n = \frac{500,000(60/50)}{0.6(90)2080(1 - 1/300)}$$
$$= 4.47 \approx 5 \text{ machines.}$$

The plant will need five machines, but that will change the utilization ratio. To calculate the new ratio, we will rearrange the formula to solve for e.

$$e = \frac{U_y t}{0.6 n T_y (1 - f_r)}$$
$$= \frac{500,000(60/50)}{0.6(5)2080(1 - 1/300)}$$
$$= 96.48\%.$$

The higher percentage utilization is due to using whole number of machines (5) instead of the fractional computation of 4.47.

Output Computations for Sequence of Machines

In a sequence of machines, the output of one set of machines will move on to the next stages for other sets of machines. The goal here is to determine how many items should be started at the first set of machines so that a given number of units will be produced from the sequence of machines. For each state,

$$\text{number of units needed at first stage} = \frac{\text{output needed}}{1 - f_r}.$$

Example

The first operation grinds a casting and 1% of the machined items are rejected instead of being passed on to the next operation. Buffing and polishing is second; 0.5% of the output from this set of machines is rejected. Problems with the third set of machines (nameplate attachment) result in a rejection rate of 2%. How many units should be started into the sequence if we are to ship 30,000 if all rejects are scrapped?

Apply the formula to each stage, starting at the last one:

$$\text{Start}_3 = \frac{30,000}{1 - 0.02}$$
$$= 30,613.$$

Then determine how many units should be started at the second stage in order to have 30,613 units at the input to the third stage:

$$\text{Start}_2 = \frac{30,613}{1 - .005}$$
$$= 30,767.$$

Finally, determine how many units should be started at the first stage:

$$\text{Start}_1 = \frac{30,767}{1 - 0.01}$$
$$= 31,078.$$

Therefore, at least 31,078 units should be started at the first stage of this manufacturing process.

Calculating Forces on Cutting Tools

This section gives formulas for finding forces required by milling machines, drills, and other tools that cut or cut into material (Moffat, 1987). Machine setting for minimum cost is also included. The following notations are used in this section:

a   cross-sectional area of chip or piece (in.^{2})
     depth of cut times feed (one spindle revolution)

c_l cost of rejecting piece outside lower limit
c_u cost of rejecting piece outside upper limit
d diameter of work or tool
g goal for machine setting (in.)
k constant, determined from tables by the formulas
L_l lower limit for accepting piece
L_u upper limit for accepting piece
p pressure at point of cut (pounds)
s cutting speed (ft/min)
T tolerance (in.)
w horsepower at tool
σ standard deviation

Calculating Pressure

A cutting tool is subjected to pressure at the point of cut, as calculated from the following:

$$p = 80{,}300(1.33)^k a,$$

where k is a constant, provided in Table 6.1.

Example

A cylinder of cast iron is being cut by a lathe to a depth of 0.06 in., with a feed of 0.0286 in. per revolution. How much pressure is at the cut?

$$p = 80{,}300(1.33)^{1.69}(0.06)(0.0286)$$
$$= 223 \text{ pounds.}$$

Finding Required Horsepower

A formula similar to the formula for pressure finds the horsepower required at the tool's cutting edge.

$$w = 2433(1.33)^k as,$$

where k is a constant given in Table 6.1.

TABLE 6.1   Constant for Pressure and Horsepower Formulas

Type of Material	Constant k
Bronze	1.26
Cast iron	1.69
Cast steel	2.93
Mild steel	4.12
High carbon steel	5.06

Example

A tool cuts mild steel to a depth of 0.045 in., feeding at 0.020 in. What horsepower is required at the cutting edge if the cutting speed is 70 ft/min?

$$w = 2.433(1.33)^{4.12}(0.045)(0.020)(70)$$
$$= 0.4963 \text{ horsepower.}$$

Machine Accuracy

The dimensional tolerance to which a piece can be held is a function of both the machining operation and the piece's dimension.

$$T = kd^{0.371}.$$

Here the coefficient k is a function of the machining operation, as shown in Table 6.2.

Example

A bar with 0.875-in. diameter is being turned on a turret lathe. What is a reasonable tolerance?

$$T = 0.003182(0.875)^{0.371}$$
$$= 0.003028 \text{ in.}$$

Calculating the Goal Dimension

The formula below finds nominal measurement for machining. This may not be the midpoint of the tolerance band. This setting will minimize costs under the following conditions:

a. If the part has not been machined enough, it can be reworked.
b. If the part has been machined too far, it must be scrapped.

TABLE 6.2 Constant k for Machining Accuracy Formula

Machining Operation	k
Drilling, rough turning	0.007816
Finish turning, milling	0.004812
Turning on turret lathe	0.003182
Automatic turning	0.002417
Broaching	0.001667
Reaming	0.001642
Precision turning	0.001378
Machine grinding	0.001008
Honing	0.000684

Work coming from an automatic machine can be expected to have a normal distribution about the center value to which the machine is set. However, it is not always desirable for the distribution to be centered about the design value because the cost of correcting rejects may not be the same on each side. This formula applies when it costs more to reject a piece because of too-small dimensions than because of too-large dimensions. It will, therefore, tell us to set the machine for a goal dimension on the large side of center, so that most of the rejects will be due to too-large dimensions (Moffat, 1987).

$$g = \frac{1}{2(L_u - L_l)} \left[L_u^2 - L_l^2 + \sigma^2 \ln\left(\frac{c_l}{c_u}\right) \right].$$

Example

A solid cylinder is to be turned down to 1.125 in. diameter by a machine under robotic control. If a piece comes off the machine measuring 1.128 in. or larger, a new shop order must be cut to have the piece machined further. That procedure costs an additional $10.80. When a piece measures 1.122 in. or smaller, it must be scrapped at a total cost of $23.35. Standard deviation of the machined measurements is 0.002 in. To what dimension should the machine be set?

$$g = \frac{1}{2(1.128 - 1.122)} \left[1.128^2 - 1.122^2 + 0.002^2 \ln\left(\frac{23.35}{10.80}\right) \right]$$
$$= 1.1253 \text{ in.}$$

It costs more to scrap a piece than to rework it, so the calculation above suggests aiming on the high side to avoid undercutting a piece. This is a reasonable practice.

The following notations are used in the next section.

d diameter of drill (in.)
f feed per revolution (in.)
f_{hp} horsepower equivalent of torque
f_{th} thrust (pounds)
f_{tq} torque (in. pounds)
v rotational velocity (rpm)

Drill Thrust

The amount of thrust required for drilling is found from the following formula:

$$f_{th} = 57.5 f^{0.8} d^{1.8} + 625 d^2.$$

Example

A 3/8-in. drill is feeding at 0.004 in. per revolution. What is the thrust?

$$f_{th} = 57.5(0.004)^{0.8}(0.375)^{1.8} + 625(0.375)^2$$
$$= 88.01 \text{ pounds.}$$

Drill Torque

This formula calculates the torque for a drilling operation.

$$f_{tq} = 25.2 f^{0.8} d^{1.8}.$$

Example

Calculate the torque for the preceding example.

$$f_{tq} = 25.2(0.004)^{0.8}(0.375)^{1.8}$$
$$= 0.052 \text{ in. pound.}$$

Drill Horsepower

This formula finds the horsepower equivalent of a torque.

$$f_{hp} = 15.87 \times 10^{-6} f_{tq} v$$
$$= 400 \times 10^{-6} f^{0.8} d^{1.8} v.$$

Example

What is the equivalent horsepower in the above example if the drill rotates at 650 rpm?

$$f_{hp} = 15.87 \times 10^{-6}(0.052)650$$
$$= 0.0005 \text{ horsepower.}$$

Calculating Speed and Rotation of Machines

This section looks at turning speed, surface speed, and rotational speed of tools and work. The notations used here are:

a_l linear feed, or advance (in./min)
a_r feed (in. per revolution)
a_t feed (in. per tooth)
d diameter of work where tool is operating (in.)
r radius of work where tool is operating (in.)
s cutting speed (ft/min)
s_r rotational speed of work (rpm)
s_s surface speed of work (ft/min)

Shaft Speed

Cylindrical shapes passing a contact point with a tool, as in a lathe, should move at a speed recommended by the manufacturer. The following formula translates surface speed of the cylinder to revolutions per minute:

$$s_r = \frac{12s_s}{\pi d}$$
$$= \frac{6s_s}{\pi r}.$$

Example

A solid metal cylinder with $1\frac{1}{4}$-in. diameter is to be worked in a lathe. Its recommended surface speed is 240 ft/min. How fast should the lathe turn?

$$s_r = \frac{12(240)}{3.1416(1.25)}$$
$$= 733 \text{ rpm.}$$

Note that the nearest available speed to 733 rpm will bring the surface speed to its recommended value.

Example

A lathe cuts at a speed of 102 ft/min on cylindrical work with a diameter of 4.385 in. What is the spindle speed?

$$s_r = \frac{12(102)}{3.1416(4.385)}$$
$$= 88.85 \text{ rpm.}$$

Example

The material should not be cut faster than 200 ft/min. What is the fastest rotation allowed when the material's radius is $2\frac{1}{2}$ in.?

$$s_r = \frac{1.9099(200)}{2.25}$$
$$= 152.79 \text{ rpm.}$$

Thus, rotational speed should be kept below 153 ft/min.

Surface Speed

The following formula finds the rate at which the work's surface passes the tool:

$$s_s = 0.2618ds_r$$
$$= 0.5236rs_r.$$

Example

Cylindrical material of 1.12-in. diameter rotates at 460 rpm. What is the work's surface speed?

$$S_s = 0.2618(1.12)460$$
$$= 134.88 \text{ ft/min.}$$

Tool Feed per Revolution

The rate at which the tool advances into or past the work is given by the formula

$$a_r = \frac{a_l}{s_r}$$
$$= \frac{0.2618 d a_l}{s_s}$$
$$= \frac{0.5236 r a_l}{s_s}.$$

Example

The tool is to advance 40 in./min when the work has a diameter of 0.875 in. If the surface speed is 180 ft/min, at what rate should the tool advance?

$$a_r = \frac{0.2618(0.875)40}{180}$$
$$= 0.0509 \text{ in. per revolution.}$$

Tool Feed per Minute

The formula below is a rearrangement of an earlier example. This is used when the rotational speed is known.

$$a_l = a_r s_r$$
$$= \frac{3.8197 a_r s_s}{d}$$
$$= \frac{1.9099 a_r s_s}{r}.$$

Example

Assume that a tool advances 0.075 in. into a work piece for each revolution and the work is rotating 660 revolutions per minute. What is the feed rate?

$$a_l = 0.075(660)$$
$$= 49.5 \text{ in./min.}$$

Tool Feed per Tooth

The following formula is used when the cutter information is known in terms of teeth instead of time:

$$a_t = \frac{a_r}{t}$$
$$= \frac{a_l}{ts_r}.$$

Example

A cutter with 33 teeth advances 0.06 in. per revolution. At what rate does it advance?

$$a_t = \frac{0.06}{33}$$
$$= 0.0018 \text{ in. per tooth.}$$

Computation to Find the Volume of Material Removed

Formulas in this section (Moffat, 1987) find the volume of material removed by operations such as cutting, drilling, and milling. In many applications, the results of these formulas (volume) will be multiplied by unit weight of the material to find the weight of material removed. The notations used are as follows:

- d diameter of hole or tool (in.)
- e distance cut extends on surface (in.)
- f linear feed rate (in./min)
- h depth of cut (in.)
- t angle of tip of drill (°)
- w width of cut (in.)
- v volume (in.^{3})
- v_r rate of volume removal (in.^{3}/min)

Time Rate for Material Removal in a Rectangular Cut

The following formula gives the rate at which the material is removed when the cut has a rectangular shape:

$$v_r = whf.$$

Example

A milling machine is cutting a 1/8-in. deep groove with the face of a 5/8-in. diameter tool. It feeds at 28 in./min. At what rate is material removed?

$$v_r = 0.625(0.125)28$$
$$= 2.1875 \text{ in.}^3/\text{min.}$$

Calculation of Linear Feed for Rectangular Material Removal

The following formula is used to find the feed rate required for a given rate of material removal:

$$f = \frac{v_r}{wh}.$$

The above formula applies when material is removed by a drill. One version of the formula approximates the volume by assuming that the drill is a plain cylinder without a triangular tip. The second version is precise and includes a correction for the angle of the drill's tip.

Example

Let us assume that we want to remove 3 in.³/min while milling 1/8-in. from the side of a block that is $1\frac{1}{16}$ in. thick. How fast should the cutter feed if we are cutting with the circumference of a tool at least $1\frac{1}{16}$ in. long?

$$f = \frac{3}{1.0625(0.125)}$$
$$= 22.6 \text{ in./min.}$$

Takt Time for Production Planning

"Takt" is the German word referring to how an orchestra conductor regulates the speed, beat, or timing so that the orchestra plays in unison. So, the idea of *takt time* is to regulate the rate time or pace of producing a completed product. This refers to the production pace at which workstations must operate in order to meet a target production output rate. The production output rate is set based on product demand. In a simple sense, if 2000 units of a widget are to be produced within an 8-h shift to meet a market demand, then 250 units must be produced per hour. That means, a unit must be produced every 60/250 = 0.24 min (14.8 s). Thus, the takt time is 14.4 s. Lean production planning then requires that workstations be balanced such that the production line generates a product every 14.4 s. This is distinguished from the *cycle time*, which refers to the actual time required to accomplish each workstation task. Cycle time may be less than, more than, or equal to takt time. Takt is not a number that can be measured with a stopwatch. It must be calculated based on the prevailing production needs and scenario. Takt time is defined by the following equation:

$$\text{Takt time} = \frac{\text{Available work time} - \text{Breaks}}{\text{Customer demand}}$$
$$= \frac{\text{Net available time per day}}{\text{Customer demand per day}}.$$

Takt time is expressed as "seconds per piece," indicating that customers are buying a product once every so many seconds. Takt time is not expressed as "pieces per second." The objective of lean production is to bring the cycle time as close to the takt time as possible; that is, choreographed. In a balanced line design, the takt time is the reciprocal of the production rate.

Improper recognition of the role of takt time can make an analyst to overestimate the production rate capability of a line. Many manufacturers have been known to over-commit to customer deliveries without accounting for the limitations imposed by takt time. Since takt time is set based on customer demand, its setting may lead to an unrealistic expectations of workstations. For example, if the constraints of the prevailing learning curve will not allow sufficient learning time for new operators, then takt times cannot be sustained. This may lead to the need for buffers to temporarily accumulate units at some workstations. But this defeats the pursuits of lean production or just-in-time. The need for buffers is a symptom of imbalances in takt time. Some manufacturers build *takt gap* into their production planning for the purpose of absorbing nonstandard occurrences in the production line. However, if there are more nonstandard or random events than have been planned for, then production rate disruption will occur.

It is important to recognize that the maximum production rate determines the minimum takt time for a production line. When demand increases, takt time should be decreased. When demand decreases, takt time should be increased. Production crew size plays an important role in setting and meeting takt time. The equation for calculating the crew size for an assembly line doing one piece flow that is paced to takt time is presented below:

$$\text{Crew size} = \frac{\text{Sum of manual cycle time}}{\text{Takt time}}.$$

Production Crew Work Rate Analysis

When resources work concurrently at different work rates, the amount of work accomplished by each may be computed for work planning purposes. The general relationship between work, work rate, and time can be expressed as

$$w = rt,$$

where w is the amount of actual work accomplished. This is expressed in appropriate units, such as miles of road completed, lines of computer code entered, gallons of oil spill cleaned, units of widgets produced, or surface area painted. Here r is the work rate per unit time at which the assigned work is accomplished and t the total time required to accomplish the work.

It should be noted that work rate can change due to the effects of learning curves. In the discussions that follow, it is assumed that work rates remain constant for at least the duration of the work being analyzed.

Work is defined as a physical measure of accomplishment with uniform density (i.e., homogeneous). For example, one square footage of construction may be said to be homogeneous if one square footage is as complex and desirable as any other square footage. Similarly, cleaning one gallon of oil spill is as good as cleaning any other gallon of oil spill within the same work environment. The production of one unit of a product is identical to the production of any other unit of the product. If uniform work density can be

TABLE 6.3 Work Rate Table for Single Resource Unit

Resource	Work rate	Time	Work Done
Resource unit	$1/x$	1	1.0

assumed for the particular work being analyzed, then the relationship is defined as one whole unit, and the tabulated relationship in Table 6.3 will be applicable for the case of a single resource performing the work.

Here $1/x$ is the amount of work accomplished per unit time. For a single resource to perform the whole unit of work, we must have the following:

$$\left(\frac{1}{x}\right)(t) = 1.0.$$

That means the magnitude of x must equal the magnitude of t. For example, if a construction worker can build one block in 30 min, then his work rate is $1/30$ of a block per minute. If the magnitude of x is greater than the magnitude of t, then only a fraction of the required work will be performed. The information about the proportion of work completed is useful for resource planning and productivity measurement purposes.

Production Work Rate Example

A production worker can custom-build three units or a product every 4 h. At that rate how long will it take to build 5 units?

From the information given, we can write the proportion 3 units is to 4 h as 5 units is to x h, where x represents the number of hours the worker will take to build 5 units. This gives

$$\frac{3\ \text{units}}{4\ \text{h}} = \frac{5\ \text{units}}{x\ \text{h}},$$

which simplifies to yield $x = 6$ h 40 min.

Case of Multiple Resources Working Together

In the case of multiple resources performing the work simultaneously, the work relationship is as presented in Table 6.4.

For multiple resources or work crew types, we have the following expression:

$$\sum_{i=1}^{n} r_i t_i = 1.0,$$

where n is the number of different crew types, r_i the work rate of crew type i, and t_i the work time of crew type i.

The expression indicates that even though the multiple crew types may work at different rates, the sum of the total work they accomplished together must equal the required whole

TABLE 6.4 Work Rate Table for Multiple Resource Units

Resource Type i	Work Rate, r_i	Time, t_i	Work Done, w_i
RES 1	r_1	t_1	r_1/t_1
RES 2	r_2	t_2	r_2/t_2
$\vdots$	$\vdots$	$\vdots$	$\vdots$
RES n	r_n	t_n	r_n/t_n
Total		1.0	

unit (i.e., the total building). For partial completion of work, the expression becomes

$$\sum_{i=1}^{n} r_i t_i = p,$$

where p is the percent completion of the required work.

Computational Examples

Suppose RES 1, working alone, can complete a construction job in 50 days. After RES 1 has been working on the job for 10 days, RES 2 was assigned to help RES 1 in completing the job. Both resources working together finished the remaining work in 15 days. It is desired to determine the work rate of RES 2.

The amount of work to be done is 1.0 whole unit. The work rate of RES 1 is 1/50 of construction work per unit time. Therefore, the amount of work completed by RES 1 in 10 days worked alone is (1/50)(10)=1/5 of the required work. This may also be expressed in terms of percent completion or earned value using C/SCSC (cost/schedule control systems criteria). The remaining work to be done is 4/5 of the total work. The two resources working together for 15 days yield the analysis shown in Table 6.5.

Thus, we have $15/50+15\, r_2 = 45$, which yields $r_2 = 1/30$ for the work rate of RES 2. This means that RES 2, working alone, could perform the construction job in 30 days. In this example, it is assumed that both resources produce identical quality of work. If quality levels are not identical for multiple resources, then the work rates may be adjusted to account for the different quality levels or a quality factor may be introduced into the analysis.

As another example, suppose the work rate of RES 1 is such that it can perform a certain task in 30 days. It is desired to add RES 2 to the task so that the completion time of the task could be reduced. The work rate of RES 2 is such that it can perform the same task

TABLE 6.5 Work Rate Analysis for Construction Example

Resource Type i	Work Rate, r_i	Time, t_i	Work Done, w_i
RES 1	1/50	15	15/50
RES 2	r_2	15	$15r_2$
Total			4/5

TABLE 6.6 Work Rate Table for Alternate Work Example

Resource Type i	Work Rate, r_i	Time, t_i	Work Done, w_i
RES 1	1/30	T	$T/30$
RES 2	1/22	T	$T/22$
Total			4/5

alone in 22 days. If RES 1 has already worked 12 days on the task before RES 2 comes in, find the completion time of the task. It is assumed that RES 1 starts the task at time 0.

As usual, the amount of work to be done is 1.0 whole unit (i.e., the full construction work). The work rate of RES 1 is 1/30 of the task per unit time and the work rate of RES 2 is 1/22 of the task per unit time. The amount of work completed by RES 1 in the 12 days it worked alone is $(1/30)(12) = 2/5$ (or 40%) of the required work. Therefore, the remaining work to be done is 2/5 (or 60%) of the full task. Let T be the time for which both resources work together. The two resources, working together, to complete the task yield the results shown in Table 6.6.

Thus, we have $T/30 + T/22 = 3/5$ which yields $T = 7.62$ days. Consequently, the completion time of the task is $(12 + T) = 19.62$ days from time zero. It is assumed that both resources produce identical quality of work and that the respective work rates remain consistent. The relative costs of the different resource types needed to perform the required work may be incorporated into the analysis as shown in Table 6.7.

Calculation of Personnel Requirements

The following expression calculates personnel requirements in a production environment:

$$N = \sum_{i=1}^{n} \frac{T_i O_i}{eH},$$

where N is the number of production employees required, n the number of types of operations, T_i the standard time required for an average operation in O_i, O_i the aggregate number of operation type i required on all the products produced per day, e the assumed production efficiency of the production facility, and H the total production time available per day.

TABLE 6.7 Incorporation of Cost into Work Rate Table

Resource i	Work Rate, r_i	Time, t_i	Work Done, w_i	Pay Rate, p_i	Total Cost, C_i
Crew A	r_1	t_1	r_1/t_1	p_1	C_1
Crew B	r_2	t_2	r_2/t_2	p_2	C_2
$\vdots$	$\vdots$	$\vdots$	$\vdots$	$\vdots$	$\vdots$
Crew n	r_n	t_n	$\frac{r_n}{t_n}$	p_n	C_n
Total			1.0		Total cost

Calculation of Machine Requirements

The following expression calculates machine requirements for a production operation:

$$N = \frac{t_r P}{t_a e},$$

where N is the number of machines required, t_r the time required in hours to process one unit of product at the machine, t_a the time in hours for which machine is available per day, P the desired production rate in units per day, and e the efficiency of the machine.

References

Badiru, A. B., Badiru, A., and Badiru, A. 2008. *Industrial ProjectManagement: Concepts, Tools, and Techniques*. Boca Raton, FL: Taylor & Francis/CRC Press.

Badiru, A. B. and Omitaomu, O. A. 2007. *Computational Economic Analysis for Engineering and Industry*. Boca Raton, FL: Taylor & Francis/CRC Press.

Badiru, A. B. 1996. *Project Management in Manufacturing and High Technology Operations*, 2nd edn. New York:Wiley.

Heragu, S. 1997. *Facilities Design*. Boston, MA: PWS Publishing Company.

Moffat, D. W. 1987. *Handbook of Manufacturing and Production Management Formulas, Charts, and Tables*. Englewood Cliffs, NJ: Prentice-Hall.

7

Forecasting Calculations

Forecasting is an important aspect of research and practice of industrial engineering. It provides information needed to make good decisions. Some of the basic techniques for forecasting include regression analysis, time-series analysis, computer simulation, and neural networks. There are two basic types of forecasting: *intrinsic forecasting* and *extrinsic forecasting*. Intrinsic forecasting is based on the assumption that historical data can adequately describe the problem scenario to be forecasted. With intrinsic forecasting, forecasting models based on historical data use extrapolation to generate estimates for the future. Intrinsic forecasting involves the following steps:

- Collecting historical data
- Developing a quantitative forecasting model based on the data collected
- Generating forecasts recursively for the future
- Revising the forecasts as new data elements become available.

Extrinsic forecasting looks outward to external factors and assumes that internal forecasts can be correlated to external factors. For example, an internal forecast of the demand for a new product may be based on external forecasts of household incomes. Good forecasts are predicated on the availability of good data.

Forecasting Based on Averages

The most common forecasting techniques are based on averages. Sophisticated quantitative forecasting models can be formulated from basic average formulas. The traditional techniques of forecasting based on averages are presented below.

Simple Average Forecast

In this method, the forecast for the next period is computed as the arithmetic average of the preceding data points. This is often referred to as *average to date*. That is

$$f_{n+1} = \frac{\sum_{t=1}^{n} d_t}{n},$$

where f_{n+1} is the forecast for period $n + 1$, d the data for the period, and n the number of preceding periods for which data are available.

Period Moving Average Forecast

In this method, the forecast for the next period is based only on the most recent data values. Whenever a new value is included, the oldest value is dropped. Thus, the average is always computed from a fixed number of values. This is represented as follows:

$$f_{n+1} = \frac{\sum_{t=n-T+1}^{n} d_t}{T}$$

$$= \frac{d_{n-T+1} + d_{n-T+2} + \cdots + d_{n-1} + d_n}{T},$$

where f_{n+1} is the forecast for period $n + 1$, d_t the datum for period t, T the number of preceding periods included in the moving average calculation, and n the current period at which forecast of f_{n+1} is calculated.

The moving average technique is an after-the-fact approach. Since T data points are needed to generate a forecast, we cannot generate forecasts for the first $T - 1$ periods. But this shortcoming is quickly overcome as more data points become available.

Weighted Average Forecast

The weighted average forecast method is based on the assumption that some data points may be more significant than others in generating future forecasts. For example, the most recent data points may weigh more than very old data points in the calculation of future estimates. This is expressed as

$$f_{n+1} = \frac{\sum_{t=1}^{n} w_i d_t}{\sum_{t=1}^{n} w_t}$$

$$= \frac{w_1 d_1 + w_2 d_2 + \cdots + w_n d_n}{w_1 + w_2 + \cdots + w_n},$$

where f_{n+1} is the weighted average forecast for period $n + 1$, d_t the datum for period t, T the number of preceding periods included in the moving average calculation, n the current period at which forecast of f_{n+1} is calculated, and w_t the weight of data point t.

The w_t's are the respective weights of the data points such that

$$\sum_{t=1}^{n} w_t = 1.0.$$

Weighted T-Period Moving Average Forecast

In this technique, the forecast for the next period is computed as the weighted average of past data points over the last T time periods. That is

$$f_{n+1} = w_1 d_n + w_2 d_{n-1} + \cdots + w_T d_{n-T+1},$$

where w_i's are the respective weights of the data points such that the weights sum up to 1.0. That is,

$$\Sigma w_i = 1.0.$$

Exponential Smoothing Forecast

This is a special case of the weighted moving average forecast. The forecast for the next period is computed as the weighted average of the immediate past data point and the forecast of the previous period. In other words, the previous forecast is adjusted based on the deviation (forecast error) of that forecast from the actual data. That is,

$$f_{n+1} = \alpha d_n + (1 - \alpha)f_n$$
$$= f_n + \alpha(d_n - f_n),$$

where f_{n+1} is the exponentially weighted average forecast for period $n + 1$, d_n the datum for period n, f_n the forecast for period n, and α the smoothing factor (real number between 0 and 1).

A low smoothing factor gives a high degree of smoothing, whereas a high value moves the forecast closer to actual data. The trade-off is in whether or not a smooth predictive curve is desired versus a close echo of actual data.

Regression Analysis

Regression analysis is a mathematical procedure for attributing the variability of one variable to changes in one or more other variables. It is sometimes called line-fitting or curve-fitting. The primary function of regression analysis is to develop a model that expresses the relationship between a dependent variable and one or more independent variables.

The effectiveness of a regression model is often tested by analysis of variance (ANOVA), which is a technique for breaking down the variance in a statistical sample into components that can be attributed to each factor affecting that sample. One major purpose of ANOVA is model testing. Model testing is important because of the serious consequences of erroneously concluding that a regression model is good when, in fact, it has little or no significance to the data. Model inadequacy often implies an error in the assumed relationships between the variables, poor data, or both. A validated regression model can be used for the following purposes:

1. Prediction/forecasting
2. Description
3. Control

Regression Relationships

Sometimes, the desired result from a regression analysis is an equation describing the best fit to the data under investigation. The "least-squares" line drawn through the data is the line of best fit. This line may be linear or curvilinear depending on the dispersion of the data. The linear situation exists in those cases where the slope of the regression equation is a constant. The nonconstant slope indicates curvilinear relationships. A plot of the data, called a scatter plot, will usually indicate whether a linear or nonlinear model will be appropriate. The major problem with the nonlinear relationship is the necessity of assuming a functional relationship before accurately developing the model. Example of regression models (simple linear, multiple, and nonlinear) are presented below:

$$Y = \beta_0 + \beta_1 x + \varepsilon,$$
$$Y = \beta_0 + \beta_1 x_1 + \beta_2 x_2 + \varepsilon,$$
$$Y = \beta_0 + \beta_1 x_1^{\alpha_1} + \beta_2 x_2^{\alpha_2} + \varepsilon,$$
$$Y = \beta_0 + \beta_1 x_1^{\alpha_1} + \beta_2 x_2^{\alpha_2} + \beta_{12} x_1^{\alpha_3} x_2^{\alpha_4} + \varepsilon,$$

where Y is the dependent variable, the x_i's are the independent variables, the β_i's are the model parameters, and ε is the error term. The error terms are assumed to be independent and identically distributed normal random variables with mean of zero and variance with a magnitude of σ^2.

Prediction

A major use of regression analysis is prediction or forecasting. Prediction can be of two basic types: interpolation and extrapolation. Interpolation predicts values of the dependent variable over the range of the independent variable or variables. Extrapolation involves predictions outside the range of the independent variables. Extrapolation carries a risk in the sense that projections are made over a data range that is not included in the development of the regression model. There is some level of uncertainty about the nature of the relationships that may exist outside the study range. Interpolation can also create a problem when the values of the independent variables are widely spaced.

Control

Extreme care is needed in using regression for control. The difficulty lies in assuming a functional relationship when, in fact, none may exist. Suppose, for example, that regression shows a relationship between chemical content in a product and noise level in the process room. Suppose further that the real reason for this relationship is that the noise level increases as the machine speed increases and that the higher machine speed produces higher chemical content. It would be erroneous to assume a functional relationship between the noise level in the room and the chemical content in the product. If this relationship does exist, then changes in the noise level could control chemical content. In this case, the real functional relationship exists between the machine speed and the chemical content. It is often difficult to prove functional relationships outside a laboratory environment because many extraneous and intractable factors may have an influence on the dependent variable. A simple example of the use of functional relationship for control can be seen in the following familiar electrical circuit equation:

$$I = \frac{V}{R},$$

where V is the voltage, I the electric current, and R the resistance. The current can be controlled by changes in the voltage, the resistance, or both. This particular equation, which has been experimentally validated, can be used as a control design.

Procedure for Regression Analysis

Problem Definition: Failure to properly define the scope of the problem could result in useless conclusions. Time can be saved throughout all phases of a regression study by knowing, as precisely as possible, the purpose of the required model. A proper definition of the problem will facilitate the selection of the appropriate variables to include in the study.

Selection of Variables: Two very important factors in the selection of variables are ease of data collection and expense of data collection. Ease of data collection deals with the accessibility and the desired form of data. We must first determine if the data can be collected and, if so, how difficult the process will be. In addition, the economic question is of prime importance. How expensive will the data be to collect and compile in a useable form? If the expense cannot be justified, then the variable under consideration may, by necessity, be omitted from the selection process.

Test of Significance of Regression: After the selection and compilation of all possible relevant variables, the next step is a test for the significance of the regression. The test should help avoid wasted effort on the use of an invalid model. The test for the significance of regression is a test to see whether at least one of the variable coefficient(s) in the regression equation is statistically different from zero. A test indicating that none of the coefficients is significantly different from zero implies that the best approximation of the data is a straight line through the data at the average value of the dependent variable regardless of the values of the independent variables. The significance level of the data is an indication of how probable it is that one has erroneously assumed a model's validity.

Coefficient of Determination

The coefficient of multiple determination, denoted by R^2, is used to judge the effectiveness of regression models containing multiple variables (i.e., multiple regression model). It indicates the proportion of the variation in the dependent variable that is explained by the model. The coefficient of multiple determination is defined as

$$R^2 = \frac{SSR}{SST}$$
$$= 1 - \frac{SSE}{SST},$$

where SSR is the sum of squares due to the regression model, SST the sum of squares total, and SSE the sum of squares due to error.

R^2 measures the proportionate reduction of total variation in the dependent variable accounted for by a specific set of independent variables. The coefficient of multiple determination, R^2, reduces to the *coefficient of simple determination*, r^2, when there is only one independent variable in the regression model. R^2 is equal to 0 when all the coefficients,

b_k, in the model are zero. That is, there is no regression fit at all. R^2 is equal to 1 when all data points fall directly on the fitted response surface. Thus, we have

$$0.0 \leq R^2 \leq 1.0.$$

The following points should be noted about regression modeling:

1. A large R^2 does not necessarily imply that the fitted model is a useful one. For example, observations may have been taken at only a few levels of the independent variables. In such a case, the fitted model may not be useful because most predictions would require extrapolation outside the region of observations. For example, for only two data points, the regression line passes perfectly through the two points and the R^2 value will be 1.0. In that case, despite the high R^2, there will be no useful prediction capability.
2. Adding more independent variables to a regression model can only increase R^2 and never reduce it. This is because the SSE cannot become larger with more independent variables, and the SST is always the same for a given set of responses.
3. Regression models developed under conditions where the number of data points is roughly equal to the number of variables will yield high values of R^2 even though the model may not be useful. For example, for only two data points, the regression line will pass perfectly through the two points and R^2 will be 1.0, but the model would have no useful prediction capability.

The strategy for using R^2 to evaluate regression models should not entirely focus on maximizing the magnitude of R^2. Rather, the intent should be to find the point at which adding more independent variables is not worthwhile in terms of the overall effectiveness of the regression model. For most practical situations, R^2 values greater than 0.62 are considered acceptable. As R^2 can often be made larger by including a large number of independent variables, it is sometimes suggested that a modified measure which recognizes the number of independent variables in the model be used. This modified measure is referred to as the "adjusted coefficient of multiple determination," or R_a^2. It is defined mathematically as follows:

$$R_a^2 = 1 - \left(\frac{n-1}{n-p} \right) \frac{\text{SSE}}{\text{SST}},$$

where n is the number of observations used to fit the model, p the number of coefficients in the model (including the constant term), and $p - 1$ the number of independent variables in the model.

R_a^2 may actually become smaller when another independent variable is introduced into the model. This is because the decrease in SSE may be more than offset by the loss of a degree of freedom in the denominator, $n - p$.

The *coefficient of multiple correlation* is defined as the positive square root of R^2. That is,

$$R = \sqrt{R^2}.$$

Thus, the higher the value of R^2, the higher the correlation in the fitted model will be.

Residual Analysis

A residual is the difference between the predicted value computed from the fitted model and the actual value from the data. The ith residual is defined as

$$e_i = Y_i - \hat{Y}_i,$$

where Y_i is the actual value and $\hat{Y}_i$ the predicted value. The SSE and the MSE are computed as

$$\text{SSE} = \sum_i e_i^2$$

$$\sigma^2 \approx \frac{\sum_i e_i^2}{n-2} = \text{MSE},$$

where n is the number of data points. A plot of residuals versus predicted values of the dependent variable can be very revealing. The plot for a good regression model will have a random pattern. A noticeable trend in the residual pattern indicates a problem with the model. Some possible reasons for an invalid regression model include the following:

- Insufficient data
- Important factors not included in model
- Inconsistency in data
- Nonexistence of any functional relationship.

Graphical analysis of residuals is important for assessing the appropriateness of regression models. The different possible residual patterns are shown in Figures 7.1 through 7.6. When we plot the residuals versus the independent variable, the result should ideally appear as shown in Figure 7.1. Figure 7.2 shows a residual pattern indicating nonlinearity of the regression function. Figure 7.3 shows a pattern suggesting nonconstant variance (i.e., variation in σ_2). Figure 7.4 presents a residual pattern implying interdependence of the error terms. Figure 7.5 shows a pattern depicting the presence of outliers. Figure 7.6 represents a pattern suggesting the omission of independent variables.

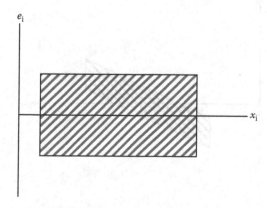

FIGURE 7.1 Ideal residual pattern.

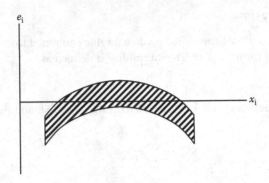

FIGURE 7.2   Residual pattern for nonlinearity.

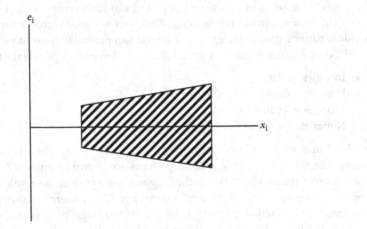

FIGURE 7.3   Residual pattern for nonconstant variance.

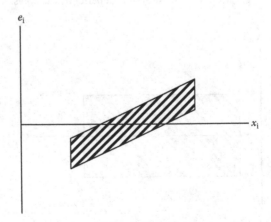

FIGURE 7.4   Residual pattern for interdependence of error terms.

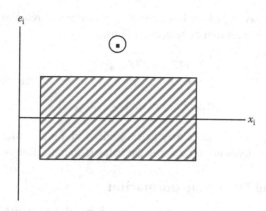

FIGURE 7.5   Residual pattern for presence of outliers.

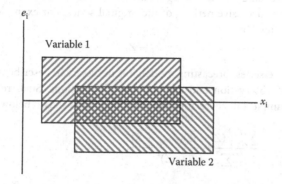

FIGURE 7.6   Residual pattern for omission of independent variables.

Time-Series Analysis

Time-series analysis is a technique that attempts to predict the future by using historical data. The basic principle of time-series analysis is that the sequence of observations is based on jointly distributed random variables. The time-series observations denoted by $Z_1, Z_2, \ldots, Z_T$ are assumed to be drawn from some joint probability density function of the form

$$f_{1,\ldots,T}(Z_{1,\ldots,T}).$$

The objective of time-series analysis is to use the joint density to make probability inferences about future observations. The concept of stationarity implies that the distribution of the time series is invariant with regard to any time displacement. That is,

$$f(Z_t, \ldots, Z_{t+k}) = f(Z_{t+m}, \ldots, Z_{t+m+k}),$$

where t is any point in time and k and m are any pair of positive integers.

A stationary time-series process has a constant variance and remains stable around a constant mean with respect to time reference. Thus,

$$E(Z_t) = E(Z_{t+m}),$$
$$V(Z_t) = V(Z_{t+m}),$$
$$\text{cov}(Z_t, Z_{t+1}) = \text{cov}(Z_{t+k}, Z_{t+k+1}).$$

Nonstationarity in a time series may be recognized in a plot of the series. A widely scattered plot with no tendency for a particular value is an indication of nonstationarity.

Stationarity and Data Transformation

In some cases where nonstationarity exists, some form of data transformation may be used to achieve stationarity. For most time-series data, the usual transformation that is employed is "differencing." Differencing involves the creation of a new series by taking differences between successive periods of the original series. For example, first, regular differences are obtained by

$$w_t = Z_t - Z_{t-1}.$$

To develop a time-series forecasting model, it is necessary to describe the relationship between a current observation and the previous observations. Such relationships are described by the sample autocorrelation function defined as shown below:

$$r_j = \frac{\sum_{t=1}^{T-j} (Z_t - \bar{Z})(Z_{t+j} - \bar{Z})}{\sum_{t=1}^{T} (Z_t - \bar{Z})^2}, \quad j = 0, 1, \ldots, T-1,$$

where T is the number of observations, Z_t the observation for time t, $\bar{Z}$ the sample mean of the series, j the number of periods separating pairs of observations, and r_j the sample estimate of the theoretical correlation coefficient.

The coefficient of correlation between two variables Y_1 and Y_2 is defined as

$$\rho_{12} = \frac{\sigma_{12}}{\sigma_1 \sigma_2},$$

where σ_1 and σ_2 are the standard deviations of Y_1 and Y_2, respectively, and σ_{12} is the covariance between Y_1 and Y_2.

The standard deviations are the positive square roots of the variances, defined as follows:

$$\sigma_1^2 = E[(Y_1 - \mu_1)^2] \quad \text{and} \quad \sigma_2^2 = E[(Y_2 - \mu_2)^2].$$

The covariance, σ_{12}, is defined as

$$\sigma_{12} = E[(Y_1 - \mu_1)(Y_2 - \mu_2)],$$

which will be zero if Y_1 and Y_2 are independent. Thus, when $\sigma_{12} = 0$, we also have $\rho_{12} = 0$. If Y_1 and Y_2 are positively related, then both σ_{12} and ρ_{12} are positive. If Y_1 and Y_2 are negatively related, then both ρ_{12} and ρ_{12} are negative. The correlation coefficient is a real number between -1 and $+1$:

$$1.0 \le \rho_j \le +1.0.$$

A time-series modeling procedure involves the development of a discrete linear stochastic process in which each observation, Z_t, may be expressed as

$$Z_t = \mu + \mu_t + \Psi_1 u_{t-1} + \Psi_2 u_{t-2} + \cdots ,$$

where μ is the mean of the process and the Ψ_i's are the model parameters, which are functions of the autocorrelations. Note that this is an infinite sum, indicating that the current observation at time t can be expressed in terms of all previous observations from the past. In a practical sense, some of the coefficients will be zero after some finite point q in the past. The u_t's form the sequence of independently and identically distributed random disturbances with mean zero and variance sigma sub u_2. The expected value of the series is obtained by

$$E(Z_t) = \mu + E(u_t + \Psi_1 u_{t-1} + \Psi_2 u_{t-2} + \cdots + \cdots + \cdots)$$
$$= \mu + E(u_t)[1 + \Psi_1 + \Psi_2 + \cdots].$$

Stationarity of the time series requires that the expected value be stable. That is, the infinite sum of the coefficients should be convergent to a fixed value, c, as shown below:

$$\sum_{i=0}^{\infty} \Psi_i = c,$$

where $\Psi_0 = 1$ and c is a constant. The theoretical variance of the process, denoted by γ_0, can be derived as

$$\gamma_0 = E[Z_t - E(Z_t)]^2$$
$$= E[(\mu + u_t + \Psi_1 y_{t-1} + \Psi_2 u_{t-2} + \cdots) - \mu]^2$$
$$= E[u_t + \Psi_1 y_{t-1} + \Psi_2 u_{t-2} + \cdots]^2$$
$$= E[u_t^2 + \Psi_1^2 u_{t-1}^2 + \Psi_2^2 u_{t-2}^2 + \cdots] + E[\text{cross-products}]$$
$$= E[u_t^2] + \Psi_1^2 E[u_{t-1}^2] + \Psi_2^2 E[u_{t-2}^2] + \cdots$$
$$= \sigma_u^2 + \Psi_1^2 \sigma_u^2 + \Psi_2^2 \sigma_u^2 + \cdots$$
$$= \sigma_u^2(1 + \Psi_1^2 + \Psi_2^2 + \cdots)$$
$$= \sigma_u^2 \sum_{i=0}^{\infty} \Psi_i^2 ,$$

where σ_u^2 represents the variance of the u_t's. The theoretical covariance between Z_t and Z_{t+j}, denoted by γ_j, can be derived in a similar manner to obtain the following:

$$\gamma_j = E\{[Z_t - E(Z_t)][Z_{t+j} - E(Z_{t+j})]\}$$
$$= \sigma_u^2(\Psi_j + \Psi_1 \Psi_{j+1} + \Psi_2 \Psi_{j+2} + \cdots)$$
$$= \sigma_u^2 \sum_{i=0}^{\infty} \Psi_i \Psi_{i+j}.$$

Sample estimates of the variances and covariances are obtained by

$$c_j = \frac{1}{T} \sum_{t=1}^{T-j} (Z_t - \bar{Z})(Z_{t+j} - \bar{Z}), \quad j = 0, 1, 2, \ldots.$$

The theoretical autocorrelations are obtained by dividing each of the autocovariances, γ_j, by γ_0. Thus, we have

$$\rho_j = \frac{\gamma_j}{\gamma_0}, \quad j = 0, 1, 2, \ldots$$

and the sample autocorrelation is obtained by

$$r_j = \frac{c_j}{c_0}, \quad j = 0, 1, 2, \ldots.$$

Moving Average Processes

If it can be assumed that $\Psi_i = 0$ for some $i > q$, where q is an integer, then our time-series model can be represented as

$$z_t = \mu + u_t + \Psi_1 u_{t-1} + \Psi_2 u_{t-2} + \cdots + \cdots + \Psi_q u_{t-q},$$

which is referred to as a moving-average process of order q, usually denoted as MA(q). For notational convenience, we will denote the truncated series as presented below:

$$Z_t = \mu + u_t - \theta_1 u_{t-1} - \theta_2 u_{t-2} - \cdots - \theta_q u_{t-q},$$

where $\theta_0 = 1$. Any MA(q) process is stationary because the condition of convergence for the Ψ_i's becomes

$$(1 + \Psi_1 + \Psi_2 + \cdots) = (1 - \theta_1 - \theta_2 - \cdots - \theta_q)$$

$$= 1 - \sum_{i=0}^{q} \theta_i,$$

which converges since q is finite. The variance of the process now reduces to

$$\gamma_0 = \sigma_u^2 \sum_{i=0}^{q} \theta_i.$$

We now have the autocovariances and autocorrelations defined, respectively, as shown below:

$$\gamma_j = \sigma_u^2 (-\theta_j + \theta_1 \theta_{j+1} + \cdots + \theta_{q-j} \theta_q), \quad j = 1, \ldots, q,$$

where $\gamma_j = 0$ for $j > q$.

$$\rho_j = \frac{(-\theta_j + \theta_1 \theta_{j+1} + \cdots + \theta_{q-j} \theta_q)}{(1 + \theta_1^2 + \cdots + \theta_q^2)}, \quad j = 1, \ldots, q,$$

where $\rho_j = 0$ for $j > q$.

Autoregressive Processes

In the preceding section, the time series, Z_t, is expressed in terms of the current disturbance, u_t, and past disturbances, u_{t-i}. An alternative is to express Z_t, in terms of the current and past observations, Z_{t-i}. This is achieved by rewriting the time-series expression as

$$u_t = Z_t - \mu - \Psi_1 u_{t-1} - \Psi_2 u_{t-2} - \cdots,$$

$$u_{t-1} = Z_{t-1} - \mu - \Psi_1 u_{t-2} - \Psi_2 u_{t-3} - \cdots,$$

$$u_{t-2} = Z_{t-2} - \mu - \Psi_1 u_{t-3} - \Psi_2 u_{t-4} - \cdots.$$

Successive back-substitutions for the u_{t-i}'s yield the following:

$$u_t = \pi_1 Z_{t-1} - \pi_2 Z_{t-2} - \cdots - \delta,$$

where π_i's and d are the model parameters and are functions of Ψ_i's and μ. We can then rewrite the model as

$$Z_t = \pi_1 Z_{t-1} + \pi_2 Z_{t-2} + \cdots + \pi_p Z_{t-p} + \delta + u_t,$$

which is referred to as an *autoregressive process of order p*, usually denoted as AR(p). For notational convenience, we will denote the autoregressive process as shown below:

$$Z_t = \varphi_1 Z_{t-1} + \varphi_2 Z_{t-2} + \cdots + \varphi_p Z_{t-p} + \delta + u.$$

Thus, AR processes are equivalent to MA processes of infinite order. Stationarity of AR processes is confirmed if the roots of the characteristic equation below lie outside the unit circle in the complex plane:

$$1 - \varphi_1 x - \varphi_2 x^2 - \cdots - \varphi_p x^p = 0,$$

where x is a dummy algebraic symbol. If the process is stationary, then we should have

$$\begin{aligned} E(Z_t) &= \varphi_1 E(Z_{t-1}) + \varphi_2 E(Z_{t-2}) + \cdots + \varphi_p E(Z_{t-p}) + \delta + E(u_t) \\ &= \varphi_1 E(Z_t) + \varphi_2 E(Z_t) + \cdots + \varphi_p E(Z_t) + \delta \\ &= E(Z_t)(\varphi_1 + \varphi_2 + \cdots + \varphi_p) + \delta, \end{aligned}$$

which yields

$$E(Z_t) = \frac{\delta}{1 - \varphi_1 - \varphi_2 - \cdots - \varphi_p}.$$

Denoting the deviation of the process from its mean by Z_t^d, the following is obtained:

$$Z_t^d = Z_t - E(Z_t) = Z_t - \frac{\delta}{1 - \varphi_1 - \varphi_2 - \cdots - \varphi_p},$$

$$Z_{t-1}^d = Z_{t-1} - \frac{\delta}{1 - \varphi_1 - \varphi_2 - \cdots - \varphi_p}.$$

Rewriting the above expression yields

$$Z_{t-1} = Z_{t-1}^d + \frac{\delta}{1 - \varphi_1 - \varphi_2 - \cdots - \varphi_p},$$

$$\vdots$$

$$Z_{t-k} = Z_{t-k}^d + \frac{\delta}{1 - \varphi_1 - \varphi_2 - \cdots - \varphi_p}.$$

If we substitute the AR(p) expression into the expression for Z_t^d, we will obtain

$$Z_t^d = \varphi_1 Z_{t-1} + \varphi_2 Z_{t-2} + \cdots + \varphi_p Z_{t-p} + \delta + u_t - \frac{\delta}{1 - \varphi_1 - \varphi_2 - \cdots - \varphi_p}.$$

Successive back-substitutions of Z_{t-j} into the above expression yields

$$Z_t^d = \varphi_1 Z_{t-1}^d + \varphi_2 Z_{t-2}^d + \cdots + \varphi_p Z_{t-p}^d + u_t.$$

Thus, the deviation series follows the same AR process without a constant term. The tools for identifying and constructing time-series models are the sample autocorrelations, r_j. For the model identification procedure, a visual assessment of the plot of r_j against j, called the sample correlogram, is used. Table 7.1 presents standard examples of *sample correlograms* and the corresponding time-series models, as commonly used in industrial time- series analysis.

A wide variety of sample correlogram patterns can be encountered in time-series analysis. It is the responsibility of the analyst to choose an appropriate model to fit the prevailing time-series data. Fortunately, several commercial statistical computer tools are available for performing time-series analysis. Analysts should consult current web postings to find the right tools for their specific application cases.

TABLE 7.1 Identification of Some Time-Series Models

Correlogram Profile	Model Type	Model
a) 1.0 — Spikes at lags 1 to q — 0 1 2	MA(2)	$Z_t = u_t - \theta_1 u_{t-1} - \theta_2 u_{t-2}$
b) 1.0 — Exponential decay	AR(1)	$Z_t = u_t + \theta_1 u_{t-1}$
c) 1.0 — Damped sine wave form	AR(2)	$Z_t = u_t + \theta_1 u_{t-1} + \theta_2 u_{t-2}$

Forecasting for Inventory Control

The important aspects of inventory control include the following:

- Ability to satisfy work demands promptly by supplying materials from stock
- Availability of bulk rates for purchases and shipping
- Possibility of maintaining more stable and level resource or workforce.

Economic Order Quantity Model

The economic order quantity (EOQ) model determines the optimal order quantity based on purchase cost, inventory carrying cost, demand rate, and ordering cost. The objective is to minimize the total relevant costs of inventory. For the formulation of the model, the following notations are used: Q is the replenishment order quantity (in units), A the fixed cost of placing an order, v the variable cost per unit of the item to be inventoried, r the inventory carrying charge per dollar of inventory per unit time, D the demand rate of the item, and TRC the total relevant costs per unit time.

Figure 7.7 shows the basic inventory pattern with respect to time. One complete cycle starts from a level of Q and ends at zero inventory.

The total relevant cost for order quantity Q is given by the expression below. Figure 7.8 shows the costs as functions of replenishment quantity:

$$\mathrm{TRC}(Q) = \frac{Qvr}{2} + \frac{AD}{Q}.$$

When the TRC(Q) function is optimized with respect to Q, we obtain the expression for the EOQ:

$$\mathrm{EOQ} = \sqrt{\frac{2AD}{vr}},$$

which represents the minimum total relevant costs of inventory. The above formulation assumes that the cost per unit is constant regardless of the order quantity. In some cases,

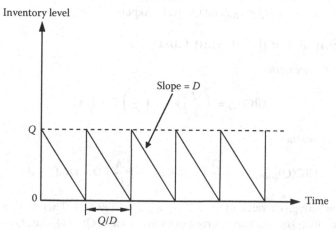

FIGURE 7.7 Basic inventory pattern.

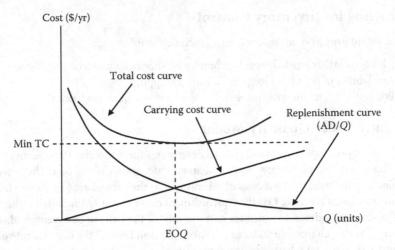

FIGURE 7.8 Inventory costs as functions of replenishment quantity.

quantity discounts may be applicable to the inventory item. The formulation for quantity discount situation is presented below.

Quantity Discount

A quantity discount may be available if the order quantity exceeds a certain level. This is referred to as the single breakpoint discount. The unit cost is represented as shown below. Figure 7.9 presents the price breakpoint for a quantity discount:

$$v = \begin{cases} v_0, & 0 \le Q < Q_b, \\ v_0(1-d), & Q_b \le Q, \end{cases}$$

where v_0 is the basic unit cost without discount, d the discount (in decimals), and it is applied to all units when $Q \le Q_b$, and Q_b the breakpoint.

Calculation of Total Relevant Cost

For $0 \le Q < Q_b$, we obtain

$$\text{TRC}(Q) = \left(\frac{Q}{2}\right) v_0 r + \left(\frac{A}{Q}\right) D + D v_0.$$

For $Q_b \le Q$, we have

$$\text{TRC}(Q)_{\text{discount}} = \left(\frac{Q}{2}\right) v_0(1-d)r + \left(\frac{A}{Q}\right) D + D v_0(1-d).$$

Note that for any given value of Q, $\text{TRC}(Q)_{\text{discount}} < \text{TRC}(Q)$. Therefore, if the lowest point on the $\text{TRC}(Q)_{\text{discount}}$ curve corresponds to a value of $Q^* > Q_b$ (i.e., Q is valid), then set $Q_{\text{opt}} = Q^*$.

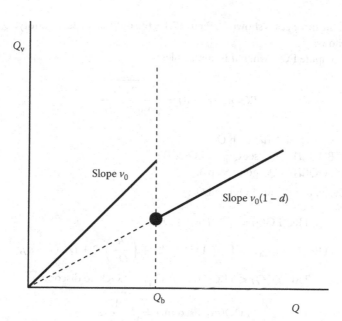

FIGURE 7.9 Price breakpoint for quantity discount.

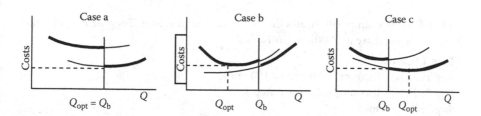

FIGURE 7.10 Cost curves for discount options.

Evaluation of the Discount Option

The trade-off between extra carrying costs and the reduction in replenishment costs should be evaluated to see if the discount option is cost-justified. A reduction in replenishment costs can be achieved by two strategies:

1. Reduction in unit value
2. Fewer replenishments per unit time

Case A: If reduction in acquisition costs is greater than extra carrying costs, then set $Q_{opt} = Q_b$.

Case B: If reduction in acquisition costs is lesser than extra carrying costs, then set $Q_{opt} = $ EOQ with no discount.

Case C: If Q_b is relatively small, then set $Q_{opt} = $ EOQ with discount. The three cases are illustrated in Figure 7.10.

Based on the three cases shown in Figure 7.10, the optimal order quantity, Q_{opt}, can be found as follows:

Step 1: Compute EOQ when d is applicable:

$$EOQ(\text{discount}) = \sqrt{\frac{2AD}{v_0(1-d)r}}.$$

Step 2: Compare EOQ(d) with Q_b:

If EOQ(d) $\geq Q_b$, set $Q_{opt} = $ EOQ(d)

If EOQ(d) $< Q_b$, go to Step 3.

Step 3: Evaluate TRC for EOQ and Q_b:

$$TRC(EOQ) = \sqrt{2ADv_0r} + Dv_0,$$

$$TRC(Q_b)_{discount} = \left(\frac{Q_b}{2}\right)v(1-d)r + \left(\frac{A}{Q_b}\right)D + Dv_0(1-d),$$

If TRC(EOQ) $<$ TRC(Q_b), set $Q_{opt} = $ EOQ(no discount) :

$$EOQ(\text{no discount}) = \sqrt{\frac{2AD}{v_0r}},$$

If TRC(EOQ) $>$ TRC(Q_b), set $Q_{opt} = Q_b$.

The following example illustrates the use of quantity discount. Suppose $d = 0.02$ and $Q_b = 100$ for the three items shown in Table 7.2.

Item 1 (Case a):

Step 1: EOQ (discount) = 19 units $<$ 100 units.

Step 2: EOQ (discount) $< Q_b$, go to step 3.

Step 3: TRC values.

$$TRC(EOQ) = \sqrt{2(1.50)(416)(14.20)(0.24)} + 416(14.20)$$

$$= \$5972.42/\text{year}$$

$$TRC(Q_b) = \frac{100(14.20)(0.98)(0.24)}{2} + \frac{(1.50)(416)}{100} + 416(14.20)(0.98)$$

$$= \$5962.29/\text{year}.$$

Since TRC(EOQ)>TRC(Q_b), set $Q_{opt} = 100$ units.

Item 2 (Case b):

TABLE 7.2 Items subject to Quantity Discount

Item	D(units/year)	v_0($/unit)	A ($)	r($/$/year)
Item 1	416	14.20	1.50	0.24
Item 2	104	3.10	1.50	0.24
Item 3	4160	2.40	1.50	0.24

Step 1: EOQ (discount) = 21 units < 100 units.
Step 2: EOQ (discount) < Q_b, go to step 3.
Step 3: TRC values.

$$\text{TRC(EOQ)} = \sqrt{2(1.50)(104)(3.10)(0.24)} + 104(3.10)$$
$$= \$337.64/\text{year}.$$

$$\text{TRC}(Q_b) = \frac{100(3.10)(0.98)(0.24)}{2} + \frac{(1.50)(104)}{100} + 104(3.10)(0.98)$$
$$= \$\,353.97/\text{year}.$$

$\text{TRC(EOQ)} < \text{TRC}(Q_b)$, set Q_{opt} = EOQ(without discount):

$$\text{EOQ} = \sqrt{\frac{2(1.50)(104)}{3.10(0.24)}}$$
$$= 20 \text{ units}.$$

Item 3 (Case c):
Step 1: Compute EOQ (discount)

$$\text{EOQ(discount)} = \sqrt{\frac{2(1.50)(4160)}{2.40(0.98)(0.24)}}$$
$$= 149 \text{ units} > 100 \text{ units}.$$

Step 2: EOQ (discount) > Q_b. Set Q_{opt} = 149 units.

Sensitivity Analysis

Sensitivity analysis involves a determination of the changes in the values of a parameter that will lead to a change in a dependent variable. It is a process for determining how wrong a decision will be if some or any of the assumptions on which the decision is based prove to be incorrect. For example, a decision may depend on the changes in the values of a particular parameter, such as inventory cost. The cost itself may in turn depend on the values of other parameters, as shown below:

Sub-parameter → Main parameter → Decision

It is of interest to determine what changes in parameter values can lead to changes in a decision. With respect to inventory management, we may be interested in the cost impact of the deviation of actual order quantity from the EOQ. The sensitivity of cost to departures from EOQ is analyzed as presented below.

Let p represent the level of change from EOQ:

$$|p| \leq 1.0,$$
$$Q' = (1 - p)\text{EOQ}.$$

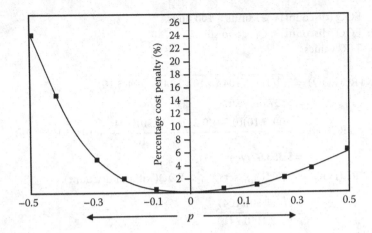

FIGURE 7.11 Sensitivity analysis based on PCP.

Percentage cost penalty (PCP) is defined as follows:

$$PCP = \frac{TRC(Q') - TRC(EOQ)}{TRC(EOQ)} \times 100$$

$$= 50 \left(\frac{p^2}{1+p} \right).$$

A plot of the cost penalty is shown in Figure 7.11. It is seen that the cost is not very sensitive to minor departures from EOQ. We can conclude that changes within 10% of EOQ will not significantly affect the total relevant cost. Two special inventory control algorithms are discussed below.

Wagner–Whitin Algorithm

The Wagner–Whitin (W–W) algorithm is an approach to deterministic inventory modeling. It is based on dynamic programming, which is a mathematical procedure for solving sequential decision problems. The assumptions of the W–W algorithm are as follows:

- Expected demand is known for N periods into the future.
- The periods are equal in length.
- No stock-out or backordering is allowed.
- All forecast demands will be met.
- Ordering cost A may vary from period to period.
- Orders are placed at the beginning of a period.
- Order lead time is zero.
- Inventory carrying cost is vr/unit/period.
- Inventory carrying cost is charged at the beginning of a period for the units carried forward from the previous period.

Notation and Variables

Let N be the number of periods in the planning horizon ($j = 1, 2, \ldots, N$), t the number of periods considered when calculating first replenishment quantity ($j = 1, 2, \ldots, t$), where $t = N$, vr the inventory carrying cost/unit/period of inventory carried forward from period j to period $j + 1$, A the ordering cost per order, I_j the inventory brought into period j, D_j the demand for period j, and Q_j the order quantity in period j.

$$\delta_j = \begin{cases} 0 & \text{if } Q_j = 0, \\ 1 & \text{if } Q_j > 1, \end{cases}$$

$$I_{j+1} = I_j + \delta_j Q_j - D_j.$$

The optimality properties of W–W algorithm are as follows:

Property 1. Replenishment takes place only when inventory level is zero.

Property 2. There is an upper limit to how far back before period j we would include D_j in a replenishment quantity.

If the requirements for period j are so large that

$$D_j(vr) > A,$$

that is,

$$D_j > \frac{A}{vr},$$

then the optimal solution will have a replenishment at the beginning of period j. That is, inventory must go to zero at the beginning of period j. The earliest value of j where $D_j > A/vr$ is used to determine the horizon (or number of periods) to be considered when calculating the first replenishment quantity.

Let $F(t)$ is the total inventory cost to satisfy demand for periods $1, 2, \ldots, t$. So, we have

$$\begin{aligned} F(t) =&\, [vrI_1 + A\delta_1 Q_1] \\ &+ [vr(I_1 + \delta_1 Q_1 - D_1) + A\delta_2 Q_2] \\ &+ [vr(I_1 + \delta_1 Q_1 - D_1 + \delta_2 Q_2 - D_2) + A\delta_3 Q_3] \\ &+ \cdots + [\cdots + A\delta_t Q_t]. \end{aligned}$$

The expression for $F(t)$ assumes that the initial inventory is known. The objective is to minimize the total inventory cost. The applicable propositions for the W–W Algorithm are summarized below.

Proposition 1:

$I_j Q_j = 0$. This is because if we are planning to place an order in period j, we will not carry an inventory I_j into period j. As stock-out is not permitted, an order must be placed in period j if the inventory carried into period j is zero. Either one of I_j or Q_j must be zero.

Proposition 2:

An optional policy exists such that for all periods, we have either of the following:

$$Q_j = 0$$

or

$$Q_j = \sum_{i=1}^{t} Q_i, \quad j \leq t \leq N.$$

This means that order quantity in period j is either zero or the total demand of some (or all) of the future periods.

Proposition 3:

There exists an optimal policy such that if demand D_{j+t} is satisfied by Q_j^*, then demands $D_j, D_{j+1}, D_{j+2}, \ldots, D_{j+t-1}$ are also satisfied by Q_j^*.

Proposition 4:

If an optimal inventory plan for the first t periods is given for an N-period model in which $I_j = 0$, for $j < t$, then it is possible to determine the optimal plan for periods $t + 1$ through N.

Proposition 5:

If an optimal inventory plan for the first t periods of an N- period model is known in which $Q_t > 0$, it is not necessary to consider periods 1 through $t - 1$ in formulating the optimal plan for the rest of the periods. That is, we re-initialize the problem to start at time t and proceed forward.

8

Six Sigma and Lean

Concept of Six Sigma

The six sigma approach, which was originally introduced by Motorola's Government Electronics Group, has caught on quickly in industry. Many major companies now embrace the approach as the key to high-quality industrial productivity. Six sigma means six standard deviations from a statistical performance average. The six sigma approach allows for no more than 3.4 defects per million parts in manufactured goods or 3.4 mistakes per million activities in a service operation. To appreciate the effect of the six sigma approach, consider a process that is 99% perfect. That process will produce 10,000 defects per million parts. With six sigma, the process will need to be 99.99966% perfect in order to produce only 3.4 defects per million. Thus, six sigma is an approach that pushes the limit of perfection.

Taguchi Loss Function

The philosophy of Taguchi loss function defines the concept of how deviation from an intended target creates a loss in the production process. Taguchi's idea of product quality analytically models the loss to the society from the time a product is shipped to customers. Taguchi loss function measures this conjectured loss with a quadratic function known as quality loss function (QLF), which is mathematically represented as shown below:

$$L(y) = k(y - m)^2,$$

where k is a proportionality constant, m the target value, and y the observed value of the quality characteristic of the product in question. The quantity $(y - m)$ represents the deviation from the target. The larger the deviation, the larger is the loss to the society. The constant k can be determined if $L(y), y$, and m are known. Loss, in the QLF concept, can be defined to consist of several components. Examples of loss are provided below:

- *Opportunity cost* of not having the service of the product due to its quality deficiency. The loss of service implies that something that should have been done to serve the society could not be done.
- *Time lost* in the search to find (or troubleshoot) the quality problem.

- *Time lost* (after finding the problem) in the attempt to solve the quality problem. The problem identification effort takes away some of the time that could have been productively used to serve the society. Thus, the society incurs a loss.
- *Productivity loss* that is incurred due to the reduced effectiveness of the product. The decreased productivity deprives the society of a certain level of service and, thereby, constitutes a loss.
- *Actual cost* of correcting the quality problem. This is, perhaps, the only direct loss that is easily recognized, but there are other subtle losses that the Taguchi method can help identify.
- *Actual loss* (e.g., loss of life) due to a failure of the product resulting from its low quality. For example, a defective automobile tire creates a potential for a traffic fatality.
- *Waste* that is generated as a result of lost time and materials due to rework and other nonproductive activities associated with low quality of work.

Identification and Elimination of Sources of Defects

The approach uses statistical methods to find problems that cause defects. For example, the total yield (number of nondefective units) from a process is determined by a combination of the performance levels of all the steps making up the process. If a process consists of 20 steps and each step is 98% perfect, then the performance of the overall process will be

$$(0.98)^{20} = 0.667608 \quad (\text{i.e., } 66.7608\%).$$

Thus the process will produce 332,392 defects per million parts. If each step of the process is pushed to the six sigma limit, then the process performance will be

$$(0.9999966)^{20} = 0.999932 \quad (\text{i.e., } 99.9932\%).$$

Thus the six sigma process will produce only 68 defects per million parts. This is a significant improvement over the original process performance. In many cases, it is not realistic to expect to achieve the six sigma level of production, but the approach helps to set a quality standard and provides a mechanism for striving to reach the goal. In effect, the six sigma process means changing the way workers perform their tasks so as to minimize the potential for defects.

The success of six sigma in industry ultimately depends on industry's ability to initiate and execute six sigma projects effectively. Thus, the project management approaches presented in this book are essential for realizing the benefits of six sigma. Project planning, organizing, team building, resource allocation, employee training, optimal scheduling, superior leadership, shared vision, and project control are all complementarily essential to implementing six sigma successfully. These success factors are not mutually exclusive. In many organizations, far too much focus is directed toward the statistical training for six sigma at the expense of proper project management development. This explains why many organizations have not been able to achieve the much- touted benefits of six sigma.

The success of the Toyota production system is not due to any special properties of the approach, but rather due to the consistency, persistence, and dedication of Toyota organizations in building their projects around all the essential success factors. Toyota focuses on changing the organizational mindset that is required in initiating and coordinating the success factors throughout the organization. Six sigma requires the management of multiple projects with an identical mindset throughout the organization. The success of this requirement depends on proper application of project management tools and techniques, as presented in the preceding chapters of this book.

Roles and Responsibilities for Six Sigma

Human roles and responsibilities are crucial in executing six sigma projects. The different categories of team players are explained below:

Executive leadership: Develops and promulgates vision and direction. Leads change and maintains accountability for organizational results (on a full-time basis).

Employee group: Includes all employees, supports organizational vision, receives and implements six sigma specs, serves as points of total process improvement, exports mission statement to functional tasks, and deploys improvement practices (on a full-time basis).

Six sigma champion: Advocates improvement projects, leads business direction, and coordinates improvement projects (on a full-time basis).

Six sigma project sponsor: Develops requirements, engages project teams, leads project scoping, and identifies resource requirements (on a part-time basis).

Master belt: Trains and coaches black belts and green belts, leads large projects, and provides leadership (on a full-time basis).

Black belt: Leads specific projects, facilitates troubleshooting, coordinates improvement groups, and trains and coaches project team members (on a full-time basis).

Green belt: Participates on black belt teams and leads small projects (on a part-time project-specific basis).

Six sigma project team members: Provides specific operational support, facilitates inward knowledge transfer, and provides links to functional areas (on a part-time basis).

Statistical Techniques for Six Sigma

Statistical process control (SPC) means controlling a process statistically. SPC originated from the efforts of the early quality control researchers. The techniques of SPC are based on basic statistical concepts normally used for statistical quality control. In a manufacturing environment, it is known that not all products are made exactly alike. There are always some inherent variations in units of the same product. The variation in the characteristics of a product provides the basis for using SPC for quality improvement. With the help of statistical approaches, individual items can be studied and general inferences can be drawn about the process or batches of products from the process. Since 100% inspection is difficult or impractical in many processes, SPC provides a mechanism to generalize with regard to process performance. SPC uses random samples generated consecutively over time. The random samples should be representative of the general process. SPC can be accomplished through the following steps:

- control charts ($\overline{X}$-chart, R-chart)
- process capability analysis (nested design, C_p, C_{p_k})
- process control (factorial design, response surface)

Control Charts

Two of the most commonly used control charts in industry are the $\overline{X}$- charts and the range charts (R-charts). The type of chart to be used normally depends on the kind of data collected. Data collected can be of two types: variable data and attribute data. The success of quality improvement depends on two major factors:

1. The quality of data available
2. The effectiveness of the techniques used for analyzing the data

Types of data for control charts

Variable data: The control charts for variable data are listed below:

- Control charts for individual data elements (X)
- Moving range chart (MR-chart)
- Average chart ($\overline{X}$-chart)
- Range chart (R-chart)
- Median chart
- Standard deviation chart (σ-chart)
- Cumulative sum chart (CUSUM)
- Exponentially weighted moving average (EWMA)

Attribute data: The control charts for attribute data are listed below:

- Proportion or fraction defective chart (p-chart) (subgroup sample size can vary)
- Percent defective chart (100 p-chart) (subgroup sample size can vary)
- Number defective chart (np-chart) (subgroup sample size is constant)
- Number defective (c-chart) (subgroup sample size = 1)
- Defective per inspection unit (u-chart) (subgroup sample size can vary)

We are using the same statistical theory to generate control limits for all the above charts except for EWMA and CUSUM.

$\overline{X}$ and R-Charts

The R-chart is a time plot useful in monitoring short-term process variations, whereas the $\overline{X}$-chart monitors the long-term variations where the likelihood of special causes is greater over time. Both charts have control lines called upper and lower control limits, as well as the central lines. The central line and control limits are calculated from the process measurements. The control limits are not specification limits. Therefore, they represent what the process is capable of doing when only common cause variation exists. If only common cause variation exists, then the data will continue to fall in a random fashion within the control limits. In this case, we say the process is in a state of statistical control.

However, if a special cause acts on the process, one or more data points will be outside the control limits, so the process is not in a state of statistical control.

Data Collection Strategies

One strategy for data collection requires that about 20–25 subgroups be collected. Twenty to twenty-five subgroups should adequately show the location and spread of a distribution in a state of statistical control. Another approach is to use run charts to monitor the process until such time as 20–25 subgroups are made available. Then, control charts can be applied with control limits included on the charts. Other data collection strategies should consider the subgroup sample size as well as the sampling frequency.

Subgroup Sample Size

The subgroup samples of size n should be taken as n consecutive readings from the process and not random samples. This is necessary in order to have an accurate estimate of the process common cause variation. Each subgroup should be selected from some small period of time or small region of space or product in order to assure homogeneous conditions within the subgroup. This is necessary because the variation within the subgroup is used in generating the control limits. The subgroup sample size n can be between four or five samples. This is a good size that balances the pros and cons of using large or small sample size for a control chart as provided below.

Advantages of using small subgroup sample size:

- Estimates of process standard deviation based on the range are as good and accurate as the estimates obtained from using the standard deviation equation which is a complex hand calculation method.
- The probability of introducing special cause variations within a subgroup is very small.
- Range chart calculation is simple and easier to compute by hand on the shop floor by operators.

Advantages of using large subgroup sample size:

- The central limit theorem supports the fact that the process average will be more normally distributed with larger sample size.
- If the process is stable, the larger the subgroup size, the better are the estimates of process variability.
- A control chart based on a larger subgroup sample size will be more sensitive to process changes.

The choice of a proper subgroup is very critical to the usefulness of any control chart. The following paragraphs explain the importance of subgroup characteristics:

- If we fail to incorporate all common cause variations within our subgroups, the process variation will be underestimated, leading to very tight control limits. Then the process will appear to go out of control too frequently even when there is no special cause.

- If we incorporate special causes within our subgroups, then we will fail to detect special causes as frequently as expected.

Frequency of Sampling

The problem of determining how frequently one should sample depends on several factors. These factors include, but are not limited to, the following.

- *Cost of collecting and testing samples:* The greater the cost of taking and testing samples, the less frequently we should sample.
- *Changes in process conditions:* The larger the frequency of changes to the process, the larger is the sampling frequency. For example, if process conditions tend to change every 15 min, then sample every 15 min. If conditions change every 2 h, then sample every 2 h.
- *Importance of quality characteristics:* The more important the quality characteristic being charted is to the customer, the more frequently the characteristic will need to be sampled.
- *Process control and capability:* The more the history of process control and capability, the less frequently the process needs to be sampled.

Stable Process

A process is said to be in a state of statistical control if the distribution of measurement data from the process has the same shape, location, and spread over time. In other words, a process is stable when the effects of all special causes have been removed from a process, so that the remaining variability is only due to common causes. Figure 8.1 shows an example of a stable distribution.

Out-of-Control Patterns

A process is said to be unstable (i.e., not in a state of statistical control) if it changes from time to time because of a shifting average, or shifting variability, or a combination of

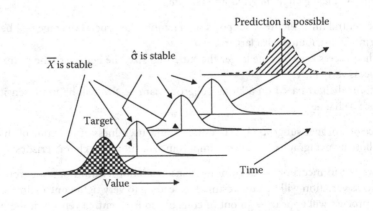

FIGURE 8.1 Stable distribution with no special causes.

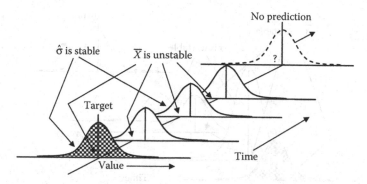

FIGURE 8.2 Unstable process average.

shifting averages and variation. Figures 8.2 through 8.4 show examples of distributions for unstable processes.

Calculation of Control Limits

- Range (R): This is the difference between the highest and lowest observations:

$$R = X_{\text{highest}} - X_{\text{lowest}}.$$

- Center lines: Calculate $\overline{X}$ and $\overline{R}$:

$$\overline{X} = \frac{\sum X_i}{m}, \quad \overline{R} = \frac{\sum R_i}{m},$$

where $\overline{X}$ is the overall process average, $\overline{R}$ the average range, m the total number of subgroups, and n the within subgroup sample size.
- Control limits based on the R-chart:

$$\text{UCL}_R = D_4\overline{R}, \quad \text{LCL}_R = D_3\overline{R}.$$

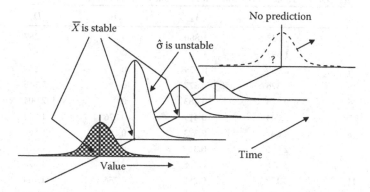

FIGURE 8.3 Unstable process variation.

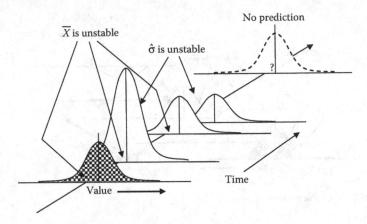

FIGURE 8.4 Unstable process average and variation.

- Estimate of process variation:

$$\hat{\sigma} = \frac{\overline{R}}{d_2}.$$

- Control limits based on $\overline{X}$-chart: Calculate the upper and lower control limits for the process average as follows:

$$\text{UCL} = \overline{X} + A_2 \overline{R}, \qquad \text{LCL} = \overline{X} - A_2 \overline{R}.$$

Table 8.1 shows the values of $d_2, A_2, D_3,$ and D_4 for different values of n. These constants are used for developing variable control charts.

Plotting Control Charts for Range and Average Charts

- Plot the range chart (R-chart) first.
- If the R-chart is in control, then plot $\overline{X}$-chart.

TABLE 8.1 Table of Constants for Variable Control Charts

n	d_2	A_2	D_3	D_4
2	1.128	1.880	0	3.267
3	1.693	1.023	0	2.575
4	2.059	0.729	0	2.282
5	2.326	0.577	0	2.115
6	0.534	0.483	0	2.004
7	2.704	0.419	0.076	1.924
8	2.847	0.373	0.136	1.864
9	2.970	0.337	0.184	1.816
10	3.078	0.308	0.223	1.777
11	3.173	0.285	0.256	1.744
12	3.258	0.266	0.284	1.716

- If the R-chart is not in control, identify and eliminate special causes, then delete points that are due to special causes, and recompute the control limits for the range chart. If the process is in control, then plot the $\overline{X}$-chart.
- Check to see if the $\overline{X}$-chart is in control; if not, search for special causes and eliminate them permanently.
- Remember to perform the eight trend tests.

Plotting Control Charts for Moving Range and Individual Control Charts

- Plot the moving range chart (MR-chart) first.
- If the MR-chart is in control, then plot the individual chart (X).
- If the MR-chart is not in control, identify and eliminate special causes, then delete special-causes points, and recompute the control limits for the moving range chart. If the MR-chart is in control, then plot the individual chart.
- Check to see if the individual chart is in control; if not, search for special causes from out-of-control points.
- Perform the eight trend tests.

Case Example: Plotting of Control Chart

An industrial engineer in a manufacturing company was trying to study a machining process for producing a smooth surface on a torque converter clutch. The quality characteristic of interest is the surface smoothness of the clutch. The engineer then collected four clutches every hour for 30 h and recorded the smoothness measurements in micro-inches. Acceptable values of smoothness lie between 0 (perfectly smooth) and 45 micro-inches. The data collected by the industrial engineer are provided in Table 8.2a. Tables 8.2b and 8.2c present the descriptive statistics calculations. Histograms of the individual and average measurements are presented in Figure 8.5.

The two histograms in Figure 8.5 show that the hourly smoothness average ranges from 27 to 32 micro-inches, much narrower than the histogram of hourly individual smoothness which ranges from 24 to 37 micro-inches. This is due to the fact that averages have less variability than individual measurements. Therefore, whenever we plot subgroup averages on an $\overline{X}$-chart, there will always exist some individual measurements that will plot outside the control limits of an $\overline{X}$- chart. The dot-plots of the surface smoothness for individual and average measurements are shown in Figure 8.6.

Calculations

I. Natural limit of the process = $\overline{X} \pm 3s$ (based on empirical rule), where s is the estimated standard deviation of all individual samples; standard deviation (special and common) $s = 2.822$; process average $\overline{X} = 29.367$. Hence, natural limit of the process = $29.367 \pm 3(2.822) = 29.367 \pm 8.466$. The natural limit of the process is between 20.90 and 37.83.

II. Inherent (common cause) process variability, $\hat{\sigma} = \overline{R}/d_2$, where $\overline{R}$ from the range chart is 5.83 and d_2 (for $n = 4$) from Table 8.1 is 2.059. Hence, $\hat{\sigma} = \overline{R}/d_2 = $

5.83/2.059 = 2.83. Thus, the total process variation s is about the same as the inherent process variability. This is because the process is in control. If the process is out of control, the total standard deviation of all the numbers will be larger than $\overline{R}/d_2$.

III. Control limits for the range chart. Obtain the constants D_3 and D_4 from Table 8.1 for $n = 4$, with $D_3 = 0$, $D_4 = 2.282$, and $\overline{R} = 172/30 = 5.73$, such that

$$\text{UCL} = D_4\overline{R} = 2.282(5.73) = 16.16, \tag{8.1}$$

$$\text{LCL} = D_3\overline{R} = 0(5.73) = 0.0. \tag{8.2}$$

TABLE 8.2a Data for Control Chart Example

	Smoothness (micro-inches)					
Subgroup No.	I	II	III	IV	Average	Range
1	34	33	24	28	29.75	10
2	33	33	33	29	32.00	4
3	32	31	25	28	29.00	7
4	33	28	27	36	31.00	9
5	26	34	29	29	29.50	8
6	30	31	32	28	30.25	4
7	25	30	27	29	27.75	5
8	32	28	32	29	30.25	4
9	29	29	28	28	28.50	1
10	31	31	27	29	29.50	4
11	27	36	28	29	30.00	9
12	28	27	31	31	29.25	4
13	29	31	32	29	30.25	3
14	30	31	31	34	31.50	4
15	30	33	28	31	30.50	5
16	27	28	30	29	28.50	3
17	28	30	33	26	29.25	7
18	31	32	28	26	29.25	6
19	28	28	37	27	30.00	10
20	30	29	34	26	29.75	8
21	28	32	30	24	28.50	8
22	29	28	28	29	28.50	1
23	27	35	30	30	30.50	8
24	31	27	28	29	28.75	4
25	32	36	26	35	32.25	10
26	27	31	28	29	28.75	4
27	27	29	24	28	27.00	5
28	28	25	26	28	26.75	3
29	25	25	32	27	27.25	7
30	31	25	24	28	27.00	7
Total					881.00	172

TABLE 8.2b   Descriptive Statistics for Individual Smoothness

N	120
Mean	29.367
Median	29.00
Tr_{mean}	29.287
SD	2.822
SE_{mean}	0.258
Min.	24.00
Max.	37.00
Q_1	28.00
Q_3	31.00

TABLE 8.2c   Descriptive Statistics for Average Smoothness

N	30
Mean	29.367
Median	29.375
Tr_{mean}	29.246
SD	1.409
SE_{mean}	0.257
Min.	26.75
Max.	32.25
Q_1	28.50
Q_3	30.25

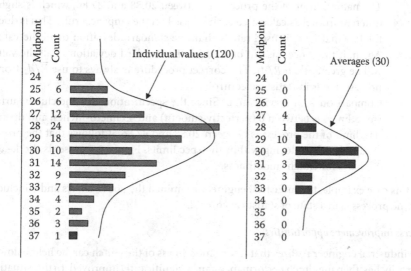

FIGURE 8.5   Histograms of individual measurements and averages for clutch smoothness.

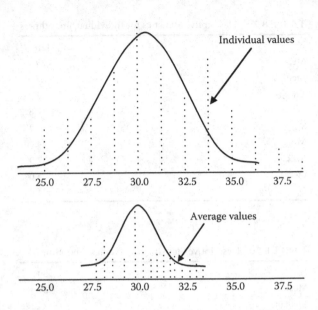

FIGURE 8.6 Dot-plots of individual measurements and averages for clutch smoothness.

IV. Control limits for the averages. Obtain the constant A_2 from Table 8.1 for $n = 4$, with $A_2 = 0.729$, such that

$$\mathrm{UCL} = \overline{X} + A2(\overline{R}) = 29.367 + 0.729(5.73) = 33.54,$$

$$\mathrm{LCL} = \overline{X} - A2(\overline{R}) = 29.367 - 0.729(5.73) = 25.19.$$

V. Natural limit of the process $= \overline{X} \pm 3(\overline{R})/d_2 = 29.367 \pm 3(2.83) = 29.367 \pm 8.49$. The natural limit of the process is between 20.88 and 37.86, which is slightly different from $\pm 3s$ calculated earlier based on the empirical rule. This is due to the fact that $\overline{R}/d_2$ is used rather than the standard deviation of all the values. Again, if the process is out of control, the standard deviation of all the values will be greater than $\overline{R}/d_2$. The correct procedure is always to use $\overline{R}/d_2$ from a process that is in statistical control.

VI. Comparison with specification. Since the specifications for the clutch surface smoothness is between 0 (perfectly smooth) and 45 micro-inches, and the natural limit of the process is between 20.88 and 37.86, it follows that the process is capable of producing within the spec limits. Figure 8.7 presents the R- and $\overline{X}$-charts for clutch smoothness.

For this case example, the industrial engineer examined the above charts and concluded that the process is in a state of statistical control.

Process improvement opportunities

The industrial engineer realizes that if the smoothness of the clutch can be held below 15 micro-inches, then the clutch performance can be significantly improved. In this situation,

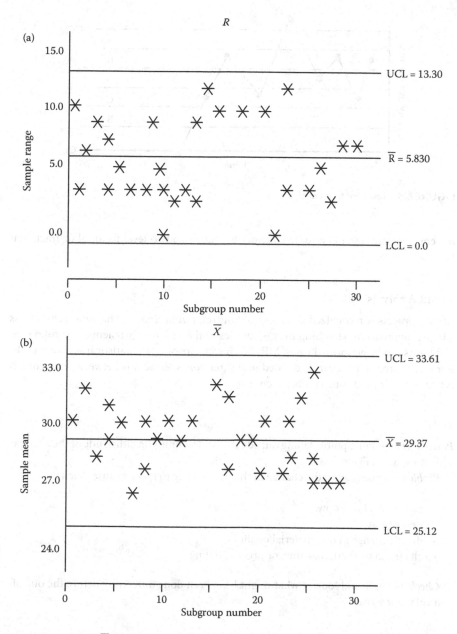

FIGURE 8.7 R and $\overline{X}$-charts for clutch smoothness. Plots (a) and (b) show alternate data dispersion patterns.

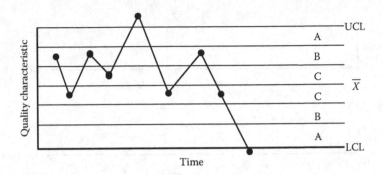

FIGURE 8.8 Test 1 for trend analysis.

the engineer can select key control factors to study in a two-level factorial or fractional factorial design.

Trend Analysis

After a process is recognized to be out of control, a zone control charting technique is a logical approach to searching for the sources of the variation problems. The following eight tests can be performed using MINITAB software or other statistical software tools. For this approach, the chart is divided into three zones. Zone A is between $\pm 3\sigma$, zone B is between $\pm 2\sigma$, and zone C is between $\pm 1\sigma$.

Test 1

Pattern: One or more points falling outside the control limits on either side of the average. This is shown in Figure 8.8.

Problem source: A sporadic change in the process due to special causes such as:

- equipment breakdown
- new operator
- drastic change in raw material quality
- change in method, machine, or process setting.

Check: Go back and look at what might have been done differently before the out-of-control point signals.

Test 2

Pattern: A run of nine points on one side of the average (Figure 8.9).

Problem source: This may be due to a small change in the level of process average. This change may be permanent at the new level.

Check: Go back to the beginning of the run and determine what was done differently at that time or prior to that time.

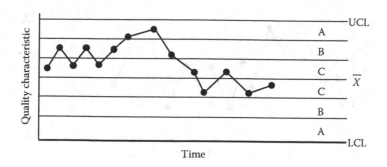

FIGURE 8.9 Test 2 for trend analysis.

Test 3

Pattern: A trend of six points in a row either increasing or decreasing as shown in Figure 8.10.

Problem source: This may be due to the following:

- gradual tool wear;
- change in characteristic such as gradual deterioration in the mixing or concentration of a chemical;
- deterioration of the plating or etching solution in electronics or chemical industries.

Check: Go back to the beginning of the run and search for the source of the run.

The above three tests are useful in providing good control of a process. However, in addition to the above three tests, some advanced tests for detecting out-of-control patterns can also be used. These tests are based on the zone control chart.

Test 4

Pattern: Fourteen points in a row alternating up and down within or outside the control limits as shown in Figure 8.11.

Problem source: This can be due to sampling variation from two different sources such as sampling systematically from high and low temperatures, or lots with two different

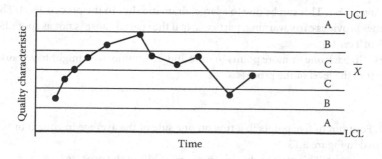

FIGURE 8.10 Test 3 for trend analysis.

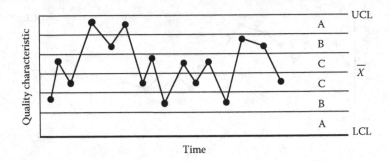

FIGURE 8.11 Test 4 for trend analysis.

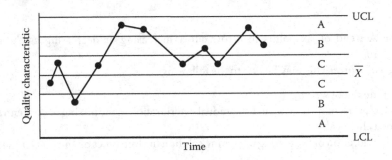

FIGURE 8.12 Test 5 for trend analysis.

averages. This pattern can also occur if adjustment is being made all the time (over control).

Check: Look for cycles in the process, such as humidity or temperature cycles, or operator over control of process.

Test 5

Pattern: Two out of three points in a row on one side of the average in zone A or beyond. An example of this is presented in Figure 8.12.

Problem source: This can be due to a large, dramatic shift in the process level. This test sometimes provides early warning, particularly if the special cause is not as sporadic as in the case of Test 1.

Check: Go back one or more points in time and determine what might have caused the large shift in the level of the process.

Test 6

Pattern: Four out of five points in a row on one side of the average in zone B or beyond, as depicted in Figure 8.13.

Problem source: This may be due to a moderate shift in the process.

Check: Go back three or four points in time.

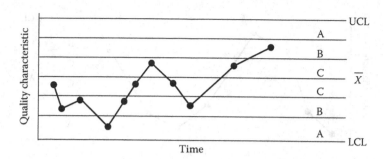

FIGURE 8.13 Test 6 for trend analysis.

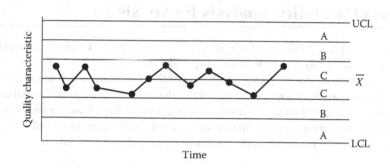

FIGURE 8.14 Test 7 for trend analysis.

Test 7

Pattern: Fifteen points in a row on either side of the average in zone C as shown in Figure 8.14.

Problem source: This is due to the following.

- Unnatural small fluctuations or absence of points near the control limits.
- At first glance may appear to be a good situation, but this is not a good control.
- Incorrect selection of subgroups. May be sampling from various subpopulations and combining them into a single subgroup for charting.
- Incorrect calculation of control limits.

Check: Look very close at the beginning of the pattern.

Test 8

Pattern: Eight points in a row on both sides of the centerline with none in zone C. An example is shown in Figure 8.15.

Problem source: No sufficient resolution on the measurement system (see the section on measurement system).

Check: Look at the range chart and see if it is in control.

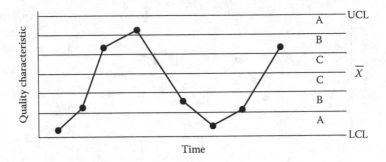

FIGURE 8.15 Test for trend analysis.

Process Capability Analysis for Six Sigma

Industrial process capability analysis is an important aspect of managing industrial projects. The capability of a process is the spread that contains almost all values of the process distribution. It is very important to note that capability is defined in terms of a distribution. Therefore, capability can only be defined for a process that is stable (has distribution) with common cause variation (inherent variability). It cannot be defined for an out-of-control process (which has no distribution) with variation special to specific causes (total variability). Figure 8.16 shows a process capability distribution.

Capable Process (C_p)

A process is capable ($C_p \geq 1$) if its natural tolerance lies within the engineering tolerance or specifications. The measure of process capability of a stable process is $6\hat{\sigma}$, where $\hat{\sigma}$ is the

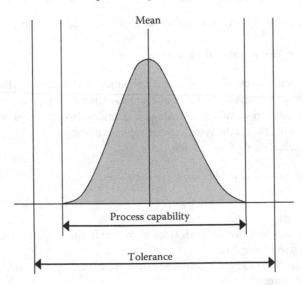

FIGURE 8.16 Process capability distribution.

inherent process variability estimated from the process. A minimum value of $C_p = 1.33$ is generally used for an ongoing process. This ensures a very low reject rate of 0.007% and therefore is an effective strategy for prevention of nonconforming items. Here C_p is defined mathematically as

$$C_p = \frac{USL - LSL}{6\hat{\sigma}}$$
$$= \frac{\text{allowable process spread}}{\text{actual process spread}},$$

where USL is the upper specification limit, LSL the lower specification limit, and C_p measures the effect of the inherent variability only.

The analyst should use $\overline{R}/d_2$ to estimate $\hat{\sigma}$ from an R-chart that is in a state of statistical control, where $\overline{R}$ is the average of the subgroup ranges and d_2 is a normalizing factor that is tabulated for different subgroup sizes (n).

We do not have to verify control before performing a capability study. We can perform the study and then verify control after the study with the use of control charts. If the process is in control during the study, then our estimates of capabilities are correct and valid. However, if the process was not in control, we would have gained useful information as well as proper insights as to the corrective actions to be pursued.

Capability Index (C_{p_k})

Process centering can be assessed when a two-sided specification is available. If the capability index (C_{p_k}) is equal to or greater than 1.33, then the process may be adequately centered. The capability index C_{p_k} can also be employed when there is only a one-sided specification. For a two-sided specification, it can be mathematically defined as

$$C_{p_k} = \min\left\{\frac{USL - \overline{X}}{3\hat{\sigma}}, \frac{\overline{X} - LSL}{3\hat{\sigma}}\right\},$$

where $\overline{X}$ is the overall process average.

However, for a one-sided specification, the actual C_{p_k} obtained is reported. This can be used to determine the percentage of observations out of specification. The overall long-term objective is to make C_p and C_{p_k} as large as possible by continuously improving or reducing the process variability, $\hat{\sigma}$, for every iteration, so that a greater percentage of the product is near the key quality characteristic target value. The ideal is to center the process with zero variability.

If a process is centered but not capable, one or several courses of action may be necessary. One of the actions may be that of integrating a designed experiment to gain additional knowledge on the process and in designing control strategies. If excessive variability is demonstrated, one may conduct a nested design with the objective of estimating the various sources of variability. These sources of variability can then be evaluated to determine what strategies to use in order to reduce or permanently eliminate them. Another action may be that of changing the specifications or continuing production and then sorting the items. Three characteristics of a process can be observed with respect to capability, as summarized below. Figures 8.17 through 8.19 present the alternative characteristics.

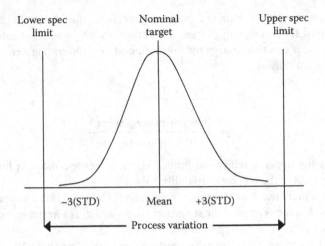

FIGURE 8.17 A process that is centered and capable.

1. The process may be centered and capable.
2. The process may be capable but not centered.
3. The process may be centered but not capable.

Process Capability Example

Step 1: Using data for the specific process, determine if the process is capable. Let us assume that the analyst has determined that the process is in a state of statistical control. For this example, the specification limits are set at 0 (lower limit) and 45 (upper limit).

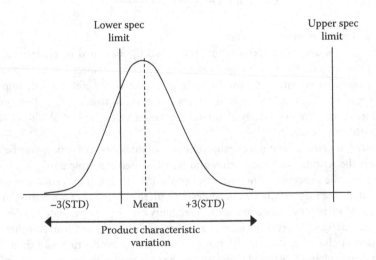

FIGURE 8.18 A process that is capable but not centered.

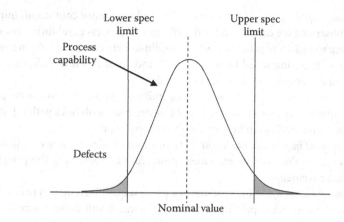

FIGURE 8.19 A process that is centered but not capable.

The inherent process variability as determined from the control chart is

$$\hat{\sigma} = \frac{\overline{R}}{d_2} = \frac{5.83}{2.059} = 2.83.$$

The capability of this process to produce within the specifications can be determined as

$$C_p = \frac{USL - LSL}{6\hat{\sigma}} = \frac{45 - 0}{6(2.83)} = 2.650.$$

The capability of the process $C_p = 2.65 > 1.0$, indicating that the process is capable of producing clutches that will meet the specifications of between 0 and 45. The process average is 29.367.

Step 2: Determine if the process can be adequately centered. Here $C_{p_k} = \min(C_l, C_u)$ can be used to determine if a process can be centered.

$$C_u = \frac{USL - \overline{X}}{3\hat{\sigma}} = \frac{45 - 29.367}{3(2.83)} = 1.84,$$

$$C_l = \frac{\overline{X} - LSL}{3\hat{\sigma}} = \frac{29.367 - 0}{3(2.83)} = 3.46.$$

Therefore, the capability index, C_{p_k}, for this process is 1.84. Since $C_{p_k} = 1.84$ is greater than 1.33, it follows that the process can be adequately centered.

Possible Applications of Process Capability Index

The potential applications of the process capability index are summarized below:

- *Communication:* C_p and C_{p_k} have been used in industry to establish a dimensionless common language useful for assessing the performance of production processes. Engineering, quality, manufacturing, and so on. can communicate and understand processes with high capabilities.

- *Continuous improvement:* The indices can be used to monitor continuous improvement by observing the changes in the distribution of process capabilities. For example, if there were 20% of processes with capabilities between 1 and 1.67 in a month, and some of these improved to between 1.33 and 2.0 the next month, then this is an indication that improvement has occurred.
- *Audits:* There are so many various kinds of audits in use today to assess the performance of quality systems. A comparison of in-process capabilities with capabilities determined from audits can help establish problem areas.
- *Prioritization of improvement:* A complete printout of all processes with unacceptable C_p or C_{p_k} values can be extremely powerful in establishing the priority for process improvements.
- *Prevention of non-conforming product:* For process qualification, it is reasonable to establish a benchmark capability of $C_{p_k} = 1.33$, which will make nonconforming products unlikely in most cases.

Potential Abuse of C_p and C_{p_k}

In spite of its several possible applications, the process capability index has some potential sources of abuse as summarized below:

- *Problems and drawbacks:* C_{p_k} can increase without process improvement even though repeated testing reduces test variability; the wider the specifications, the larger the C_p or C_{p_k}, but the action does not improve the process.
- Analysts tend to focus on numbers rather than on processes.
- *Process control:* Analysts tend to determine process capability before statistical control has been established. Most people are not aware that capability determination is based on the process common cause variation and what can be expected in the future. The presence of special causes of variation makes prediction impossible and the capability index unclear.
- *Non-normality:* Some processes result in non-normal distribution for some characteristics. As capability indices are very sensitive to departures from normality, data transformation may be used to achieve approximate normality.
- *Computation:* Most computer-based tools do not use $\overline{R}/d_2$ to calculate σ.

When analytical and statistical tools are coupled with sound managerial approaches, an organization can benefit from a robust implementation of improvement strategies. One approach that has emerged as a sound managerial principle is lean, which has been successfully applied to many industrial operations.

Lean Principles and Applications

What is "lean"? Lean means the identification and elimination of sources of *waste* in operations. Recall that six sigma involves the identification and elimination of the source

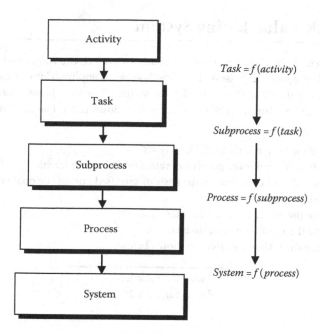

FIGURE 8.20 Hierarchy of process components.

of *defects*. When Lean and six sigma are coupled, an organization can derive the double benefit of reducing waste and defects in operations, which leads to what is known as Lean–Six-Sigma. Consequently, the organization can achieve higher product quality, better employee morale, better satisfaction of customer requirements, and more effective utilization of limited resources. The basic principle of "Lean" is to take a close look at the elemental compositions of a process so that non-value-adding elements can be located and eliminated.

Kaizen of a Process

By applying the Japanese concept of "Kaizen," which means "take apart and make better," an organization can redesign its processes to be lean and devoid of excesses. In a mechanical design sense, this can be likened to finite-element analysis, which identifies how the component parts of a mechanical system fit together. It is by identifying these basic elements that improvement opportunities can be easily and quickly recognized. It should be recalled that the process of work breakdown structure in project management facilitates the identification of task- level components of an endeavor. Consequently, using a project management approach facilitates the achievement of the objectives of "lean." Figure 8.20 shows a process decomposition hierarchy that may help to identify elemental characteristics that may harbor waste.

Lean Task Value Rating System

In order to identify value-adding elements of a Lean project, the component tasks must be ranked and comparatively assessed. The method below applies relative ratings to tasks. It is based on the distribution of a total point system. The total points available to the composite process or project are allocated across individual tasks. The steps are explained below:

1. Let T be the total points available to tasks.
2. $T = 100n$, where n is the number of raters on the rating team.
3. Rate the value of each task on the basis of specified output (or quality) criteria on a scale of 0 to 100.
4. Let x_{ij} be the rating for task i by rater j.
5. Let m be the number of tasks to be rated.
6. Organize the ratings by rater j as shown below:

Rating for Task 1:	x_{ij}
Rating for Task 2:	x_{2j}
$\vdots$	$\vdots$
Rating for Task m:	x_{mj}
Total rating points	100

7. Tabulate the ratings by the raters as shown in Table 8.3 and calculate the overall weighted score for each Task i from the expression

$$w_i = \frac{1}{n} \sum_{j=1}^{n} x_{ij}.$$

The w_i are used to rank order the tasks to determine the relative value-added contributions of each. Subsequently, using a preferred cut-off margin, the low or noncontributing activities can be slated for elimination.

In terms of activity prioritization, a comprehensive Lean analysis can identify the important versus unimportant, and urgent versus not urgent tasks. It is within the

TABLE 8.3 Lean Task Rating Matrix

	Rating by Rater ($j = 1$)	Rating by Rater ($j = 2$)	$\cdots$	Rating by Rater, n	Total Points from Task i	w_i
Rating for Task $i = 1$						
Rating for Task $i = 2$						
$\vdots$						
Rating for Task m						
Total points from rater j	100	100		100	$100n$	

TABLE 8.4 Pareto Analysis of Unimportant Process Task Elements

	Urgent	Not urgent
Important	20%	80%
Unimportant	80%	20%

unimportant and not urgent quadrant that one will find "waste" task elements that should be eliminated. Using the familiar Pareto distribution format, Table 8.4 presents an example of task elements within a 20% waste elimination zone.

It is conjectured that activities that fall in the "unimportant" and "not urgent" zone run the risk of generating points of waste in any productive undertaking. That zone should be the first target of review for tasks that can be eliminated. Granted that there may be some "sacred cow" activities that an organization must retain for political, cultural, or regulatory reasons, attempts should still be made to categorize all task elements of a project. The long-established industrial engineering principle of time-and-motion studies is making a comeback due to the increased interest in eliminating waste in Lean initiatives.

9

Risk Computations

Uncertainty is a reality in any decision-making process. Industrial engineers take risk and uncertainty into consideration in any decision scenario. Industrial engineers identify, analyze, quantify, evaluate, track, and develop mitigation strategies for risks. Traditional decision theory classifies decisions under three different influences:

- *Decision under certainty:* Made when possible event(s) or outcome(s) of a decision can be positively determined.
- *Decisions under risk:* Made using information on the probability that a possible event or outcome will occur.
- *Decisions under uncertainty:* Made by evaluating possible event(s) or outcome(s) without information on the probability that the event(s) or outcome(s) will occur.

Cost Uncertainties

In an inflationary economy, costs can become very dynamic and intractable. Cost estimates include various tangible and intangible components of a project, such as machines, inventory, training, raw materials, design, and personnel wages. Costs can change during a project for a number of reasons including:

- External inflationary trends
- Internal cost adjustment procedures
- Modification of work process
- Design adjustments
- Changes in cost of raw materials
- Changes in labor costs
- Adjustment of work breakdown structure
- Cash flow limitations
- Effects of tax obligations

These cost changes and others combine to create uncertainties in the project's cost. Even if the cost of some of the parameters can be accurately estimated, the overall project cost may still be uncertain due to the few parameters that cannot be accurately estimated.

Schedule Uncertainties

Unexpected engineering change orders and other changes in a project environment may necessitate schedule changes, which introduce uncertainties into the project. The following are some of the reasons for changes in project schedules:

- Task adjustments
- Changes in scope of the work
- Changes in delivery arrangements
- Changes in project specification
- Introduction of new technology

Performance Uncertainties

Performance measurement involves observing the value of parameter(s) during a project and comparing the actual performance, based on the observed parameter(s), to the expected performance. Performance control then takes appropriate actions to minimize the deviations between the actual performance and expected performance. Project plans are based on the expected performance of the project parameters. Performance uncertainties exist when the expected performance cannot be defined in definite terms. As a result, project plans require a frequent review.

The project management team must have a good understanding of the factors that can have a negative impact on the expected project performance. If at least some of the sources of deficient performance can be controlled, then the detrimental effects of uncertainties can be alleviated. The most common factors that can influence the project performance include the following:

- Redefinition of project priorities
- Changes in management control
- Changes in resource availability
- Changes in work ethic
- Changes in organizational policies and procedures
- Changes in personnel productivity
- Changes in quality standards

To minimize the effect of uncertainties in project management, a good control must be maintained over the various sources of uncertainty discussed above. The same analytical tools that are effective for one category of uncertainties should also work for other categories.

Decision Tables and Trees

Decision tree analysis is used to evaluate sequential decision problems. In project management, a *decision tree* may be useful for evaluating sequential project milestones. A decision problem under certainty has two elements: *action* and *consequence*. The decision-maker's choices are the actions, whereas the results of those actions are the consequences. For example, in a Critical Path Method (CPM) network planning, the choice of one task among three potential tasks in a given time slot represents a potential action. The consequences of choosing one task over another may be characterized in terms of the slack time created in the network, the cost of performing the selected task, the resulting effect on the project completion time, or the degree to which a specified performance criterion is satisfied.

If the decision is made under uncertainty, as in Program Evaluation and Review Technique (PERT) network analysis, a third element, called an *event*, is introduced into the decision problem. We extended the CPM task selection example to a PERT analysis, and the actions may be defined as Select Task 1, Select Task 2, and Select Task 3. The durations associated with the three possible actions can be categorized as *long task duration*, *medium task duration*, and *short task duration*. The actual duration of each task is uncertain. Thus, each task has some probability of exhibiting long, medium, or short durations.

The events can be identified as weather incidents: rain or no rain. The incidents of rain or no rain are uncertain. The consequences may be defined as *increased project completion time, decreased project completion time*, and *unchanged project completion time*. These consequences are also uncertain due to the probable durations of the tasks and the variable choices of the decision-maker. That is, the consequences are determined partly by choice and partly by chance. The consequences also depend on which event occurs—rain or no rain.

To simplify the decision analysis, the decision elements may be summarized by using a decision table. A decision table shows the relationship between pairs of decision elements. Table 9.1 shows the decision table for the preceding example. In Table 9.1, each row corresponds to an event and each column corresponds to an action. The consequences appear as entries in the body of Table 9.1. The consequences have been coded as I (increased), D (decreased), and U (unchanged). Each event–action combination has a specific consequence associated with it.

In some decision problems, the consequences may not be unique. Thus, a consequence, which is associated with a particular event–action pair, may also be associated with another event–action pair. The actions included in the decision table are the only ones that the decision-maker wishes to consider. Subcontracting or task elimination, for example, are other possible choices for the decision-maker. The actions included in the decision problem are mutually exclusive and collectively exhaustive, so that exactly one will be selected. The events are also mutually exclusive and collectively exhaustive.

TABLE 9.1 Decision Table for Task Selection

	Actions								
	Task 1			Task 2			Task 3		
Event	Long	Medium	Short	Long	Medium	Short	Long	Medium	Short
Rain	I	I	U	I	U	D	I	I	U
No rain	I	D	D	U	D	D	U	U	U

I, increased project duration; D, decreased project duration; U, unchanged project duration.

The decision problem can also be conveniently represented as a decision tree as shown in Figure 9.1. The tree representation is particularly effective for decision problems with choices that must be made at different times over an extended period. Resource-allocation decisions, for example, must be made several times during a project's life cycle. The choice of actions is shown as a fork with a separate branch for each action. The events are also represented by branches in separate fields.

To avoid confusion in very complex decision trees, the nodes for action forks are represented by squares, whereas the nodes for event forks are represented by circles. The basic convention for constructing a tree diagram is that the flow should be chronological from left to right. The actions are shown on the initial fork because the decision must be made before the actual event is known. The events are shown as branches in the third-stage forks. The consequence resulting from an event–action combination is shown as the endpoint of the corresponding path from the root of the tree.

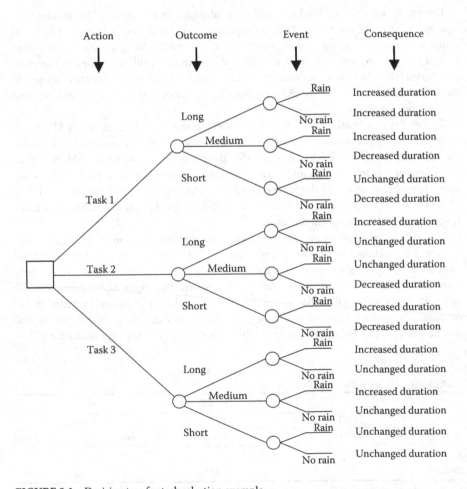

FIGURE 9.1 Decision tree for task selection example.

Figure 9.1 shows six paths leading to an increase in the project duration, five paths leading to a decrease in project duration, and seven paths leading to an unchanged project duration. The total number of paths is given by

$$P = \prod_{i=1}^{N} n_i,$$

where P is the total number of paths in the decision tree, N the number of decision stages in the tree, and n_i the number of branches emanating from each node in stage i.

Thus, for the sample given in Figure 9.1, the number of paths is $P = (3)(3)(2) = 18$ paths. As mentioned previously, some of the paths, even though they are distinct, lead to identical consequences.

Probability values can be incorporated into the decision structure as shown in Figure 9.2. Note that the selection of a task at the decision node is based on choice rather than probability. In this example, we assume that the probability of having a particular task duration is independent of whether or not it rains. In some cases, the weather

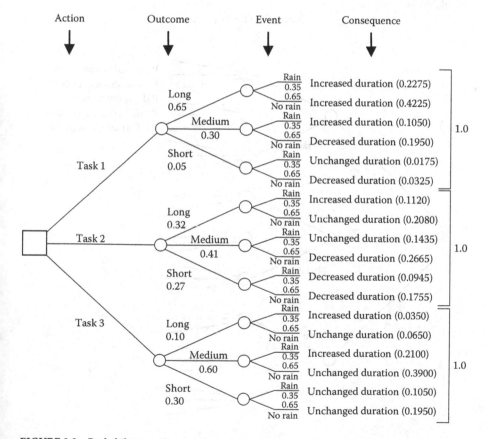

FIGURE 9.2 Probability tree diagram for task selection example.

sensitivity of a task may influence the duration of the task. Also, the probability of rain or no rain is independent of any other element in the decision structure.

If the items in the probability tree are interdependent, then the appropriate conditional probabilities would need to be computed. This will be the case if the duration of a task is influenced by whether or not it rains. In such a case, the probability tree should be redrawn as shown in Figure 9.3, which indicates that the weather event will need to be observed first before the task duration event can be determined. In Figure 9.3, the conditional probability of each type of duration, given that it rains or it does not rain, will need to be calculated.

The respective probabilities of the three possible consequences are shown in Figure 9.2. The probability at the end of each path is computed by multiplying the individual probabilities along the path. For example, the probability of having an increased project completion time along the first path (Task 1, Long Duration, and Rain) is calculated as

$$(0.65)(0.35) = 0.2275.$$

Similarly, the probability for the second path (Task 1, Long Duration, and No Rain) is calculated as

$$(0.65)(0.65) = 0.4225.$$

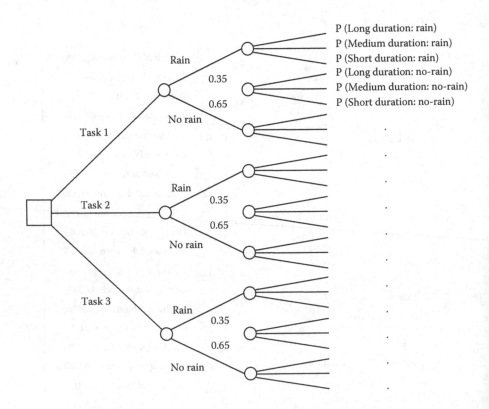

FIGURE 9.3 Probability tree for weather-dependent task durations.

TABLE 9.2 Probability Summary for Project Completion Time

Consequence	Selected Task					
	Task 1		Task 2		Task 3	
Increased duration	0.2275 + 0.4225 + 0.105	0.755	0.112	0.112	0.035 + 0.21	0.245
Decreased duration	0.195 + 0.0325	0.2275	0.2665 + 0.0945 + 0.1755	0.5635	0.0	0.0
Unchanged duration	0.0175	0.0175	0.208 + 0.1435	0.3515	0.065 + 0.39 + 0.105 + 0.195	0.755
Sum of probabilities		1.0		1.0		1.0

The sum of the probabilities at the end of the paths associated with each action (choice) is equal to 1 as expected. Table 9.2 presents a summary of the respective probabilities of the three consequences based on the selection of each task. The probability of having an increased project duration when Task 1 is selected is calculated as

$$\text{Probability} = 0.2275 + 0.4225 + 0.105 = 0.755.$$

Likewise, the probability of having an increased project duration when Task 3 is selected is calculated as

$$\text{Probability} = 0.035 + 0.21 = 0.245.$$

If the selection of tasks at the first node is probable in nature, then the respective probabilities would be included in the calculation procedure. For example, Figure 9.4 shows a case where Task 1 is selected 25% of the time, Task 2 is selected 45% of the time, and Task 3 is selected 30% of the time. The resulting end probabilities for the three possible consequences have been revised accordingly. Note that all probabilities at the end of all the paths add up to 1 in this case. Table 9.3 presents the summary of the probabilities of the three consequences for the case of weather-dependent task durations.

The examples presented above can be extended to other decision problems in project management, which can be represented in terms of decision tables and trees. For example, resource-allocation decision problems under uncertainty can be handled by appropriate decision tree models.

Reliability Calculations

$$\text{Variance:} \quad \sigma^2 = \frac{1}{\lambda^2}.$$

$$\text{Standard deviation:} \quad \sigma = \sqrt{\frac{1}{\lambda^2}} = \frac{1}{\lambda}.$$

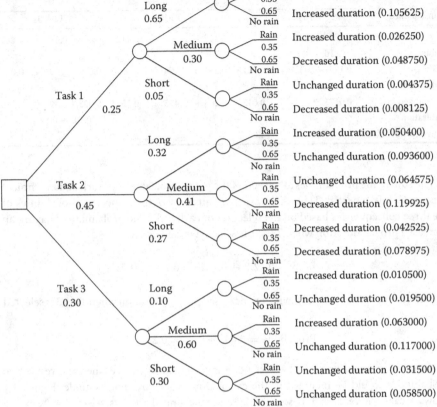

FIGURE 9.4 Modified probability tree for task selection example.

General Reliability Definitions

a. Reliability function:
The reliability function $R(t)$, also known as the survival function $S(t)$, is defined by

$$R(t) = S(t) = 1 - F(t).$$

TABLE 9.3 Summary for Weather-Dependent Task Durations

Consequence	Path Probabilities	Row Total
Increased duration	$0.056875 + 0.105625 + 0.02625 + 0.0504 + 0.0105 + 0.063$	0.312650
Decreased duration	$0.04875 + 0.119925 + 0.042525 + 0.078975$	0.290175
Unchanged duration	$0.004375 + 0.008125 + 0.0936 + 0.064575 + 0.0195 +$	0.397175
	$0.117 + 0.0315 + 0.0585$	
Column total		1.0

b. Failure distribution function:

The failure distribution function is the probability of an item failing in the time interval $0 \le \tau \le t$:

$$F(t) = \int_0^t f(\tau) \, d\tau, \quad t \ge 0.$$

c. Failure rate:

The failure rate of the unit is

$$z(t) = \lim_{\Delta t \to 0} \frac{F(t - \Delta t)}{R(t)} = \frac{f(t)}{R(t)}.$$

d. Mean time to failure:

The mean time to failure (MTTF) of a unit is

$$\text{MTTF} = \int_0^\infty f(t)t \, dt = \int_0^\infty R(t) \, dt.$$

e. Reliability of the system:

The reliability of the system is the product of the reliability functions of the components $R_1 \cdots R_n$:

$$R_s(t) = R_1 \cdot R_2 \cdots R_n = \prod_{i-1}^n R_i(t).$$

Exponential Distribution Used as Reliability Function

a. Reliability function:

$$R(t) = e^{-\lambda t} \quad (\lambda = \text{constant}).$$

b. Failure distribution function:

$$F(t) = 1 - e^{-\lambda t}.$$

c. Density function of failure:

$$f(t) = \lambda \, e^{-\lambda t}.$$

d. Failure rate:

$$z(t) = \frac{f(t)}{R(t)} = \lambda.$$

e. Mean time to failure:

$$\text{MTTF} = \int_0^\infty e^{-\lambda t} \, dt = \frac{1}{\lambda}.$$

f. System reliability:

$$R_s(t) = e^{-k}, \quad \text{where } k = t \sum_{i=1}^n \lambda_i.$$

g. Cumulative failure rate:

$$z_s = \lambda_1 + \lambda_2 + \cdots + \lambda_n = \sum_{i=1}^{n} \lambda_i = \frac{1}{\text{MTBF}}.$$

MTBF = Mean Time Between Failures

Reliability Formulas

$$\text{Variance:} \quad \sigma^2 = \frac{1}{\lambda^2}.$$

$$\text{Standard deviation:} \quad \sigma = \sqrt{\frac{1}{\lambda^2}} = \frac{1}{\lambda}.$$

General Reliability Definitions

a. Reliability function:
   The reliability function $R(t)$, also known as the survival function $S(t)$, is defined by

$$R(t) = S(t) = 1 - F(t).$$

b. Failure distribution function:
   The failure distribution function is the probability of an item failing in the time interval $0 \leq \tau \leq t$:

$$F(t) = \int_0^t f(\tau)\, d\tau, \quad t \geq 0.$$

c. Failure rate:
   The failure rate of the unit is

$$z(t) = \lim_{\Delta t \to 0} \frac{F(t - \Delta t)}{R(t)} = \frac{f(t)}{R(t)}.$$

d. Mean time to failure:
   The MTTF of a unit is

$$\text{MTTF} = \int_0^\infty f(t)t\, dt = \int_0^\infty R(t)\, dt.$$

e. Reliability of the system:
   The reliability of the system is the product of the reliability functions of the components $R_1 \cdots R_n$:

$$R_s(t) = R_1 \cdot R_2 \cdots R_n = \prod_{i-1}^{n} R_i(t).$$

Exponential Distribution Used as Reliability Function

a. Reliability function:
$$R(t) = e^{-\lambda t} \quad (\lambda = \text{constant})$$

b. Failure distribution function:
$$F(t) = 1 - e^{-\lambda t}.$$

c. Density function of failure:
$$f(t) = \lambda\, e^{-\lambda t}.$$

d. Failure rate:
$$z(t) = \frac{f(t)}{R(t)} = \lambda.$$

e. Mean time to failure:
$$\text{MTTF} = \int_0^\infty e^{-\lambda t}\, dt = \frac{1}{\lambda}.$$

f. System reliability:
$$R_s(t) = e^{-k}, \quad \text{where } k = t \sum_{i=1}^{n} \lambda_i.$$

g. Cumulative failure rate:
$$z_s = \lambda_1 + \lambda_2 + \cdots + \lambda_n = \sum_{i=1}^{n} \lambda_i = \frac{1}{\text{MTBF}}.$$

10

Computations for Project Analysis

Computations for project analysis focus on cost, schedule, and performance. Project planning, organizing, scheduling, and control all have a time basis that can benefit from computational analysis. The examples in this chapter illustrate typical computations for project analysis. Some of the questions that drive the computation needs are summarized below.

Planning

How long will engineering and design take?
How long will R&D analysis take?
How long will it take to build up inventory?
What is the installation time?
How long will equipment order take?
Will work occur during any special times (holidays, plant shutdowns)?

Organizing

When will the necessary administrative resources be available?
How much time can resources commit to the project?
When are deliverables expected?

Scheduling

When can project be executed? Is there enough time between design and execution?
When will manufacturing equipment be available?
When will production resources be available?
Is there a special promotion that most be worked around? How does this effect execution timing?

Control

Can execution time be controlled?
In terms of budgeting, what are the time constraints?
What is the time frame for feedback? Productivity analysis? Quality analysis?
What is the time frame to make a GO or NO-GO decision?

CPM Scheduling

Project scheduling is often the most visible step in the sequence of steps of project analysis. The two most common techniques of basic project scheduling are the Critical Path Method (CPM) and Program Evaluation and Review Technique (PERT). The network of activities contained in a project provides the basis for scheduling the project and can be represented graphically to show both the contents and objectives of the project. Extensions to CPM and PERT include Precedence Diagramming Method (PDM) and Critical Resource Diagramming (CRD). These extensions were developed to take care of specialized needs in a particular project scenario. PDM permits the relaxation of the precedence structures in a project so that the project duration can be compressed. CRD handles the project scheduling process by using activity-resource assignment as the primary scheduling focus. This approach facilitates resource-based scheduling rather than activity- based scheduling so that resources can be more effectively utilized.

CPM network analysis procedures originated from the traditional Gantt chart, or bar chart, developed by Henry L. Gantt during World War I. There are several mathematical techniques for scheduling activities, especially where resource constraints are a major factor. Unfortunately, the mathematical formulations are not generally practical due to the complexity involved in implementing them for realistically large projects. Even computer implementations of the complex mathematical techniques often become too cumbersome for real-time managerial applications. A basic CPM project network analysis is typically implemented in three phases:

- Network planning phase
- Network scheduling phase
- Network control phase

Network planning: In network planning phase, the required activities and their precedence relationships are determined. Precedence requirements may be determined on the basis of the following:

- Technological constraints
- Procedural requirements
- Imposed limitations

The project activities are represented in the form of a network diagram. The two popular models for network drawing are the activity-on-arrow (AOA) and the activity-on-node (AON). In the AOA approach, arrows are used to represent activities, whereas nodes represent starting and end points of activities. In the AON approach, conversely, nodes represent activities, whereas arrows represent precedence relationships. Time, cost, and resource requirement estimates are developed for each activity during the network-planning phase and are usually based on historical records, time standards, forecasting, regression functions, or other quantitative models.

Network scheduling: Network scheduling phase is performed by using forward-pass and backward-pass computations. These computations give the earliest and latest starting and finishing times for each activity. The amount of "slacks" or "floats" associated with each activity is determined. The activity path that includes the least slack in the network is used to determine the critical activities. This path also determines the duration of the project. Resource allocation and time-cost trade-offs are other functions performed during network scheduling.

Network control: Network control involves tracking the progress of a project on the basis of the network schedule and taking corrective actions when needed. An evaluation of actual performance versus expected performance determines deficiencies in the project progress. The advantages of project network analysis are presented below.

Advantages for Communication

- It clarifies project objectives.
- It establishes the specifications for project performance.
- It provides a starting point for more detailed task analysis.
- It presents a documentation of the project plan.
- It serves as a visual communication tool.

Advantages for Control

- It presents a measure for evaluating project performance.
- It helps determine what corrective actions are needed.
- It gives a clear message of what is expected.
- It encourages team interaction.

Advantages for Team Interaction

- It offers a mechanism for a quick introduction to the project.
- It specifies functional interfaces on the project.
- It facilitates ease of task coordination.

Figure 10.1 shows the graphical representation for the AON network. The network components are as follow:

- *Node*: A node is a circular representation of an activity.
- *Arrow*: An arrow is a line connecting two nodes and having an arrowhead at one end. The arrow implies that the activity at the tail of the arrow precedes the one at the head of the arrow.
- *Activity*: An activity is a time-consuming effort required to perform a part of the overall project. It is represented by a node in the AON system or by an arrow in the AOA system. The job the activity represents may be indicated by a short phrase or symbol inside the node or along the arrow.

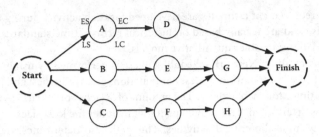

FIGURE 10.1 Graphical representation of AON network.

- *Restriction*: A restriction is a precedence relationship that establishes the sequence of activities. When one activity must be completed before another activity can begin, the first is said to be a predecessor of the second.
- *Dummy*: A dummy is used to indicate one event of a significant nature (e.g., milestone). It is denoted by a dashed circle and treated as an activity with zero time duration. It is not required in the AON method. However, it may be included for convenience, network clarification, or to represent a milestone in the progress of the project.
- *Predecessor activity*: A predecessor activity is one which immediately precedes the one being considered.
- *Successor activity*: A successor activity is one that immediately follows the one being considered.
- *Descendent activity*: A descendent activity is any activity restricted by the one under consideration.
- *Antecedent activity*: An antecedent activity is any activity that must precede the one being considered. Activities A and B are antecedents of D. Activity A is antecedent of B, and A has no antecedent.
- *Merge point*: A merge point exists when two or more activities are predecessors to a single activity. All activities preceding the merge point must be completed before the merge activity can commence.
- *Burst point*: A burst point exists when two or more activities have a common predecessor. None of the activities emanating from the same predecessor activity can be started until the burst-point activity is completed.
- *Precedence diagram*: A precedence diagram is a graphical representation of the activities making up a project and the precedence requirements needed to complete the project. Time is conventionally shown to be from left to right, but no attempt is made to make the size of the nodes or arrows proportional to the duration of time.

Activity Precedence Relationships

Network Notation

A: Activity identification
ES: Earliest starting time

EC: Earliest completion time
LS: Latest starting time
LC: Latest completion time
t: Activity duration

Forward-Pass Calculations

Step 1: Unless otherwise stated, the starting time of a project is set equal to time zero. That is, the first node, *node 1*, in the network diagram has an ES of zero. Thus,

$$ES(1) = 0.$$

If a desired starting time, t_0, is specified, then $ES(1) = t_0$.

Step 2: The ES time for any node (activity j) is equal to the maximum of the ECs of the immediate predecessors of the node. That is,

$$ES(i) = \max\{EC(j)\}, \quad j \in P(i),$$

where $P(i) = \{$set of immediate predecessors of activity $i\}$.

Step 3: The EC of activity i is the activity's ES plus its estimated time t_i. That is,

$$EC(i) = ES(i) + t_i.$$

Step 4: The EC of a project is equal to the EC of the last node, n, in the project network. That is,

$$EC(\text{Project}) = EC(n).$$

Step 5: Unless the LC of a project is explicitly specified, it is set equal to the EC of the project. This is called the zero project slack convention. That is,

$$LC(\text{Project}) = EC(\text{Project}).$$

Step 6: If a desired deadline, T_p, is specified for the project, then

$$LC(\text{Project}) = T_p.$$

It should be noted that an LC or deadline may sometimes be specified for a project on the basis of contractual agreements.

Step 7: The LC for activity j is the smallest of the LSs of the activity's immediate successors. That is,

$$LC(j) = \min, \quad i \in S(j),$$

where $S(j) = \{$immediate successors of activity $j\}$.

Step 8: The LS for activity j is the LC minus the activity time. That is,

$$LS(j) = LC(j) - t_i.$$

CPM Example

Table 10.1 presents the data for a simple project network. The AON network for the example is given in Figure 10.2. Dummy activities are included in the network to designate single starting and end points for the network.

The forward-pass calculation results are shown in Figure 10.3. Zero is entered as the ES for the initial node. As the initial node for the example is a dummy node, its duration is zero. Thus, EC for the starting node is equal to its ES. The ES values for the immediate successors of the starting node are set equal to the EC of the START node and the resulting EC values are computed. Each node is treated as the "start" node for its successor or successors. However, if an activity has more than one predecessor, the maximum of the ECS of the preceding activities is used as the activity's starting time. This happens in the case of activity G, whose ES is determined as max$\{6, 5, 9\} = 9$. The earliest project completion time for the example is 11 days. Note that this is the maximum of the immediately preceding ECs: max$\{6, 11\} = 11$. As the dummy-ending node has no duration, its EC is set equal to its ES of 11 days.

TABLE 10.1 Data for Sample Project for CPM Analysis

Activity	Predecessor	Duration (days)
A	—	2
B	—	6
C	—	4
D	A	3
E	C	5
F	A	4
G	B, D, E	2

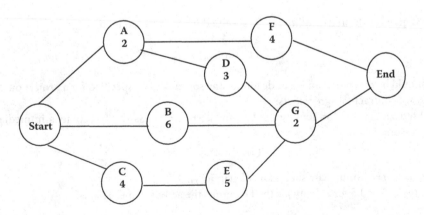

FIGURE 10.2 Example of activity network.

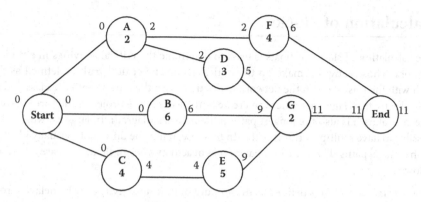

FIGURE 10.3 Forward-pass analysis for CPM example.

Backward-Pass Calculations

The backward-pass computations establish the LS and LC for each node in the network. The results of the backward-pass computations are shown in Figure 10.4. As no deadline is specified, the LC of the project is set equal to the EC. By backtracking and using the network analysis rules presented earlier, the LCs and LSs are determined for each node. Note that in the case of activity A with two immediate successors, the LC is determined as the minimum of the immediately succeeding LSs. That is, min{6, 7} = 6. A similar situation occurs for the dummy-starting node. In that case, the LC of the dummy start node is min{0, 3, 4} = 0. As this dummy node has no duration, the LS of the project is set equal to the node's LC. Thus, the project starts at time 0 and is expected to be completed by time 11.

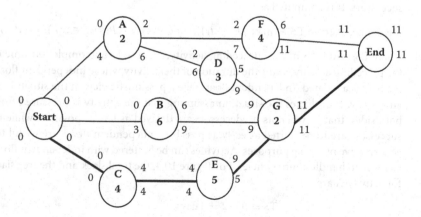

FIGURE 10.4 Backward-pass analysis for CPM example.

Calculation of Slacks

The calculation of slacks or floats is used to determine the critical activities in a project network. Those activities make up the critical path. The critical path is defined as the path with the least slack in the network. All activities on the critical path are classified as critical activities. These activities can create bottlenecks in the project if they are delayed. The critical path is also the longest path in the network diagram. In large networks, it is possible to have multiple critical paths. In this case, it may be difficult to visually identify all the critical paths. There are four basic types of activity slack or float. They are described below:

1. *Total slack* (TS) is defined as the amount of time an activity may be delayed from its ES without delaying the LC of the project. The total slack of activity j is the difference between the LC and the EC of the activity, or the difference between the LS and the ES of the activity:

$$TS(j) = LC(j) - EC(j) \quad \text{or} \quad TS(j) = LS(j) - ES(j).$$

2. *Free slack* (FS) is the amount of time an activity may be delayed from its ES without delaying the starting time of any of its immediate successors. An activity's free slack is calculated as the difference between the minimum ES of the activity's successors and the EC of the activity:

$$FS(j) = \min\{ES(i)\} - EC(j), \quad j \in S(j).$$

3. *Interfering slack* (IS) is the amount of time by which an activity interferes with (or obstructs) its successors when its total slack is fully used. It is computed as the difference between the total slack and free slack:

$$IS(j) = TS(j) - FS(j).$$

4. *Independent float* (IF) is the amount of float that an activity will always have regardless of the completion times of its predecessors or the starting times of its successors. It is computed as

$$IF = \max\{0, (\min ES_j - \max LC_i - t_k)\}, \quad j \in S(k), \ i \in P(k), \ j \in S(k), \ i \in P(k),$$

where ES_j is the ES of the succeeding activity, LC_i the latest completion time of the preceding activity, and t the duration of the activity whose independent floats are being calculated. Independent floats take a pessimistic view of the situation of an activity. It evaluates the situation assuming that the activity is pressured from both sides, that is, when its predecessors are delayed as late as possible, while its successors are to be started as early as possible. Independent floats are useful for conservative planning purposes. Activities can be buffered with independent floats as a way to handle contingencies. For Figure 10.4, the total slack and the free slack for activity A are

$$TS = 6 - 2 = 4 \text{ days},$$
$$FS = \min\{2, 2\} - 2 = 2 - 2 = 0.$$

TABLE 10.2   Result of CPM Analysis for Sample Project

Activity	Duration (days)	ES	EC	LS	LC	TS	FS	Critical
A	2	0	2	4	6	4	0	—
B	6	0	6	3	9	3	3	—
C	4	0	4	0	4	0	0	Critical
D	3	2	5	6	9	4	4	—
E	5	4	9	4	9	0	0	Critical
F	4	2	6	7	11	5	5	—
G	2	9	11	9	11	0	0	Critical

Similarly, the total slack and the free slack for activity F are

$$TS = 11 - 6 = 5 \text{ days},$$
$$FS = \min\{11\} - 6 = 11 - 6 = 5 \text{ days}.$$

Table 10.2 presents a tabulation of the results of the CPM example. Table 10.2 contains the earliest and latest times for each activity, as well as the total and free slacks. The results indicate that the minimum total slack in the network is zero. Thus, activities C, E, and G are identified as the critical activities. The critical path is highlighted in Figure 10.4 and consists of the following sequence of activities:

$$\text{START} \longrightarrow C \longrightarrow E \longrightarrow G \longrightarrow \text{END}.$$

The total slack for the overall project itself is equal to the total slack observed on the critical path. The minimum slack in most networks will be zero because the ending LC is set equal to the ending EC. If a deadline is specified for a project, then the project's LC should be set to the specified deadline. In that case, the minimum total slack in the network will be given by

$$TS_{min} = (\text{Project deadline}) - EC \text{ of the last node}.$$

This minimum total slack will then appear as the total slack for each activity on the critical path. If a specified deadline is lower than the EC at the finish node, then the project will start out with a negative slack. This means that it will be behind schedule before it even starts. It may then become necessary to expedite some activities (i.e., crashing) in order to overcome the negative slack. Figure 10.5 shows an example with a specified project deadline. In this case, the deadline of 18 days occurs after the EC of the last node in the network.

Calculations for Subcritical Paths

In a large project network, there may be paths that are near critical. Such paths require almost as much attention as the critical path because they have a high risk of becoming critical when changes occur in the network. Analysis of subcritical paths may help in

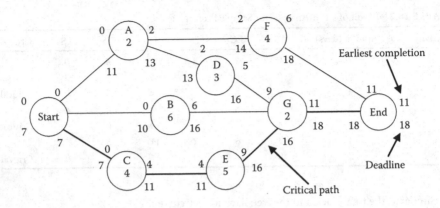

FIGURE 10.5 CPM network with deadline.

the classification of tasks into ABC categories on the basis of Pareto analysis, which separates the most important activities from the less important ones. This can be used for more targeted allocation of resources. With subcritical analysis, attention can shift from focusing only on the critical path to managing critical and near-critical tasks. Steps for identifying the subcritical paths are as follow:

Step 1: Sort activities in increasing order of total slacks.

Step 2: Partition the sorted activities into groups based on the magnitudes of total slacks.

Step 3: Sort the activities within each group in increasing order of their ESs.

Step 4: Assign the highest level of criticality to the first group of activities (e.g., 100%). This first group represents the usual critical path.

Step 5: Calculate the relative criticality indices for the other groups in decreasing order of criticality.

The path's criticality level is obtained from the formula

$$\lambda = \frac{\alpha_2 - \beta}{\alpha_2 - \alpha_1} \times 100,$$

where α_1 is the minimum total slack in the network, α_2 the maximum total slack in the network, and β the total slack for the path whose criticality is to be calculated.

The above procedure yields relative criticality levels between 0% and 100%. Table 10.3 presents an example of path criticality levels. The criticality level may be converted to a scale between 1 (least critical) and 10 (most critical) by the scaling factor below:

$$\lambda' = 1 + 0.09\lambda.$$

Plotting of Gantt Charts

A project schedule is developed by mapping the results of CPM analysis to a calendar timeline. The Gantt chart is one of the most widely used tools for presenting project

TABLE 10.3 Analysis of Subcritical Paths

Path number	Activities on path	Total slacks	λ(%)	λ'(%)
1	A, C, G, H	0	100	10
2	B, D, E	1	97.56	9.78
3	F, I	5	87.81	8.90
4	J, K, L	9	78.05	8.03
5	O, P, Q, R	10	75.61	7.81
6	M, S, T	25	39.02	4.51
7	N, AA, BB, U	30	26.83	3.42
8	V, W, X	32	21.95	2.98
9	Y, CC, EE	35	17.14	2.54
10	DD, Z, FF	41	0	1.00

schedules. It can show planned and actual progress of activities. As a project progresses, markers are made on the activity bars to indicate actual work accomplished. Figure 10.6 presents the Gantt chart for our illustrative example, using the earliest starting (ES) times from Table 10.2. Figure 10.7 presents the Gantt chart for the example based on the latest starting (LS) times. Critical activities are indicated by the shaded bars.

Figure 10.6 shows that the starting time of activity F can be delayed from day 2 until day 7 (i.e., TS = 5) without delaying the overall project. Likewise, A, D, or both may be delayed by a combined total of 4 days (TS = 4) without delaying the overall project. If all the 4 days of slacks are used up by A, then D cannot be delayed. If A is delayed by 1 day, then D can be delayed by up to 3 days without, however, causing a delay of G,

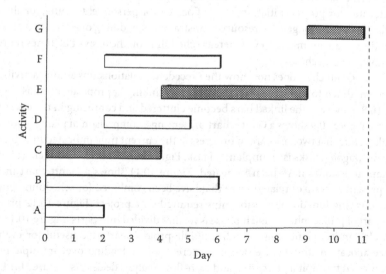

FIGURE 10.6 Gantt chart based on ESs.

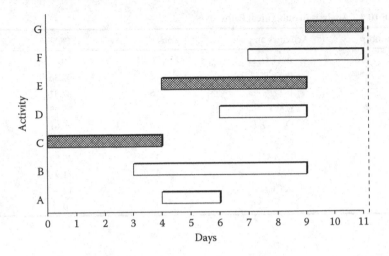

FIGURE 10.7   Gantt chart based on LSs.

which determines project completion. The Gantt chart also indicates that activity B may be delayed by up to 3 days without affecting the project's completion time.

In Figure 10.7, the activities are shown scheduled by their LCs. This represents an extreme case where activity slack times are fully used. No activity in this schedule can be delayed without delaying the project. In Figure 10.7, only one activity is scheduled over the first 3 days. This may be compared to the schedule in Figure 10.6, which has three starting activities. The schedule in Figure 10.7 may be useful if there is a situational constraint that permits only a few activities to be scheduled in the early stages of the project. Such constraints may involve shortage of project personnel, lack of initial budget, time allocated for project initiation, time allocated for personnel training, an allowance for a learning period, or general resource constraints. Scheduling of activities based on ES times indicates an optimistic view, whereas scheduling on the basis of LS times represents a pessimistic approach.

The basic Gantt chart does not show the precedence relationships among activities, but it can be modified to show these relationships by linking appropriate bars, as shown in Figure 10.8. However, the linked bars become cluttered and confusing in the case of large networks. Figure 10.9 shows a Gantt chart presenting a comparison of planned and actual schedules. Note that two tasks are in progress at the current time indicated in Figure 10.7. One of the ongoing tasks is an unplanned task. Figure 10.10 shows a Gantt chart in which important milestones have been indicated. Figure 10.11 shows a Gantt chart in which bars represent a group of related tasks that have been combined for scheduling purposes or for conveying functional relationships required on a project. Figure 10.12 presents a Gantt chart of project phases. Each phase is further divided into parts. Figure 10.13 shows a multiple- project Gantt chart. Such multiple-project charts are useful for evaluating resource-allocation strategies, especially where resource loading over multiple projects may be needed for capital budgeting and cash-flow analysis decisions. Figure 10.14 shows a cumulative slippage (CumSlip) chart that is useful for project tracking and control. The

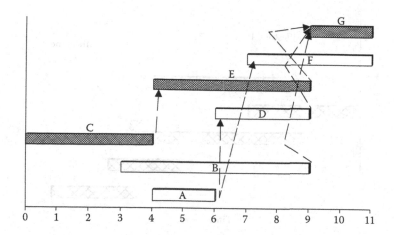

FIGURE 10.8 Linked bars in Gantt chart.

chart shows the accumulation of slippage over a time span, either for individual activities or for the total project. A tracking bar is drawn to show the duration (starting and end points) as of the last time of review.

Calculations for Project Crashing

Crashing is the expediting or compression of activity duration. Crashing is done as a trade-off between a shorter task duration and a higher task cost. It must be determined whether the total cost savings realized from reducing the project duration is enough to justify the higher costs associated with reducing individual task durations. If there is a

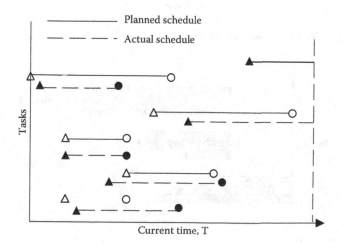

FIGURE 10.9 Progress monitoring Gantt chart.

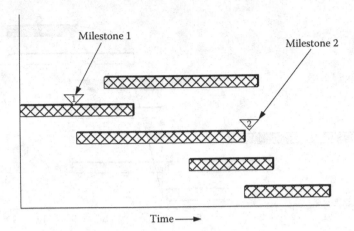

FIGURE 10.10 Milestone Gantt chart.

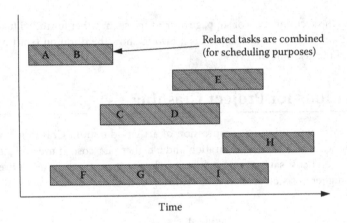

FIGURE 10.11 Task combination Gantt chart.

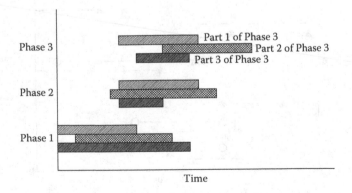

FIGURE 10.12 Phase-based Gantt chart.

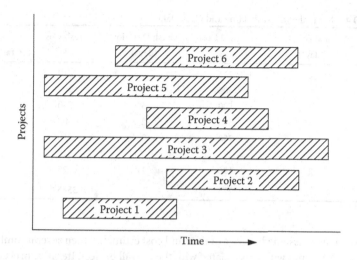

FIGURE 10.13 Multiple-project Gantt chart.

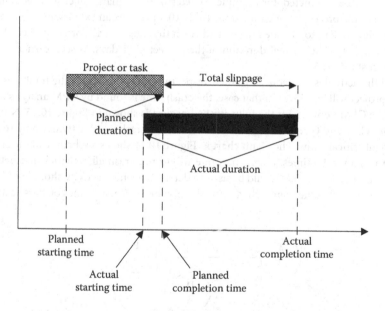

FIGURE 10.14 Cumulative slippage (CumSlip) chart.

delay penalty associated with a project, it may be possible to reduce the total project cost even though crashing increases individual task costs. If the cost savings on the delay penalty is higher than the incremental cost of reducing the project duration, then crashing is justified. Normal task duration refers to the time required to perform a task under normal circumstances. *Crash task duration* refers to the reduced time required to perform a task when additional resources are allocated to it.

TABLE 10.4 Normal and Crash Time and Cost Data

Activity	Normal Duration (days)	Normal Cost ($)	Crash Duration (days)	Crash Cost ($)	Crashing Ratio
A	2	210	2	210	0
B	6	400	4	600	100
C	4	500	3	750	250
D	3	540	2	600	60
E	5	750	3	950	100
F	4	275	3	310	35
G	2	100	1	125	25
		2775		3545	

If each activity is assigned a range of time and cost estimates, then several combinations of time and cost values will be associated with the overall project. Iterative procedures are used to determine the best time or cost combination for a project. Time-cost trade-off analysis may be conducted, for example, to determine the marginal cost of reducing the duration of the project by one time unit. Table 10.4 presents an extension of the data for the sample problem to include normal and crash times as well as normal and crash costs for each activity. The normal duration of the project is 11 days, as seen earlier, and the normal cost is $2775.

If all the activities are reduced to their respective crash durations, the total crash cost of the project will be $3545. In that case, the crash time is found by CPM analysis to be 7 days. The CPM network for the fully crashed project is shown in Figure 10.15. Note that activities C, E, and G remain critical. Sometimes, the crashing of activities may result in additional critical paths. The Gantt chart in Figure 10.16 shows a schedule of the crashed project using the ES times. In practice, one would not crash all activities in a network. Rather, some selection rule would be used to determine which activity should be crashed and by how much. One approach is to crash only the critical activities or those activities

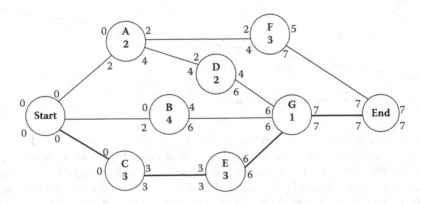

FIGURE 10.15 Example of fully crashed CPM network.

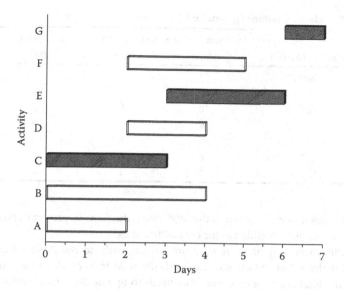

FIGURE 10.16 Gantt chart of fully crashed CPM network.

with the best ratios of incremental cost versus time reduction. The last column in Table 10.4 presents the respective ratios for the activities in our sample. The crashing ratios are computed as

$$r = \frac{\text{Crash cost} - \text{Normal cost}}{\text{Normal duration} - \text{Crash duration}}.$$

Activity G offers the lowest cost per unit time reduction of $25. If the preferred approach is to crash only one activity at a time, we may decide to crash activity G first and evaluate the increase in project cost versus the reduction in project duration. The process can then be repeated for the next best candidate for crashing, which is activity F in this case. The project completion time is not reduced any further because activity F is not a critical activity. After F has been crashed, activity D can then be crashed. This approach is repeated iteratively in order of activity preference until no further reduction in project duration can be achieved or until the total project cost exceeds a specified limit.

A more comprehensive analysis is to evaluate all possible combinations of the activities that can be crashed. However, such a complete enumeration would be prohibitive, because there would be a total of 2^c crashed networks to evaluate, where c is the number of activities that can be crashed out of the n activities in the network ($c \leq n$). For our sample, only six out of the seven activities in the network can be crashed. Thus, a complete enumeration will involve $2^6 = 64$ alternative networks. Table 10.5 shows seven of the 64 crashing options. Activity G, which offers the best crashing ratio, reduces the project duration by only 1 day. Even though activities F, D, and B are crashed by a total of 4 days at an incremental cost of $295, they do not generate any reduction in project duration. Activity E is crashed by 2 days and generates a reduction of 2 days in project duration. Activity C, which is crashed by 1 day, generates a further reduction of 1 day in the project duration.

TABLE 10.5 Selected Crashing Options for CPM Example

Option Number	Activities Crashed	Network Duration (days)	Time Reduction (days)	Incremental Cost ($)	Total Cost ($)
1	None	11	–	–	2775
2	G	10	1	25	2800
3	G, F	10	0	35	2835
4	G, F, D	10	0	60	2895
5	G, F, D, B	10	0	200	3095
6	G, F, D, B, E	8	2	200	3295
7	G, F, D, B, E, C	7	1	250	3545

It should be noted that the activities that generate reductions in project duration are the ones that were earlier identified as the critical activities.

In general, there may be more than one critical path, so the project analyst needs to check for the set of critical activities with the least total crashing ratio in order to minimize the total crashing cost. Also, one needs to update the critical paths every time a set of activities is crashed because new activities may become critical in the meantime. For the network given in Figure 10.15, the path C → E → G is the only critical path. Therefore, we do not need to consider crashing other jobs because the incurred cost will not affect the project completion time. There are 12 possible ways one can crash activities C, G, and E in order to reduce the project time.

Several other approaches exist in determining which activities to crash in a project network. Two alternative approaches are presented below for computing the crashing ratio, r. The first one directly uses the criticality of an activity to determine its crashing ratio, whereas the second one uses the calculation shown below:

$$r = \text{Criticality index,}$$

$$r = \frac{\text{Crash cost} - \text{Normal cost}}{(\text{Normal duration} - \text{Crash duration}) \, (\text{Criticality index})}.$$

The first approach gives crashing priority to the activity with the highest probability of being on the critical path. In deterministic networks, this refers to the critical activities. In stochastic networks, an activity is expected to fall on the critical path only a percentage of the time. The second approach is a combination of the approach used for the illustrative sample and the criticality index approach. It reflects the process of selecting the least-cost expected value. The denominator of the expression represents the expected number of days by which the critical path can be shortened.

Calculations for Project Duration Diagnostics

PERT incorporates variability in activity durations into project duration analysis and diagnostics. Real-life activities are often prone to uncertainties, which in turn determine the actual duration of the activities. In CPM, activity durations are assumed to be fixed.

In PERT, the uncertainties in activity durations are accounted for by using three time estimates for each activity. The three time estimates represent the spread of the estimated activity duration. The greater the uncertainty of an activity, the wider its range of estimates will be. Predicting project completion time is essential for avoiding the proverbial "90% complete, but 90% remains" syndrome of managing large projects. The diagnostic techniques presented in this chapter are useful for having a better handle on predicting project durations.

PERT Formulas

PERT uses three time estimates (a, m, b) and the following equations, respectively, to compute expected duration and variance for each activity:

$$t_e = \frac{a + 4m + b}{6}$$

and

$$s^2 = \frac{(b - a)^2}{36},$$

where a is the optimistic time estimate, m the most likely time estimate, b the pessimistic time estimate $(a < m < b)$, t_e the expected time for the activity, and s^2 the variance of the duration of the activity.

After obtaining the estimate of the duration for each activity, the network analysis is carried out in the same manner as CPM analysis.

Activity Time Distributions

PERT analysis assumes that the probabilistic properties of activity duration can be modeled by the beta probability density function, which is shown in Figure 10.17 with alternate shapes. The uniform distribution between 0 and 1 is a special case of the beta distribution, with both shape parameters equal to 1.

The triangular probability density function has been used as an alternative to the beta distribution for modeling activity times. The triangular density has three essential parameters: minimum value (a), mode (m), and maximum (b). Figure 10.18 presents a graphical representation of the triangular density function.

For cases where only two time estimates, instead of three, are to be used for network analysis, the uniform density function may be assumed for the activity times. This is acceptable for situations for which the extreme limits of an activity's duration can be estimated and it can be assumed that the intermediate values are equally likely to occur. Figure 10.19 presents a graphical representation of the uniform distribution. In this case, the expected activity duration is computed as the average of the upper and lower limits of the distribution. The appeal of using only two time estimates a and b is that the estimation error due to subjectivity can be reduced and the estimation task is simplified. Even when

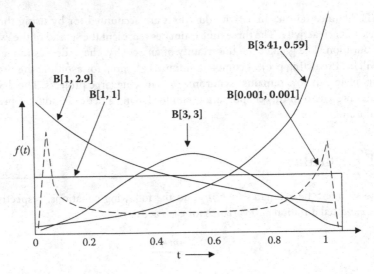

FIGURE 10.17   Beta distribution.

a uniform distribution is not assumed, other statistical distributions can be modeled over the range from *a* to *b*.

Regardless of whichever distribution is used, once the expected activity durations have been computed, the analysis of the activity network is carried out just as in the case of CPM.

Project Duration Analysis

If the project duration can be assumed to be approximately normally distributed, then the probability of meeting a specified deadline can be computed by finding the area under

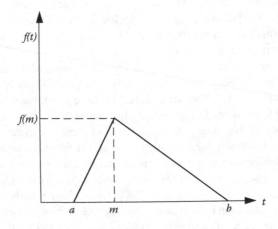

FIGURE 10.18   Triangular probability density function.

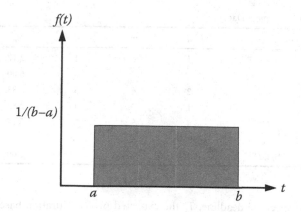

FIGURE 10.19 Uniform probability density function.

the standard normal curve to the left of the deadline. It is important to do an analytical analysis of the project duration because project contracts often come with clauses dealing with the following:

1. No-excuse completion deadline agreement.
2. Late-day penalty clause for not completing within deadline.
3. On-time completion incentive award for completing on or before the deadline.
4. Early finish incentive bonus for completing ahead of schedule.

Figure 10.20 shows an example of a normal distribution describing the project duration. The variable T_d represents the specified deadline.

A relationship between the standard normal random variable z and the project duration variable can be obtained by using the following familiar transformation formula:

$$z = \frac{T_d - T_e}{S},$$

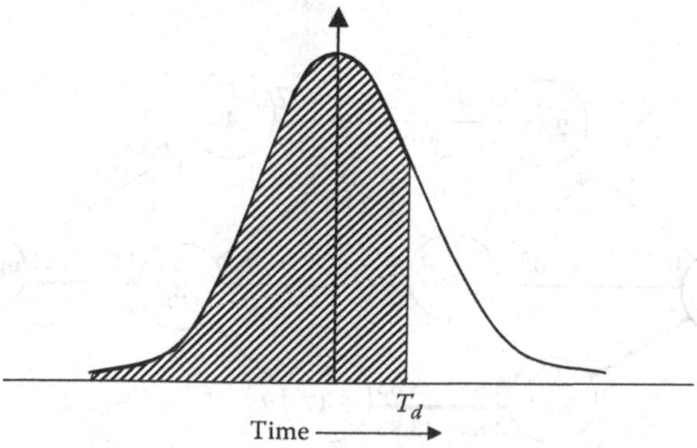

FIGURE 10.20 Project deadline under the normal curve.

TABLE 10.6 PERT Project Data

Activity	Predecessors	a	m	b	t_e	s^2
A	—	1	2	4	2.17	0.2500
B	—	5	6	7	6.00	0.1111
C	—	2	4	5	3.83	0.2500
D	A	1	3	4	2.83	0.2500
E	C	4	5	7	5.17	0.2500
F	A	3	4	5	4.00	0.1111
G	B, D, E	1	2	3	2.00	0.1111

where T_d is the specified deadline, T_e the expected project duration based on network analysis, and S the standard deviation of the project duration.

The probability of completing a project by the deadline T_d is then computed as

$$P(T \leq T_d) = P\left(z \leq \frac{T_d - T_e}{S}\right).$$

This probability is obtained from the standard normal table available in most statistical books. Examples presented below illustrate the procedure for probability calculations in PERT. Suppose we have the project data presented in Table 10.6. The expected activity durations and variances as calculated by the PERT formulas are shown in the last two columns of Table 10.6. Figure 10.21 shows the PERT network. Activities C, E, and G are shown to be critical, and the project completion time is determined to be 11 time units.

The probability of completing the project on or before a deadline of 10 time units (i.e., $T_d = 10$) is calculated as shown below:

$$T_e = 11,$$
$$S^2 = V[C] + V[E] + V[G]$$
$$= 0.25 + 0.25 + 0.1111$$

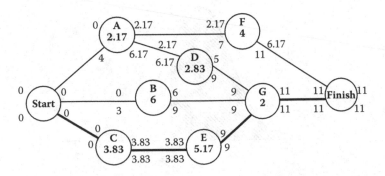

FIGURE 10.21 PERT network example.

$$= 0.6111,$$

$$S = \sqrt{0.6111}$$

$$= 0.7817,$$

and

$$P(T \leq T_d) = P(T \leq 10)$$

$$= P\left(z \leq \frac{10 - T_e}{S}\right)$$

$$= P\left(z \leq \frac{10 - 11}{0.7817}\right)$$

$$= P(z \leq -1.2793)$$

$$= 1 - P(z \leq 1.2793)$$

$$= 1 - 0.8997$$

$$= 0.1003.$$

Thus, there is just over 10% probability of finishing the project within 10 days. By contrast, the probability of finishing the project in 13 days is calculated as

$$P(T \leq 13) = P\left(z \leq \frac{13 - 11}{0.7817}\right)$$

$$= P(z \leq 2.5585)$$

$$= 0.9948.$$

This implies that there is over 99% probability of finishing the project within 13 days. Note that the probability of finishing the project in exactly 13 days will be zero. If we desire the probability that the project can be completed within a certain lower limit (T_L) and a certain upper limit (T_U), the computation will proceed as follows.

Let $T_L = 9$ and $T_U = 11.5$, then

$$P(T_L \leq T \leq T_U) = P(9 \leq T \leq 11.5)$$

$$= P(T \leq 11.5) - P(T \leq 9)$$

$$= P\left(z \leq \frac{11.5 - 11}{0.7817}\right) - P\left(z \leq \frac{9 - 11}{0.7817}\right)$$

$$= P(z \leq 0.6396) - P(z \leq -2.5585)$$

$$= P(z \leq 0.6396) - [1 - P(z \leq 2.5585)]$$

$$= 0.7389 - [1 - 0.9948]$$

$$= 0.7389 - 0.0052$$

$$= 0.7337.$$

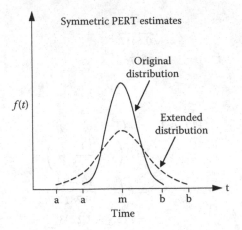

FIGURE 10.22 Interval extension for symmetric distribution.

Simulation of Project Networks

Computer simulation is a tool that can be effectively utilized to enhance project planning, scheduling, and control. At any given time, only a small segment of a project network will be available for direct observation and analysis. The major portion of the project will either have been in the past or will be expected in the future. Such unobservable portions of the project can be analyzed through simulation analysis. Using the historical information from previous segments of a project and the prevailing events in the project environment, projections can be made about future expectations of the project. Outputs of simulation can alert management to real and potential problems. The information provided by simulation can be very helpful in project selection decisions also. Simulation-based project analysis may involve the following:

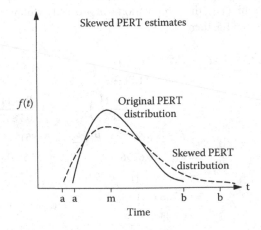

FIGURE 10.23 Interval extension for skewed distribution.

- Analytical modeling of activity times
- Simulation of project schedule
- What-if analysis and statistical modeling
- Management decisions and sensitivity analysis

Activity time modeling. The true distribution of activity times will rarely be known. Even when known, the distribution may change from one type to another depending on who is performing the activity, where the activity is performed, and when the activity is performed. Simulation permits a project analyst to experiment with different activity time distributions. Commonly used activity time distributions are beta, normal, uniform, and triangular distributions. The results of simulation experiments can guide the analyst in developing definite action plans. Using the three PERT estimates a (optimistic time), m (most likely time), and b (pessimistic time), a beta distribution with appropriate shape parameters can be modeled for each activity time. Figure 10.22 shows plots for a symmetric spread of PERT estimates, whereas Figure 10.23 shows plots for s skewed spread.

11

Product Shape and Geometrical Calculations

Product development is one of the major functions of industrial engineers. Industrial engineers work closely with mechanical, electrical, aerospace, and civil engineers to determine the best product development decisions with particular reference to product shape, geometry, cost, material composition, material handling, safety, and ergonomics. These decisions usually involve specific product shape and geometric calculations. Basic and complex examples are presented in this chapter.

Equation of a Straight Line

Lines and curves form the basic shapes of products. Concatenation lines and curves are used to form contour surfaces found in products. The general form of the equation of a straight line is

$$Ax + By + c = 0.$$

The standard form of the equation is

$$y = mx + b.$$

This is known as the *slope–intercept* form.
The point–slope form is

$$y - y_1 = m(x - x_1).$$

Given two points, then

$$\text{slope:} \quad m = \frac{(y_2 - y_1)}{(x_2 - x_1)},$$

the angle between lines with slopes m_1 and m_2 is

$$\alpha = \arctan\left[\frac{(m_2 - m_1)}{(1 + m_2 m_1)}\right].$$

Two lines are perpendicular if $m_1 = -1/m_2$.

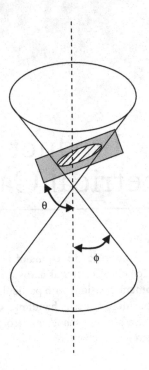

FIGURE 11.1   Plot of conic sections.

The distance between two points is

$$d = \sqrt{(y_2 - y_1)^2 + (x_2 - x_1)^2}.$$

Quadratic Equation

$$ax^2 + bx + c = 0,$$

$$x = \text{Roots} = \frac{-b \pm \sqrt{b^2 - 4ac}}{2a}.$$

Conic Sections

A conic section is a curve obtained by intersecting a cone (more precisely, a right circular conical surface) with a plane, as shown in Figure 11.1, and given by the formula

$$\text{Eccentricity:} \quad e = \frac{\cos \theta}{\cos \phi}.$$

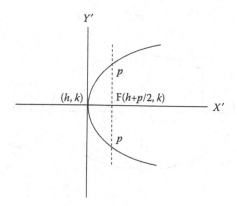

FIGURE 11.2 Conic section for parabola.

Case 1 of Conic Section: $e = 1$ (Parabola)

Standard form:

$$(y - k)^2 = 2p(x - h)$$

with center located at (h, k). X' and Y' are translated axes (see Figure 11.2).
When $h = k = 0$, we have the following:

$$\text{Focus:} \quad \left(\frac{p}{2}, 0\right).$$

$$\text{Directrix:} \quad x = -\frac{p}{2}.$$

Case 2 of Conic Section: $e < 1$ (Ellipse)

Standard form:

$$\frac{(x - h)^2}{a^2} + \frac{(y - k)^2}{b^2} = 1$$

with center at (h, k) (see Figure 11.3).
When $h = k = 0$,

$$\text{Eccentricity:} \quad e = \sqrt{1 - \frac{b^2}{a^2}} = \frac{c}{a}.$$

$$b = a\sqrt{1 - e^2}$$

$$\text{Focus:} \quad (\pm ae, 0).$$

$$\text{Directrix:} \quad x = \pm\frac{a}{e}.$$

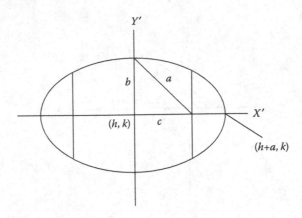

FIGURE 11.3 Conic section for ellipse.

Case 3 Hyperbola: $e > 1$

Standard form:

$$\frac{(x-h)^2}{a^2} - \frac{(y-k)^2}{b^2} = 1$$

with center at (h, k) (see Figure 11.4).

When $h = k = 0$,

$$\text{Eccentricity:} \quad e = \sqrt{1 + \frac{b^2}{a^2}} = \frac{c}{a},$$

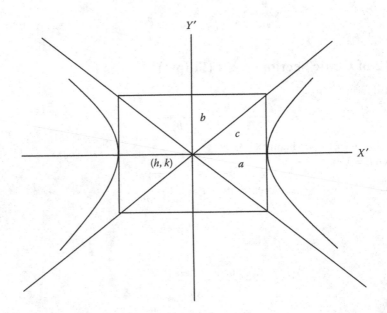

FIGURE 11.4 Conic section for hyperbola.

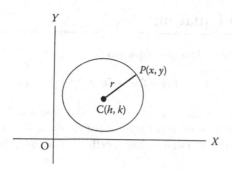

FIGURE 11.5 Conic section for circle.

where $b = a\sqrt{e^2 - 1}$.

$$\text{Focus:} \quad (\pm ae, 0).$$

$$\text{Directrix:} \quad x = \pm\frac{a}{e}.$$

Case 4 $e = 1$ (Circle)

Standard form:

$$(x - h)^2 + (y - k)^2 = r^2$$

with center at (h, k) (see Figure 11.5).

$$\text{Radius:} \quad r = \sqrt{(x - h)^2 + (y - k)^2}.$$

Length of the tangent from a point: Using the general form of the equation of a circle, the length of the tangent, t, is found by substituting the coordinates of a point $P(x', y')$ and the coordinates of the center of the circle into the standard equation (see Figure 11.6):

$$t^2 = (x' - h)^2 + (y' - k)^2 - r^2.$$

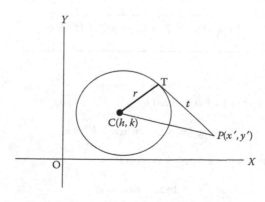

FIGURE 11.6 Circle and tangent line.

Conic Section Equation

The general form of the conic section equation is

$$Ax^2 + Bxy + Cy^2 + Dx + Ey + F = 0,$$

where A and C are not both zero. If $B^2 - AC < 0$, an ellipse is defined. If $B^2 - AC > 0$, a hyperbola is defined. If $B^2 - AC = 0$, the conic is a parabola. If $A = C$ and $B = 0$, a circle is defined. If $A = B = C = 0$, a straight line is defined. The equation,

$$x^2 + y^2 + 2ax + 2by + c = 0$$

is the normal form of the conic section equation, if that conic section has a principal axis parallel to a coordinate axis.

$$h = -a,$$
$$k = -b,$$

and

$$r = \sqrt{a^2 + b^2 - c}.$$

If $a^2 + b^2 - c$ is positive, then we have a circle with center $(-a, -b)$. If $a^2 + b^2 - c$ equals zero, we have a point at $(-a, -b)$. If $a^2 + b^2 - c$ is negative, we have a locus that is *imaginary*.

Quadric Surface (Sphere)

General form:
$$(x - h)^2 + (y - k)^2 + (z - m)^2 = r^2$$

with center at (h, k, m).

In a three-dimensional space, the distance between two points is

$$d = \sqrt{(x_2 - x_1)^2 + (y_2 - y_1)^2 + (z_2 + z_1)^2}.$$

Logarithm:

The logarithm of x to the base b is defined by

$$\log_b(x) = c, \quad \text{where} \quad b^c = x.$$

Special definitions for $b = e$ or $b = 10$ are

$$\ln x, \quad \text{base} = e.$$

$$\log x, \quad \text{base} = 10.$$

To change from one base to another:

$$\log_b(x) = \frac{\log_a x}{\log_a b}.$$

For example,

$$\ln x = \frac{\log_{10} x}{\log_{10} e} = 2.302585(\log_{10} x).$$

Identities

$$\log_b b^n = n,$$

$$\log x^c = c \log x, \quad x^c = \text{antilog}(c \log x),$$

$$\log xy = \log x + \log y,$$

$$\log_b b = 1, \quad \log 1 = 0,$$

$$\log \left(\frac{x}{y} \right) = \log x - \log y.$$

Trigonometry

Trigonometric functions are defined using a right triangle (see Figure 11.7).

$$\sin \theta = \frac{y}{r}, \quad \cos \theta = \frac{x}{r},$$

$$\tan \theta = \frac{y}{x}, \quad \cot \theta = \frac{x}{y},$$

$$\csc \theta = \frac{r}{y}, \quad \sec \theta = \frac{r}{x}.$$

Law of Sines

$$\frac{a}{\sin A} = \frac{b}{\sin B} = \frac{c}{\sin C}.$$

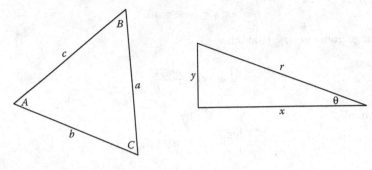

FIGURE 11.7   Triangles for trigonometric functions.

Law of Cosines

$$a^2 = b^2 + c^2 - 2bc \cos A,$$

$$b^2 = a^2 + c^2 - 2ac \cos B,$$

and

$$c^2 = a^2 + b^2 - 2ab \cos C.$$

Identities

$$\csc \theta = \frac{1}{\sin \theta},$$

$$\sec \theta = \frac{1}{\cos \theta},$$

$$\tan \theta = \frac{\sin \theta}{\cos \theta},$$

$$\cot \theta = \frac{1}{\tan \theta},$$

$$\sin^2 \theta + \cos^2 \theta = 1,$$

$$\tan^2 \theta + 1 = \sec^2 \theta,$$

$$\cot^2 \theta + 1 = \csc^2 \theta,$$

$$\sin(\alpha + \beta) = \sin \alpha \cos \beta + \cos \alpha \sin \beta,$$

$$\cos(\alpha + \beta) = \cos \alpha \cos \beta - \sin \alpha \sin \beta,$$

$$\sin 2\alpha = 2\sin\alpha\cos\alpha,$$

$$\cos 2\alpha = \cos^2\alpha - \sin^2\alpha = 1 - 2\sin^2\alpha = 2\cos^2\alpha - 1,$$

$$\tan 2\alpha = \frac{2\tan\alpha}{1 - \tan^2\alpha},$$

$$\cot 2\alpha = \frac{\cot^2\alpha - 1}{2\cot\alpha},$$

$$\tan(\alpha + \beta) = \frac{\tan\alpha + \tan\beta}{1 - \tan\alpha\tan\beta},$$

$$\cot(\alpha + \beta) = \frac{\cot\alpha + \cot\beta - 1}{\cot\alpha + \cot\beta},$$

$$\sin(\alpha - \beta) = \sin\alpha\cos\beta - \cos\alpha\sin\beta,$$

$$\cos(\alpha - \beta) = \cos\alpha\cos\beta + \sin\alpha\sin\beta,$$

$$\tan(\alpha - \beta) = \frac{\tan\alpha - \tan\beta}{1 + \tan\alpha\tan\beta},$$

$$\cot(\alpha - \beta) = \frac{\cot\alpha\cot\beta + 1}{\cot\beta - \cot\alpha},$$

$$\sin\left(\frac{\alpha}{2}\right) = \pm\sqrt{\frac{1 - \cos\alpha}{2}},$$

$$\cos\left(\frac{\alpha}{2}\right) = \pm\sqrt{\frac{1 + \cos\alpha}{2}},$$

$$\tan\left(\frac{\alpha}{2}\right) = \pm\sqrt{\frac{1 - \cos\alpha}{1 + \cos\alpha}},$$

$$\cot\left(\frac{\alpha}{2}\right) = \pm\sqrt{\frac{1 + \cos\alpha}{1 - \cos\alpha}},$$

$$\sin\alpha\sin\beta = \left(\frac{1}{2}\right)[\cos(\alpha - \beta) - \cos(\alpha + \beta)],$$

$$\cos\alpha\cos\beta = \left(\frac{1}{2}\right)[\cos(\alpha - \beta) + \cos(\alpha + \beta)],$$

$$\sin \alpha \cos \beta = \left(\frac{1}{2}\right)[\sin(\alpha + \beta) + \sin(\alpha - \beta)],$$

$$\sin \alpha + \sin \beta = 2 \sin \left(\frac{1}{2}\right)(\alpha + \beta) \cos \left(\frac{1}{2}\right)(\alpha - \beta),$$

$$\sin \alpha - \sin \beta = 2 \cos \left(\frac{1}{2}\right)(\alpha + \beta) \sin \left(\frac{1}{2}\right)(\alpha - \beta),$$

$$\cos \alpha + \cos \beta = 2 \cos \left(\frac{1}{2}\right)(\alpha + \beta) \cos \left(\frac{1}{2}\right)(\alpha - \beta),$$

$$\cos \alpha - \cos \beta = -2 \sin \left(\frac{1}{2}\right)(\alpha + \beta) \sin \left(\frac{1}{2}\right)(\alpha - \beta).$$

Complex Numbers

Definition

$$i = \sqrt{-1}$$

$$(a + ib) + (c + id) = (a + c) + i(b + d),$$

$$(a + ib) - (c + id) = (a - c) + i(b - d),$$

$$(a + ib)(c + id) = (ac - bd) + i(ad + bc),$$

$$\frac{(a + ib)}{(c + id)} = \frac{(a + ib)(c - id)}{(c + id)(c - id)} = \frac{(ac + bd) + i(bc - ad)}{c^2 + d^2}.$$

Polar Coordinates

$$x = r \cos \theta, \quad y = r \sin \theta, \quad \theta = \arctan \left(\frac{y}{x}\right),$$

$$r = |x + iy| = \sqrt{x^2 + y^2},$$

$$x + iy = r(\cos \theta + i \sin \theta) = r e^{i\theta},$$

$$r_1(\cos \theta_1 + i \sin \theta_1)][r_2(\cos \theta_2 + i \sin \theta_2)] = r_1 r_2[\cos(\theta_1 + \theta_2) + i \sin(\theta_1 + \theta_2)],$$

$$(x + iy)^n = [r(\cos \theta + i \sin \theta)]^n = r^n(\cos n\theta + i \sin n\theta),$$

$$\frac{r_1(\cos \theta_1 + i \sin \theta_1)}{r_2(\cos \theta_2 + i \sin \theta_2)} = \frac{r_1}{r_2}[\cos(\theta_1 - \theta_2) + \sin(\theta_2)].$$

Euler's Identity

$$e^{i\theta} = \cos\theta + i\sin\theta$$

$$e^{-i\theta} = \cos\theta - i\sin\theta,$$

$$\cos\theta = \frac{e^{i\theta} + e^{-i\theta}}{2}, \quad \sin\theta = \frac{e^{i\theta} + e^{-i\theta}}{2i}.$$

Roots

If k is any positive integer, any complex number (other than zero) has k distinct roots. The k roots of $r(\cos\theta + i\sin\theta)$ can be found by substituting successively $n = 0, 1, 2, \ldots,$ $(k-1)$ in the formula

$$w = \sqrt[k]{r}\left[\cos\left(\frac{\theta}{k} + n\frac{360°}{k}\right) + i\sin\left(\frac{\theta}{k} + n\frac{360°}{k}\right)\right].$$

Matrices

A matrix is an ordered rectangular array of numbers with m rows and n columns. The element a_{ij} refers to row i and column j.

Matrix Multiplication

If $A = (a_{ik})$ is an $m \times n$ matrix and $B = (b_{kj})$ is an $n \times s$ matrix, the matrix product AB is an $m \times s$ matrix, C, defined as follows:

$$C = c_{ij} = \left(\sum_{l=1}^{n} a_{il}b_{lj}\right),$$

where n is the integer representing the number of columns of A and the number of rows of B; l and $k = 1, 2, \ldots, n$.

Matrix Addition

If $A = (a_{ij})$ and $B = (b_{ij})$ are two matrices of the same size $m \times n$, the sum $A + B$ is the $m \times n$ matrix $C = (c_{ij})$, where $c_{ij} = a_{ij} + b_{ij}$.

Identity Matrix

The matrix $I = (a_{ij})$ is a square $n \times n$ identity matrix where $a_{ij} = 1$ for $i = 1, 2, \ldots, n$ and $a_{ij} = 0$ for $i \neq j$.

Matrix Transpose

The matrix B is the transpose of the matrix A if each entry b_{ji} in B is the same as the entry a_{ij} in A and conversely. This is denoted as $B = A^{T}$.

Matrix Inverse

The inverse B of a square $n \times n$ matrix A is $B = A^{-1} = \text{adj}(A)/|A|$, where $\text{adj}(A)$ is adjoint of A (obtained by replacing A^T elements with their cofactors) and $|A|$ is the determinant of A. Also, $AA^{-1} = A^{-1}A = I$, where I is the identity matrix.

Determinants

A determinant of order n consists of n^2 numbers, called the elements of the determinant, arranged in n columns and enclosed by two vertical lines. In any determinant, the minor of a given element is the determinant that remains after all of the elements are struck out that lie in the same row and in the same column as the given element. Consider an element which lies in the jth column and the ith row. The cofactor of this element is the value of the minor of the element (if $i + j$ is even), and it is the negative of the value of the minor of the element (if $i + j$ is odd).

If n is greater than 1, the value of a determinant of order n is the sum of the n products formed by multiplying each element of some specified row (or column) by its cofactor. This sum is called the *expansion of the determinant* [according to the elements of the specified row (or column)]. For a second-order determinant

$$\begin{vmatrix} a_1 & a_2 \\ b_1 & b_2 \end{vmatrix} = a_1 b_2 - a_2 b_1.$$

For a third-order determinant

$$\begin{vmatrix} a_1 & a_2 & a_3 \\ b_1 & b_2 & b_3 \\ c_1 & c_2 & c_3 \end{vmatrix} = a_1 b_2 c_3 + a_2 b_3 c_1 + a_3 b_1 c_2 - a_3 b_2 c_1 - a_2 b_1 c_3 - a_1 b_3 c_2.$$

Vectors

$$\mathbf{A} = a_x \mathbf{i} + a_y \mathbf{j} + a_z \mathbf{k}.$$

Addition and subtraction:

$$\mathbf{A} + \mathbf{B} = (a_x + b_x)\mathbf{i} + (a_y + b_y)\mathbf{j} + (a_z + b_z)\mathbf{k},$$

$$\mathbf{A} - \mathbf{B} = (a_x - b_x)\mathbf{i} + (a_y - b_y)\mathbf{j} + (a_z - b_z)\mathbf{k}.$$

The *dot product* is a *scalar product* and represents the projection of $\mathbf{B}$ onto $\mathbf{A}$ times $|\mathbf{A}|$. It is given by

$$\mathbf{A} \cdot \mathbf{B} = a_x b_x + a_y b_y + a_z b_z = |\mathbf{A}||\mathbf{B}| \cos \theta = \mathbf{B} \cdot \mathbf{A}.$$

The *cross product* is a *vector product* of magnitude $|\mathbf{B}||\mathbf{A}| \sin \theta$ which is perpendicular to the plane containing $\mathbf{A}$ and $\mathbf{B}$. The product is

$$\mathbf{A} \times \mathbf{B} = \begin{vmatrix} \mathbf{i} & \mathbf{j} & \mathbf{k} \\ a_x & a_y & a_z \\ b_x & b_y & b_z \end{vmatrix} = -\mathbf{B} \times \mathbf{A}.$$

The sense of $\mathbf{A} \times \mathbf{B}$ is determined by the right-hand rule

$$\mathbf{A} \times \mathbf{B} = |\mathbf{A}||\mathbf{B}|\mathbf{n} \sin \theta,$$

where n is the unit vector perpendicular to the plane of $\mathbf{A}$ and $\mathbf{B}$.

Gradient, Divergence, and Curl

$$\nabla \phi = \left(\frac{\delta}{\delta x}\mathbf{i} + \frac{\delta}{\delta y}\mathbf{j} + \frac{\delta}{\delta z}\mathbf{k} \right) \phi,$$

$$\nabla \cdot \mathbf{V} = \left(\frac{\delta}{\delta x}\mathbf{i} + \frac{\delta}{\delta y}\mathbf{j} + \frac{\delta}{\delta z}\mathbf{k} \right) \cdot (V_1\mathbf{i} + V_2\mathbf{j} + V_3\mathbf{k}),$$

$$\nabla \times \mathbf{V} = \left(\frac{\delta}{\delta x}\mathbf{i} + \frac{\delta}{\delta y}\mathbf{j} + \frac{\delta}{\delta z}\mathbf{k} \right) \times (V_1\mathbf{i} + V_2\mathbf{j} + V_3\mathbf{k}).$$

The Laplacian of a scalar function ϕ is

$$\nabla^2 \phi = \frac{\delta^2 \phi}{\delta x^2} + \frac{\delta^2 \phi}{\delta y^2} + \frac{\delta^2 \phi}{\delta z^2}.$$

Identities

$$\mathbf{A} \cdot \mathbf{B} = \mathbf{B} \cdot \mathbf{A}, \quad \mathbf{A} \cdot (\mathbf{B} + \mathbf{C}) = \mathbf{A} \cdot \mathbf{B} + \mathbf{A} \cdot \mathbf{C},$$

$$\mathbf{A} \cdot \mathbf{A} = |\mathbf{A}|^2,$$

$$\mathbf{i} \cdot \mathbf{i} = \mathbf{j} \cdot \mathbf{j} = \mathbf{k} \cdot \mathbf{k} = 1,$$

$$\mathbf{i} \cdot \mathbf{j} = \mathbf{j} \cdot \mathbf{k} = \mathbf{k} \cdot \mathbf{i} = 0.$$

If $\mathbf{A} \cdot \mathbf{B} = 0$, then either $\mathbf{A} = 0$, $\mathbf{B} = 0$, or $\mathbf{A}$ is perpendicular to $\mathbf{B}$.

$$\mathbf{A} \times \mathbf{B} = -\mathbf{B} \times \mathbf{A},$$

$$\mathbf{A} \times (\mathbf{B} + \mathbf{C}) = (\mathbf{A} \times \mathbf{B}) + (\mathbf{A} \times \mathbf{C}),$$

$$(\mathbf{B} + \mathbf{C}) \times \mathbf{A} = (\mathbf{B} \times \mathbf{A}) + (\mathbf{C} \times \mathbf{A}),$$

$$i \times i = j \times j = k \times k = 0,$$

$$i \times j = k = -j \times i, \quad j \times k = i = -k \times j,$$

$$k \times i = j = -i \times k.$$

If $\mathbf{A} \times \mathbf{B} = 0$, then either $\mathbf{A} = 0$, $\mathbf{B} = 0$, or $\mathbf{A}$ is parallel to $\mathbf{B}$.

$$\nabla^2 \phi = \nabla \cdot (\nabla \phi) = (\nabla \cdot \nabla)\phi,$$

$$\nabla \times \nabla \phi = 0,$$

$$\nabla \cdot (\nabla \times \mathbf{A}) = 0,$$

$$\nabla \times (\nabla \times \mathbf{A}) = \nabla(\nabla \cdot \mathbf{A}) - \nabla^2 \mathbf{A}.$$

Progressions and Series

Arithmetic Progression

To determine whether a given finite sequence of numbers is an arithmetic progression, subtract each number from the following number. If the differences are equal, the series is arithmetic.

1. The first term is a
2. The common differences is d
3. The number of terms is n
4. The last or nth term is l
5. The sum of n terms is S

$$l = a + (n - 1)d,$$

$$S = \frac{n(a + 1)}{2} = \frac{n[2a + (n - 1)d]}{2}.$$

Geometric Progression

To determine whether a given finite sequence is a geometric progression (G.P.), divide each number after the first by the preceding number. If the quotients are equal, the series is geometric.

1. The first term is a
2. The common ratio is r
3. The number of terms is n
4. The last or nth term is l
5. The sum of n terms is S

$$l = ar^{n-1},$$

$$S = \frac{a(l - r^n)}{l - r}, \quad r \neq 1$$

$$S = \frac{a - rl}{l - r}, \quad r \neq 1,$$

$$\lim_{n \to \infty} S_n = \frac{a}{l - r}, \quad r < 1.$$

A G.P. converges if $|r| < 1$ and it diverges if $|r| > 1$.

Properties of Series

$$\sum_{i-1}^{n} c = nc, \quad c = \text{constant},$$

$$\sum_{i=1}^{n} cx_i = c \sum_{i=1}^{n} x_i,$$

$$\sum_{i=1}^{n} (x_i + y_i - z_i) = \sum_{i=1}^{n} x_i + \sum_{i=1}^{n} y_i - \sum_{i=1}^{n} z_i,$$

$$\sum_{x=1}^{n} x = \frac{n + n^2}{2}.$$

Power Series

$$\sum_{i=0}^{\infty} a_i (x - a)^i.$$

1. A power series, which is convergent in the interval $-R < x < R$, defines a function of x that is continuous for all values of x within the interval and is said to represent the function in that interval.
2. A power series may be differentiated term by term within its interval of convergence. The resulting series has the same interval of convergence as the original series (except possibly at the endpoints of the series).
3. A power series may be integrated term by term provided the limits of integration are within the interval of convergence of the series.
4. Two power series may be added, subtracted, or multiplied, and the resulting series in each case is convergent, at least, in the interval common to the two series.
5. Using the process of long division (as for polynomials), two power series may be divided one by the other within their common interval of convergence.

Taylor's Series

$$f(x) = f(a) + \frac{f'(a)}{1!}(x-a) + \frac{f''(a)}{2!}(x-a)^2 + \cdots + \frac{f^{(n)}(a)}{n!}(x-a)^n + \cdots$$

is called Taylor series, and the function $f(x)$ is said to be expanded about the point a in a Taylor series.

If $a = 0$, Taylor series equation becomes a Maclaurin series.

Differential Calculus

The Derivative

For any function $y = f(x)$, the derivative is given by

$$D_x y = \frac{dy}{dx} = y',$$

$$y' = \lim_{\Delta x \to 0} \left(\frac{\Delta y}{\Delta x} \right),$$

$$= \lim_{\Delta x \to 0} \left\{ \frac{[f(x+\Delta x) = f(x)]}{\Delta x} \right\},$$

where y' is the slope of the curve $f(x)$.

Test for a Maximum

$y = f(x)$ is a maximum for
  $x = a$, if $f'(a) = 0$ and $f''(a) < 0$.

Test for a Minimum

$y = f(x)$ is a minimum for
  $x = a$, if $f'(a) = 0$ and $f''(a) > 0$..

Test for a Point of Inflection

$y = f(x)$ has a point of inflection at $x = a$,
  if $f''(a) = 0$ and if $f''(x)$ changes sign as x increases through $x = a$.

The Partial Derivative

In a function of two independent variables x and y, a derivative with respect to one of the variables may be found if the other variable is *assumed* to remain constant. If y is *kept fixed*, the function

$$z = f(x, y)$$

becomes a function of the *single variable* x, and its derivative (if it exists) can be found. This derivative is called the *partial derivative* of z with respect to x. The partial derivative with respect to x is denoted as follows:

$$\frac{\delta z}{\delta x} = \frac{\delta f(x, y)}{\delta x}.$$

The curvature K of a curve at P is the limit of its average curvature for the arc PQ as Q approaches P. This is also expressed as: the curvature of a curve at a given point is the rate of change of its inclination with respect to its arc length:

$$K = \lim_{\Delta s \to 0} \left(\frac{\Delta \alpha}{\Delta s} \right) = \frac{d\alpha}{ds}.$$

Curvature in Rectangular Coordinates

$$K = \frac{y''}{[1 + (y')^2]^{3/2}}.$$

When it may be easier to differentiate the function with respect to y rather than x, the notation x' will be used for the derivative.

$$x' = \frac{dx}{dy},$$

$$K = \frac{-x''}{[1 + (x')^2]^{3/2}}.$$

The Radius of Curvature

The *radius of curvature* R at any point on a curve is defined as the absolute value of the reciprocal of the curvature K at that point.

$$R = \frac{1}{|K|}, \qquad K \neq 0,$$

$$R = \left| \frac{[1 + (y')^2]^{3/2}}{|y''|} \right|, \qquad y'' \neq 0.$$

L'Hospital's Rule

If the fractional function $f(x)/g(x)$ assumes one of the indeterminate forms $0/0$ or ∞/∞ (where α is finite or infinite), then

$$\lim_{x \to \alpha} \frac{f(x)}{g(x)}$$

is equal to the first of the expressions

$$\lim_{x \to \alpha} \frac{f'(x)}{g'(x)}, \quad \lim_{x \to \alpha} \frac{f''(x)}{g''(x)}, \quad \lim_{x \to \alpha} \frac{f'''(x)}{g'''(x)},$$

which is not indeterminate, provided such first indicated limit exists.

Integral Calculus

The definite integral is defined as

$$\lim_{n \to \infty} \sum_{i=1}^{n} f(x_i), \quad \Delta x_i = \int_a^b f(x)dx.$$

Also, $\Delta x_i \to 0$ for all i.

A list of derivatives and integrals is available in the "Derivatives and Indefinite Integrals" section. The integral equations can be used along with the following methods of integration:

a. Integration by parts (integral equation #6)
b. Integration by substitution
c. Separation of rational fractions into partial fractions

Derivatives and Indefinite Integrals

In these formulas, u, v, and w represent functions of x. Also, a, c, and n represent constants. All arguments of the trigonometric functions are in radians. A constant of integration should be added to the integrals. To avoid terminology difficulty, the following definitions are followed:

$$\arcsin u = \sin^{-1} u, \quad (\sin u)^{-1} = \frac{1}{\sin u}.$$

Derivatives

1. $\dfrac{dc}{dx} = 0,$

2. $\dfrac{dx}{dx} = 1,$

3. $\dfrac{d(cu)}{dx} = c\dfrac{du}{dx},$

4. $\dfrac{d(u+v-w)}{dx} = \dfrac{du}{dx} + \dfrac{dv}{dx} - \dfrac{dw}{dx},$

5. $\dfrac{d(uv)}{dx} = \dfrac{udv}{dx} + \dfrac{vdu}{dx},$

6. $\dfrac{d(uvw)}{dx} = \dfrac{uvdw}{dx} + \dfrac{uwdv}{dx} + \dfrac{vwdu}{dx},$

7. $\dfrac{d(u/v)}{dx} = \dfrac{vdu/dx - udv/dx}{v^2},$

8. $\dfrac{d(u^n)}{dx} = nu^{n-1}\dfrac{du}{dx},$

9. $\dfrac{d[f(u)]}{dx} = \left\{\dfrac{d[f(u)]}{du}\right\}\dfrac{du}{dx},$

10. $\dfrac{du}{dx} = \dfrac{1}{dx/du},$

11. $\dfrac{d(\log_a u)}{dx} = (\log_a e)\dfrac{1}{u}\dfrac{du}{dx},$

12. $\dfrac{d(\ln u)}{dx} = \dfrac{1}{u}\dfrac{du}{dx},$

13. $\dfrac{d(a^u)}{dx} = (\ln a)a^u\dfrac{du}{dx},$

14. $\dfrac{d(e^u)}{dx} = e^u\dfrac{du}{dx},$

15. $\dfrac{d(u^v)}{dx} = vu^{v-1}\dfrac{du}{dx} + (\ln u)u^v\dfrac{dv}{dx},$

16. $\dfrac{d(\sin u)}{dx} = \cos u\dfrac{du}{dx},$

17. $\dfrac{d(\cos u)}{dx} = -\sin u\dfrac{du}{dx},$

18. $\dfrac{d(\tan u)}{dx} = \sec^2 u\dfrac{du}{dx},$

19. $\dfrac{d(\cot u)}{dx} = -\csc^2 u\dfrac{du}{dx},$

20. $\dfrac{d(\sec u)}{dx} = \sec u\tan u\dfrac{du}{dx},$

21. $\dfrac{d(\csc u)}{dx} = -\csc u\cot u\dfrac{du}{dx},$

22. $\dfrac{d(\sin^{-1}u)}{dx} = \dfrac{1}{\sqrt{1-u^2}}\dfrac{du}{dx}\left(\dfrac{-\pi}{2} \le \sin^{-1}u \le \dfrac{\pi}{2}\right),$

23. $\dfrac{d(\cos^{-1}u)}{dx} = -\dfrac{1}{\sqrt{1-u^2}}\dfrac{du}{dx} \quad (0 \le \cos^{-1}u \le \pi),$

24. $\dfrac{d(\tan^{-1}u)}{dx} = \dfrac{1}{1+u^2}\dfrac{du}{dx}\left(\dfrac{-\pi}{2} < \tan^{-1}u < \dfrac{\pi}{2}\right),$

25. $\dfrac{d(\cot^{-1}u)}{dx} = -\dfrac{1}{1+u^2}\dfrac{du}{dx}(0 < \cot^{-1}u < \pi),$

26. $\dfrac{d(\sec^{-1}u)}{dx} = \dfrac{1}{\sqrt[u]{u^2-1}}\dfrac{du}{dx}\left(0 < \sec^{-1}u < \dfrac{\pi}{2}\right)\left(-\pi \le \sec^{-1}u < \dfrac{-\pi}{2}\right),$

27. $\dfrac{d(\csc^{-1}u)}{dx} = -\dfrac{1}{\sqrt[u]{u^2-1}}\dfrac{du}{dx}\left(0 < \csc^{-1}u \le \dfrac{\pi}{2}\right)\left(-\pi < \csc^{-1}u \le \dfrac{-\pi}{2}\right).$

Indefinite Integrals

1. $\displaystyle\int df(x) = f(x),$

2. $\displaystyle\int dx = x,$

3. $\displaystyle\int af(x)dx = a\int f(x)dx,$

4. $\displaystyle\int [u(x) \pm v(x)]dx = \int u(x)dx \pm \int v(x)dx,$

5. $\displaystyle\int x^m dx = \frac{x^{m+1}}{m+1}, \quad m \neq -1,$

6. $\displaystyle\int u(x)dv(x) = u(x)v(x) - \int v(x)du(x),$

7. $\displaystyle\int \frac{dx}{ax+b} = \frac{1}{a}\ln|ax+b|,$

8. $\displaystyle\int \frac{dx}{\sqrt{x}} = 2\sqrt{x},$

9. $\displaystyle\int a^x dx = \frac{a^x}{\ln a},$

10. $\displaystyle\int \sin x dx = -\cos x,$

11. $\displaystyle\int \cos x dx = \sin x,$

12. $\displaystyle\int \sin^2 x dx = \frac{x}{2} - \frac{\sin 2x}{4},$

13. $\displaystyle\int \cos^2 x dx = \frac{x}{2} + \frac{\sin 2x}{4},$

14. $\displaystyle\int x \sin x dx = \sin x - x \cos x,$

15. $\displaystyle\int x \cos x dx = \cos x + x \sin x,$

16. $\displaystyle\int \sin x \cos x dx = \frac{\sin^2 x}{2},$

17. $\displaystyle\int \sin ax \cos bx dx = -\frac{\cos(a-b)x}{2(a-b)} - \frac{\cos(a+b)x}{2(a+b)}(a^2 \neq b^2),$

18. $\displaystyle\int \tan x dx = -\ln|\cos x| = \ln|\sec x|,$

19. $\displaystyle\int \cot x dx = -\ln|\csc x| = \ln|\sin x|,$

20. $\displaystyle\int \tan^2 x dx = \tan x - x,$

21. $\displaystyle\int \cot^2 x dx = -\cot x - x,$

22. $\displaystyle\int e^{ax} dx = \frac{1}{a}e^{ax},$

23. $\displaystyle\int xe^{ax} dx = \frac{e^{ax}}{a^2}(ax-1),$

24. $\int \ln x \, dx = x[\ln(x) - 1], \quad x > 0,$

25. $\int \dfrac{dx}{a^2 + x^2} = \dfrac{1}{a} \tan^{-1} \dfrac{x}{a}, \quad a \neq 0,$

26. $\int \dfrac{dx}{ax^2 + c} = \dfrac{1}{\sqrt{ac}} \tan^{-1} \left(x \sqrt{\dfrac{a}{c}} \right), \quad a > 0, c > 0,$

27a. $\int \dfrac{dx}{ax^2 + bx + c} = \dfrac{2}{\sqrt{4ac - b^2}} \tan^{-1} \dfrac{2ax + b}{\sqrt{4ac - b^2}}, \quad 4ac - b^2 > 0,$

27b. $\int \dfrac{dx}{ax^2 + bx + c} = \dfrac{1}{\sqrt{b^2 - 4ac}} \ln \left| \dfrac{2ax + b - \sqrt{b^2 - 4ac}}{2ax + b + \sqrt{b^2 - 4ac}} \right|, \quad b^2 - 4ac > 0,$

27c. $\int \dfrac{dx}{ax^2 + bx + c} = -\dfrac{2}{2ax + b}, \quad b^2 - 4ac = 0.$

Mensuration of Areas and Volumes

Nomenclature

- A total surface area
- P perimeter
- V volume

Parabola

$$A = \frac{2bh}{3}$$

and

$$A = \frac{bh}{3}$$

Ellipse

$$A = \pi ab,$$

$$P_{\text{approx}} = 2\pi \sqrt{\frac{a^2 + b^2}{2}},$$

$$P = \pi(a + b)\left[1 + \left(\frac{1}{2}\right)^2 \lambda^2 + \left(\frac{1}{2} \times \frac{1}{4}\right)^2 \lambda^4 + \left(\frac{1}{2} \times \frac{1}{4} \times \frac{3}{6}\right)^2 \lambda^6 \right.$$

$$\left. + \left(\frac{1}{2} \times \frac{1}{4} \times \frac{3}{6} \times \frac{5}{8}\right)^2 \lambda^8 + \left(\frac{1}{2} \times \frac{1}{4} \times \frac{3}{6} \times \frac{5}{8} \times \frac{7}{10}\right)^2 \lambda^{10} + \cdots \right],$$

where

$$\lambda = \frac{a - b}{a + b}.$$

Circular Segment

$$A = \frac{r^2(\phi - \sin\phi)}{2},$$

$$\phi = \frac{s}{r} = 2\left\{\arccos\left[\frac{r-d}{r}\right]\right\}.$$

Circular Sector

$$A = \frac{\phi r^2}{2} = \frac{sr}{2},$$

$$\phi = \frac{s}{r}.$$

Sphere

$$V = \frac{4\pi r^3}{3} = \frac{\pi d^3}{6},$$

$$A = 4\pi r^2 = \pi d^2.$$

Parallelogram

$$P = 2(a + b),$$

$$d_1 = \sqrt{a^2 + b^2 - 2ab(\cos\phi)},$$

$$d_2 = \sqrt{a^2 + b^2 + 2ab(\cos\phi)},$$

$$d_1^2 + d_2^2 = 2(a^2 + b^2),$$

$$A = ah = ab(\sin\phi),$$

If $a = b$, the parallelogram is a rhombus.

Regular Polygon (*n* Equal Sides)

$$\phi = \frac{2\pi}{n},$$

$$\theta = \left[\frac{\pi(n-2)}{n} \right] = \pi \left(1 - \frac{2}{n} \right),$$

$$P = ns,$$

$$s = 2r \left[\tan \left(\frac{\phi}{2} \right) \right],$$

$$A = \left(\frac{nsr}{2} \right).$$

Prismoid

$$V = \left(\frac{h}{6} \right) (A_1 + A_2 + 4A).$$

Right Circular Cone

$$V = \frac{\pi r^2 h}{3},$$

$$A = \text{side area} + \text{base area}$$
$$= \pi r(r + \sqrt{r^2 + h^2}),$$

$$A_x : A_b = x^2 : h^2.$$

Right Circular Cylinder

$$V = \pi r^2 h = \frac{\pi d^2 h}{4},$$

$$A = \text{side areas} + \text{end areas} = 2\pi r(h + r).$$

Paraboloid of Revolution

$$V = \frac{\pi d^2 h}{8}.$$

Centroids and Moments of Inertia

The *location of the centroid of an area*, bounded by the axes and the function $y = f(x)$, can be found by integration:

$$x_c = \frac{\int x \, dA}{A},$$

$$y_c = \frac{\int x \, dA}{A},$$

$$A = \int f(x) \, dx,$$

$$dA = f(x) \, dx = g(y) \, dy.$$

The *first moment of area* with respect to the y-axis and the x-axis, respectively, are

$$M_y = \int x \, dA = x_c A$$

and

$$M_x = \int y \, dA = y_c A.$$

The *moment of inertia (second moment of area)* with respect to the y-axis and the x-axis, respectively, are

$$I_y = \int x^2 \, dA$$

and

$$I_x = \int y^2 \, dA.$$

The moment of inertia taken with respect to an axis passing through the area's centroid is the *centroidal moment of inertia*. The *parallel axis theorem* for the moment of inertia with respect to another axis parallel with and located d units from the centroidal axis is expressed by

$$I_{\text{parallel axis}} = I_c + Ad^2.$$

In a plane,

$$J = \int r^2 \, dA = I_x + I_y.$$

Difference Equations

Difference equations are used to model discrete systems. Systems which can be described by difference equations include computer program variables iteratively evaluated in a loop, sequential circuits, cash flows, recursive processes, systems with time-delay components, and so on. Any system whose input $v(t)$ and output $y(t)$ are defined only at the equally spaced intervals $t = kT$ can be described by a difference equation.

First-Order Linear Difference Equation

The difference equation

$$P_k = P_{k-1}(1+i) - A$$

represents the balance P of a loan after the kth payment A. If P_k is defined as $y(k)$, the model becomes

$$y(k) - (1+i)y(k-1) = -A.$$

Second-Order Linear Difference Equation

The Fibonacci number sequence can be generated by

$$y(k) = y(k-1) + y(k-2),$$

where $y(-1) = 1$ and $y(-2) = 1)$. An alternate form for this model is

$$f(k+2) = f(k+1) + f(k)$$

with $f(0) = 1$ and $f(1) = 1$.

Numerical Methods

Newton's Method for Root Extraction

Given a function $f(x)$ which has a simple root of $f(x) = 0$ at $x = a$ an important computational task would be to find that root. If $f(x)$ has a continuous first derivative, then the $(j+1)$st estimate of the root is

$$a^{j+1} = a^j - \frac{f(x)}{df(x)/dx}\bigg|_{x=a^j}.$$

The initial estimate of the root a^0 must be near enough to the actual root to cause the algorithm to converge to the root.

Newton's Method of Minimization

Given a scalar value function

$$h(x) = h(x_1, h_2, \ldots, x_n),$$

find a vector $x^* \in R_n$ such that

$$h(x^*) \le h(x) \quad \text{for all } x.$$

Newton's algorithm is

$$x_{k+1} = x_k - \left(\frac{\partial^2 h}{\partial x^2}\bigg|_{x=x_k}\right)^{-1} \frac{\partial h}{\partial x}\bigg|_{x=x_k},$$

where

$$\frac{\partial h}{\partial x} = \begin{bmatrix} \dfrac{\partial h}{\partial x_1} \\ \dfrac{\partial h}{\partial x_2} \\ \cdots \\ \cdots \\ \dfrac{\partial h}{\partial x_n} \end{bmatrix}$$

and

$$\frac{\partial^2 h}{\partial x^2} = \begin{bmatrix} \dfrac{\partial^2 h}{\partial x_1^2} & \dfrac{\partial^2 h}{\partial x_1 \partial x_2} & \cdots & \cdots & \dfrac{\partial^2 h}{\partial x_1 \partial x_n} \\ \dfrac{\partial^2 h}{\partial x_1 \partial x_2} & \dfrac{\partial^2 h}{\partial x_2^2} & \cdots & \cdots & \dfrac{\partial^2 h}{\partial x_2 \partial x_n} \\ \cdots & \cdots & \cdots & \cdots & \cdots \\ \cdots & \cdots & \cdots & \cdots & \cdots \\ \dfrac{\partial^2 h}{\partial x_1 \partial x_n} & \dfrac{\partial^2 h}{\partial x_2 \partial x_n} & \cdots & \cdots & \dfrac{\partial^2 h}{\partial x_n^2} \end{bmatrix}.$$

Numerical Integration

Three of the more common numerical integration algorithms used to evaluate the integral

$$\int_a^b f(x)\mathrm{d}x$$

are as follows.

Euler's or Forward Rectangular Rule

$$\int_a^b f(x)\mathrm{d}x \approx \Delta x \sum_{k=0}^{n-1} f(a + k\Delta x).$$

Trapezoidal Rule

For $n = 1$

$$\int_a^b f(x)\mathrm{d}x \approx \Delta x \left[\frac{f(a) + f(b)}{2} \right]$$

and for $n > 1$

$$\int_a^b f(x)\mathrm{d}x \approx \frac{\Delta x}{2} \left[f(a) + 2\sum_{k=1}^{n-1} f(a + k\Delta x) + f(b) \right].$$

Simpson's Rule/Parabolic Rule (*n* Must Be an Even Integer)

For $n = 2$

$$\int_a^b f(x)dx \approx \left(\frac{b-a}{6}\right)\left[f(a) + 4f\left(\frac{a+b}{2}\right) + f(b)\right]$$

and for $n \geq 4$

$$\int_a^b f(x)dx \approx \frac{\Delta x}{3}\left[f(a) + 2\sum_{k=2,4,6,\ldots}^{n-2} f(a+k\Delta x) + 4\sum_{k=1,3,5,\ldots}^{n-1} f(a+k\Delta x) + f(b)\right]$$

with $\Delta x = (b-a)/n$ and n is the number of intervals between data points.

Calculation of Best-Fit Circle*

This section illustrates a good example of how to apply geometrical calculations to product shape design. Many manufacturers use coordinate measuring machines (CMMs) to check key dimensions in their parts. The CMM is used to pick a series of points on the surface of the parts and these points are used to estimate key dimensions—distances, angles, and radii by fitting curves to groups of points. There are several well-known methods for computing a *best-fit* line to a set of points (least-squares and Tchebyshev minimax are the best known), but methods for fitting a circle to a set of points are harder to find. Here is a method for computing a *best-fit circle* after the fashion of the least- squares method (see Figure 11.8).

Suppose we desire a circle, $(x-h)^2 + (y-k)^2 = r^2$, which *best-fits* an arbitrary set of $N > 3$ points. Specifically, we need to determine (h, k) and r so as to minimize some error measure. A convenient error measure is the variance of $r^2(\sigma^2)$, which will not involve carrying any radicals through the calculations. From statistics, the variance

$$\sigma^2(r^2) = \frac{\sum r^4}{N} - \left(\frac{\sum r^2}{N}\right)^2.$$

So, for (h, k) yet to be determined, and an arbitrary point (x_i, y_i), we have the following:

$$r_i^2 = x_i^2 - 2hx_i + h^2 + y_i^2 - 2ky_i + k^2$$

and

$$\begin{aligned}
r_i^4 = &\ x_i^4 - 4hx_i^3 + 6h^2x_i^2 - 4h^3x_i + h^4 + 2x_i^2y_i^2 - 4kx_i^2y_i \\
&- 4hx_iy_i^2 + 2k^2x_i^2 + 8hkx_iy_i + 2h^2y_i^2 - 4hk^2x_i - 4h^2ky_i \\
&+ 2h^2k^2 + y_i^4 - 4ky_i^3 + 6k^2y_i^2 - 4k^3y_i + k^4.
\end{aligned}$$

* From Galer, C. K. 2008. Best-fit circle. *The Bent of Tau Beta Pi.* 2008, p. 37.

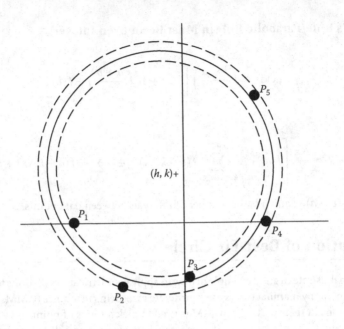

FIGURE 11.8   Best-fit circle.

Summing over the entire point set, we obtain the following (the reader can work through the intermediate steps, if desired):

$$
\begin{aligned}
\sigma^2 = \frac{1}{N}&\left[\sum x^4 + 2\sum x^2 y^2 + \sum y^4\right] \\
&- \frac{1}{N^2}\left[\left(\sum x^2\right)^2 + 2\sum x^2\sum y^2 + \left(\sum y^2\right)^2\right] \\
&- \frac{4}{N}\left[h\sum x^3 + k\sum x^2 y + h\sum xy^2 + k\sum y^3\right] \\
&+ \frac{4}{N^2}\left[h\sum x\sum x^2 + k\sum x^2\sum y + h\sum x\sum y^2 + k\sum y\sum y^2\right] \\
&+ \frac{4}{N}\left[h^2\sum x^2 + 2hk\sum xy + k^2\sum y^2\right] \\
&- \frac{4}{N^2}\left[h^2\left(\sum x\right)^2 + 2hk\sum x\sum y + k^2\left(\sum y\right)^2\right].
\end{aligned}
$$

Now we want to minimize σ^2, so we take the partial derivatives with respect to h and k and set them to zero. This yields the normal equations for (h, k):

$$
\begin{aligned}
&\left[N\sum x^2 - \left(\sum x\right)^2\right]h + \left[N\sum xy - \sum x\sum y\right]k \\
&\quad = \frac{1}{2}\left[N\left(\sum x^3 + \sum xy^2\right) - \sum x\left(\sum x^2 + \sum y^2\right)\right],
\end{aligned}
$$

$$\left[N\sum xy-\sum x\sum y\right]h+\left[N\sum y^2-\left(\sum y\right)^2\right]k$$
$$=\frac{1}{2}\left[N\left(\sum x^2y+\sum y^3\right)-\sum y\left(\sum x^2+\sum y^2\right)\right].$$

We can greatly simplify the normal equations and greatly improve computational round-off error by translating the set of points to a co-ordinate system whose origin is the centroid of the point set:

$$x_i'=x_i-\overline{x},\quad \overline{x}=\frac{\sum x}{N}$$

and

$$y_i'=y_i-\overline{y},\quad \overline{y}=\frac{\sum y}{N}.$$

In this co-ordinate system, $\sum x'=\sum y'=0$, and the normal equations become

$$h'\sum x'^2+k'\sum x'y'=\frac{1}{2}\left(\sum x'^3+\sum x'y'^2\right)$$

and

$$h'\sum x'y'+k'\sum y'^2=\frac{1}{2}\left(\sum x'^2y'+\sum y'^3\right).$$

Having thus found (h',k'), "un-translate" to the original co-ordinate system:

$$h=h'+\overline{x},\quad k=k'+\overline{y}.$$

We can now compute

$$r_i=\sqrt{(x_i-h)^2+(y_i-k)^2}$$

and

$$\overline{r}=\frac{\sum r}{N}.$$

12

General Engineering Calculations

Six Simple Machines for Materials Handling

Material handling design and implementation constitute one of the basic functions in the practice of industrial engineering. Calculations related to six simple machines are useful for assessing mechanical advantage for material handling purposes. The mechanical advantage is the ratio of the force of resistance to the force of effort given by

$$MA = \frac{F_R}{F_E},$$

where MA is the mechanical advantage, F_R the force of resistance (N), and F_E the force of effort (N).

Machine 1: The Lever

A lever consists of a rigid bar that is free to turn on a pivot, which is called a fulcrum. The law of simple machines as applied to levers is

$$F_R L_R = F_E L_E.$$

Machine 2: Wheel and Axle

A wheel-and-axle system consists of a large wheel attached to an axle so that both turn together:

$$F_R r_R = F_E r_E,$$

where F_R is the force of resistance (N), F_E the force of effort (N), r_R the radius of resistance wheel (m), and r_E the radius of effort wheel (m).

The mechanical advantage is

$$MA_{\text{wheel and axle}} = \frac{r_E}{r_R}.$$

Machine 3: The Pulley

If a pulley is fastened to a fixed object, it is called a fixed pulley. If the pulley is fastened to the resistance to be moved, it is called a moveable pulley. If one continuous cord is used, then the ratio reduces according to the number of strands holding the resistance in the pulley system.

The effort force equals the tension in each supporting stand. The mechanical advantage of the pulley is given by the formula

$$\mathrm{MA}_{\text{pulley}} = \frac{F_R}{F_E} = \frac{nT}{T} = n,$$

where T is the tension in each supporting strand, N the number of strands holding the resistance, F_R the force of resistance (N), and F_E the force of effort (N).

Machine 4: The Inclined Plane

An inclined plane is a surface set at an angle from the horizontal and used to raise objects that are too heavy to lift vertically.

The mechanical advantage of an inclined plane is

$$\mathrm{MA}_{\text{inclined plane}} = \frac{F_R}{F_E} = \frac{l}{h},$$

where F_R is the force of resistance (N), F_E the force of effort (N), l the length of the plane (m), and h the height of the plane (m).

Machine 5: The Wedge

The wedge is a modification of the inclined plane. The mechanical advantage of a wedge can be found by dividing the length of either slope by the thickness of the longer end.

As with the inclined plane, the mechanical advantage gained by using a wedge requires a corresponding increase in distance.

The mechanical advantage is

$$\mathrm{MA} = \frac{s}{T},$$

where MA is the mechanical advantage, s the length of either slope (m), and T the thickness of the longer end (m).

Machine 6: The Screw

A screw is an inclined plane wrapped around a circle. From the law of machines, we obtain

$$F_R h = F_E U_E.$$

However, for advancing a screw with a screwdriver, the mechanical advantage is

$$\mathrm{MA}_{\text{screw}} = \frac{F_R}{F_E} = \frac{U_E}{h},$$

where F_R is the force of resistance (N), F_E the effort force (N), h the pitch of screw, and U_E the circumference of the handle of the screw.

Mechanics: Kinematics

Scalars and Vectors

The mathematical quantities that are used to describe the motion of objects can be divided into two categories: scalars and vectors.

a. Scalars
  Scalars are quantities that can be fully described by a magnitude alone.
b. Vectors
  Vectors are quantities that can be fully described by both a magnitude and a direction.

Distance and Displacement

a. Distance
  Distance is a scalar quantity that refers to how far an object has gone during its motions.
b. Displacement
  Displacement is the change in position of the object. It is a vector quantity that includes the magnitude as a distance, such as 5 miles, and a direction, such as north.

Acceleration

Acceleration is the change in velocity per unit of time. Acceleration is a vector quality.

Speed and Velocity

a. Speed
  The distance traveled per unit of time is called the speed: for example, 35 miles per hour. Speed is a scalar quantity.
b. Velocity
  The quantity that combines both the speed of an object and its direction of motion is called velocity. Velocity is a vector quantity.

Frequency

Frequency is the number of complete vibrations per unit time in simple harmonic or sinusoidal motion.

Period

Period is the time required for one full cycle. It is the reciprocal of the frequency.

Angular Displacement

Angular displacement is the rotational angle through which any point on a rotating body moves.

Angular Velocity

Angular velocity is the ratio of angular displacement to time.

Angular Acceleration

Angular acceleration is the ratio of angular velocity with respect to time.

Rotational Speed

Rotational speed is the number of revolutions (a revolution is one complete rotation of a body) per unit of time.

Uniform Linear Motion

A path is a straight line. The total distance traveled corresponds with the rectangular area in the diagram $v - t$.

a. Distance:

$$S = Vt.$$

b. Speed:

$$V = \frac{S}{t},$$

where s is the distance (m), v the speed (m/s), and t the time (s).

Uniform Accelerated Linear Motion

1. If $v0 > 0$ and $a > 0$, then
   a. Distance:

$$s = v_0t + \frac{at^2}{2}.$$

   b. Speed:

$$v = v_0 + at,$$

   where s is the distance (m), v the speed (m/s), t the time (s), v_0 the initial speed (m/s), and a the acceleration (m/s^2).
2. If $v0 = 0$ and $a > 0$, then
   a. Distance:

$$s = \frac{at^2}{2}.$$

   The shaded areas in diagram $v - t$ represent the distance s traveled during the time period t.
   b. Speed:

$$v = at,$$

where s is the distance (m), v the speed (m/s), v_0 the initial speed (m/s), and a the acceleration (m/s^2).

Rotational Motion

Rotational motion occurs when the body itself is spinning. The path is a circle about the axis.

a. Distance:

$$s = r\varphi.$$

b. Velocity:

$$v = r\omega.$$

c. Tangential acceleration:

$$a_t = r\alpha.$$

d. Centripetal acceleration:

$$a_n = \omega^2 r = \frac{v^2}{r}.$$

where φ is the angle determined by s and r (rad), ω the angular velocity (s^{-1}), α the angular acceleration (s^{-2}), a_t the tangential acceleration (s^{-2}), and a_n the centripetal acceleration (s^{-2}).

Distance s, velocity v, and tangential acceleration a_t are proportional to radius r.

Uniform Rotation and a Fixed Axis

For $\omega_0 = $ constant; $\alpha = 0$:

a. Angle of rotations:

$$\varphi = \omega t.$$

b. Angular velocity:

$$\omega = \frac{\varphi}{t},$$

where φ is the angle of rotation (rad), ω the angular velocity (s^{-1}), α the angular acceleration (s^{-2}), and ω_0 the initial angular speed (s^{-1}).

The shaded area in the diagram $\omega - t$ represents the angle of rotation $\varphi = 2\pi n$ covered during time period t.

Uniform Accelerated Rotation about a Fixed Axis

1. If $\omega_0 > 0$ and $\alpha > 0$, then
   a. Angle of rotation:

$$\varphi = \frac{1}{2}(\omega_0 + \omega) = \omega_0 t + \frac{1}{2}\alpha t^2.$$

   b. Angular velocity:

$$\omega = \omega_0 + \alpha t = \sqrt{\omega_0^2 + 2\alpha\varphi},$$
$$\omega_0 = \omega - \alpha t = \sqrt{\omega^2 - 2\alpha\varphi}.$$

c. Angular acceleration:

$$\alpha = \frac{\omega - \omega_0}{t} = \frac{\omega^2 - \omega_0{}^2}{2\varphi}.$$

d. Time:

$$t = \frac{\omega - \omega_0}{\alpha} = \frac{2\varphi}{\omega_0 - \omega}.$$

2. If $\omega_0 = 0$ and $a = $ constant, then
   a. Angle of rotations:

$$\varphi = \frac{\omega t}{2} = \frac{at}{2} = \frac{\omega^2}{2a}.$$

b. Angular velocity:

$$\omega = \sqrt{2a\varphi} = \frac{2\varphi}{t} = at, \quad \omega_0 = 0.$$

c. Angular acceleration:

$$a = \frac{\omega}{t} = \frac{2\varphi}{t^2} = \frac{\omega^2}{2\varphi}.$$

d. Time:

$$t = \sqrt{\frac{2\varphi}{a}} = \frac{\omega}{a} = \frac{2\varphi}{\omega}.$$

Simple Harmonic Motion

Simple harmonic motion occurs when an object moves repeatedly over the same path in equal time intervals.

The maximum deflection from the position of rest is called "amplitude."

A mass on a spring is an example of an object in simple harmonic motion. The motion is sinusoidal in time and demonstrates a single frequency.

a. Displacement:

$$s = A \sin(\omega t + \varphi_0).$$

b. Velocity:

$$v = A\omega \cos(\omega t + \varphi_0).$$

c. Angular acceleration:

$$a = -A\alpha\omega^2 \sin(\omega t + \varphi_0).$$

where s is the displacement, A the amplitude, φ_0 the angular position at time $t = 0$, φ the angular position at time t, and T the time period.

Pendulum

A pendulum consists of an object suspended so that it swings freely back and forth about a pivot.

Period:

$$T = 2\pi\sqrt{\frac{l}{g}},$$

where T is the time period (s), l the length of the pendulum (m), and $g = 9.81$ (m/s^2) or 32.2 (ft/s^2).

Free Fall

A free-falling object is an object that is falling under the sole influence of gravity.

a. Initial speed:

$$v_0 = 0.$$

b. Distance:

$$h = -\frac{gt^2}{2} = -\frac{vt}{2} = -\frac{v^2}{2g}.$$

c. Speed:

$$v = +gt = -\frac{2h}{t} = \sqrt{-2gh}.$$

d. Time:

$$t = +\frac{v}{g} = -\frac{2h}{v} = \sqrt{-\frac{2h}{g}}.$$

Vertical Project

a. Initial speed:

$$v_0 > 0 \quad \text{(upwards)}; \qquad v_0 < 0 \quad \text{(downwards)}.$$

b. Distance:

$$h = v_0 t - \frac{gt^2}{2} = (v_0 + v)\frac{t}{2}, \quad h_{max} = \frac{v_0^2}{2g}.$$

c. Time:

$$t = \frac{v_0 - v}{g} = \frac{2h}{v_0 + v}, \quad t_{h_{max}} = \frac{v_0}{g}.$$

where v is the velocity (m/s), h the distance (m), and g the acceleration due to gravity (m/s^2).

Angled Projections

Upwards ($\alpha > 0$); downwards ($\alpha < 0$)

a. Distance:

$$s = v_0 t \cos \alpha.$$

b. Altitude:

$$h = v_0 t \sin \alpha - \frac{g t^2}{2} = s \tan \alpha - \frac{g s^2}{2 v_0^2 \cos \alpha},$$

$$h_{max} = \frac{v_0^2 \sin^2 \alpha}{2g}.$$

c. Velocity:

$$v = \sqrt{v_0^2 - 2gh} = \sqrt{v_0^2 + g^2 t^2 - 2 g v_0 t \sin \alpha}.$$

d. Time:

$$t_{h_{max}} = \frac{v_0 \sin \alpha}{g}, \qquad t_{s_1} = \frac{2 v_0 \sin \alpha}{g}.$$

Horizontal Projection ($\alpha = 0$):

a. Distance:

$$s = v_0 t = v_0 \sqrt{\frac{2h}{g}}.$$

b. Altitude:

$$h = -\frac{g t^2}{2}.$$

c. Trajectory velocity:

$$v = \sqrt{v_0^2 + g^2 t^2}.$$

where v_0 is the initial velocity (m/s), v the trajectory velocity (m/s), s the distance (m), and h the height (m).

Sliding Motion on an Inclined Plane

1. If excluding friction ($\mu = 0$), then
   a. Velocity:

$$v = at = \frac{2s}{t} = \sqrt{2as}.$$

   b. Distance:

$$s = \frac{a t^2}{2} = \frac{vt}{2} = \frac{v^2}{2a}.$$

   c. Acceleration:

$$a = g \sin \alpha.$$

2. If including friction ($\mu > 0$), then
   a. Velocity:

$$v = at = \frac{2s}{t} = \sqrt{2as}.$$

   b. Distance:

$$s = \frac{at^2}{2} = \frac{vt}{2} = \frac{v^2}{2a}.$$

   c. Accelerations:

$$s = \frac{at^2}{2} = \frac{vt}{2} = \frac{v^2}{2a},$$

where μ is the coefficient of sliding friction, g the acceleration due to gravity ($= 9.81 \text{m/s}^2$), v_0 the initial velocity (m/s), v the trajectory velocity (m/s), s the distance (m), a the acceleration (m/s^2), and α the inclined angle.

Rolling Motion on an Inclined Plane

1. If excluding friction ($f = 0$), then
   a. Velocity:

$$v = at = \frac{2s}{t} = \sqrt{2as}.$$

   b. Acceleration:

$$a = \frac{gr^2}{l^2 + k^2} \sin \alpha.$$

   c. Distance:

$$s = \frac{at^2}{2} = \frac{vt}{2} = \frac{v^2}{2a}.$$

   d. Tilting angle:

$$\tan \alpha = \mu_0 \left(\frac{r^2 + k^2}{k^2} \right).$$

2. If including friction ($f > 0$), then
   a. Distance:

$$s = \frac{at^2}{2} = \frac{vt}{2} = \frac{v^2}{2a}.$$

   b. Velocity:

$$v = at = \frac{2s}{t} = \sqrt{2as}.$$

   c. Accelerations:

$$a = gr^2 \frac{\sin \alpha - (f/r) \cos \alpha}{l^2 + k^2}.$$

   d. Tilting angle:

$$\tan \alpha_{min} = \frac{f}{r}, \qquad \tan \alpha_{max} = \mu_0 \left(\frac{r^2 + k^2 - fr}{k^2} \right).$$

The value of k can be the calculated by the formulas given below:

Ball	Solid cylinder	Pipe with low wall thickness
$k^2 = \dfrac{2r^2}{5}$	$k^2 = \dfrac{r^2}{2}$	$k^2 = \dfrac{r_i^2 + r_0^2}{2} \approx r^2$

where s is the distance (m), v the velocity (m/s), a the acceleration (m/s^2), α the tilting angle (°), f the lever arm of rolling resistance (m), k the radius of gyration (m), μ_0 the coefficient of static friction, and g the acceleration due to gravity (m/s^2).

Mechanics: Dynamics

Newton's First Law of Motion

Newton's First Law or the Law of Inertia: An object that is in motion continues in motion with the same velocity at a constant speed and in a straight line, and an object at rest continues at rest unless an unbalanced (outside) force acts upon it.

Newton's Second Law of Motion

The second law of motion, called the law of accelerations: The total force acting on an object equals the mass of the object times its acceleration. In equation form, this law is

$$F = ma,$$

where F is the total force (N), m the mass (kg), and a the acceleration (m/s^2).

Newton's Third Law of Motion

The third law of motion, called the law of action and reaction, can be stated as follows:
    For every force applied by object A to object B (action), there is a force exerted by object B on object A (the reaction) which has the same magnitude but is opposite in direction.
    In equation form, this law is

$$F_B = -F_A,$$

where F_B is the force of action (N) and F_A the force of reaction (N).

Momentum of Force

The momentum can be defined as a mass in motion. Momentum is a vector quantity; in other words, the direction is important:

$$p = mv.$$

Impulse of Force

The impulse of a force is equal to the change in momentum that the force causes in an object:

$$I = Ft,$$

where p is the momentum (N s), m the mass of the object (kg), v the velocity of the object (m/s), I the impulse of force (N s), F the force (N), and t the time (s).

Law of Conservation of Momentum

One of the most powerful laws in physics is the law of momentum conservation, which can be stated as follows: In the absence of external forces, the total momentum of the system is constant.

If two objects of masses m_1 and m_2 having velocities v_1 and v_2 collide and then separate with velocities v'_1 and v'_2, then the equation for the conservation of momentum is

$$m_1 v_1 + m_2 v_2 = m_1 v_1 + m_2 v_2.$$

Friction

Friction is a force that always acts parallel to the surface in contact and opposite to the direction of motion. Starting friction is greater than moving friction. Friction increases as the force between the surfaces increases.

The characteristics of friction can be described by the following equation:

$$F_f = \mu F_n,$$

where F_f is the frictional force (N), F_n the normal force (N), and μ the coefficient of friction ($\mu = \tan \alpha$).

General Law of Gravity

Gravity is a force that attracts bodies of matter toward each other. Gravity is the attraction between any two objects that have mass.

The general formula for gravity is

$$F = \Gamma \frac{m_A m_B}{r^2},$$

where m_A, m_B are the masses of objects A and B (kg), F the magnitude of attractive force between objects A and B (N), r the distance between objects A and B (m), Γ the gravitational constant (N m^2/kg^2), and $\Gamma = 6.67 \times 10^{-11}$ N m^2/kg^2.

Gravitational Force

The force of gravity is given by the equation

$$F_G = g \frac{R_e^2 m}{(R_e + h)^2}.$$

On the earth surface, $h = 0$, so

$$F_G = mg,$$

where F_G is the force of gravity (N), R_e the radius of the earth ($R_e = 6.37 \times 10^6$ m), m the mass (kg), g the acceleration due to gravity (m/s^2), and $g = 9.81$m/s^2 or 32.2 ft/s^2.

The acceleration of a falling body is independent of the mass of the object.

The weight F_w on an object is actually the force of gravity on that object:

$$F_w = mg.$$

Centrifugal Force

Centrifugal force is the apparent force drawing a rotating body away from the center of rotation and it is caused by the inertia of the body. Centrifugal force can be calculated by the following formula:

$$F_c = \frac{mv^2}{r} = m\omega^2 r.$$

Centripetal Force

Centripetal force is defined as the force acting on a body in curvilinear motion that is directed toward the center of curvature or axis of rotation. Centripetal force is equal in magnitude to centrifugal force but in the opposite direction.

$$F_{cp} = -F_c = \frac{mv^2}{r},$$

where F_c is the centrifugal force (N), F_{cp} the centripetal force (N), m the mass of the body (kg), v velocity of the body (m/s), r the radius of curvature of the path of the body (m), and ω the angular velocity (s^{-1}).

Torque

Torque is the ability of a force to cause a body to rotate about a particular axis. Torque can have either a clockwise or an anticlockwise direction. To distinguish between the two possible directions of rotation, we adopt the convention that an anticlockwise torque is positive and that a clockwise torque is negative. One way to quantify a torque is

$$T = Fl,$$

where T is the torque (N m or lb ft), F the applied force (N or lb), and l the length of the torque arm (m or ft).

Work

Work is the product of a force in the direction of the motion and the displacement.

a. Work done by a constant force:

$$W = F_s s = Fs \cos \alpha,$$

where W is the work (N m = J), F_s the component of force along the direction of movement (N), and s the distance the system is displaced (m).

b. Work done by a variable force:
    If the force is not constant along the path of the object, we need to calculate the force over very tiny intervals and then add them up. This is exactly what the integration over differential small intervals of a line can accomplish:

$$W = \int_{s_i}^{s_f} F_s(s)\, ds = \int_{s_i}^{s_f} F(s) \cos \alpha \, ds,$$

where $F_s(s)$ is the component of the force function along the direction of movement (N), $F(s)$ the function of the magnitude of the force vector along the displacement curve (N), s_i the initial location of the body (m), s_f the final location of the body (m), and α the angle between the displacement and the force.

Energy

Energy is defined as the ability to do work. The quantitative relationship between work and mechanical energy is expressed by

$$\text{TME}_i + W_{ext} = \text{TME}_f,$$

where TME_i is the initial amount of total mechanical energy (J), W_{ext} the work done by external forces (J), and TME_f the final amount of total mechanical energy (J).

There are two kinds of mechanical energy: kinetic and potential.

a. Kinetic energy:
    Kinetic energy is the energy of motion. The following equation is used to represent the kinetic energy of an object:

$$E_k = \frac{1}{2}mv^2,$$

where m is the mass of the moving object (kg) and v the velocity of the moving object (m/s).

b. Potential energy:
    Potential energy is the stored energy of a body and is due to its internal characteristics or its position. Gravitational potential energy is defined by the formula

$$E_{pg} = mgh,$$

where E_{pg} is the gravitational potential energy (J), m the mass of the object (kg), h the height above the reference level (m), and g the acceleration due to gravity (m/s²).

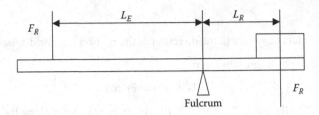

FIGURE 12.1 The lever.

Conservation of Energy

In any isolated system, energy can be transformed from one kind to another, but the total amount of energy is constant (conserved), that is,

$$E = E_k + E_p + E_e + \cdots = \text{constant.}$$

Conservation of mechanical energy is given by

$$E_k + E_p = \text{constant.}$$

Power

Power is the rate at which work is done, or the rate at which energy is transformed from one form to another. Mathematically, it is computed using the following equation:

$$P = \frac{W}{t},$$

where P is the power (W), W the work done (J), and t the time (s).

The standard metric unit of power is watt (W). As is implied by the equation for power, a unit of power is equivalent to a unit of work divided by a unit of time. Thus, a watt is equivalent to Joule/second (J/s). As the expression for work is

$$W = Fs,$$

the expression for power can be rewritten as

$$P = Fv,$$

where s is the displacement (m) and v the speed (m/s).

Appendix A: Mathematical Patterns, Series, and Formulae

Number Sequence and Patterns

Numbers can have interesting patterns. Here we list the most common patterns and how they are made.

$$1 \times 8 + 1 = 9$$
$$12 \times 8 + 2 = 98$$
$$123 \times 8 + 3 = 987$$
$$1234 \times 8 + 4 = 9876$$
$$12345 \times 8 + 5 = 98765$$
$$123456 \times 8 + 6 = 987654$$
$$1234567 \times 8 + 7 = 9876543$$
$$12345678 \times 8 + 8 = 98765432$$
$$123456789 \times 8 + 9 = 987654321$$

$$1 \times 9 + 2 = 11$$
$$12 \times 9 + 3 = 111$$
$$123 \times 9 + 4 = 1111$$
$$1234 \times 9 + 5 = 11111$$
$$12345 \times 9 + 6 = 111111$$
$$123456 \times 9 + 7 = 1111111$$
$$1234567 \times 9 + 8 = 11111111$$
$$12345678 \times 9 + 9 = 111111111$$
$$123456789 \times 9 + 10 = 1111111111$$

$$9 \times 9 + 7 = 88$$
$$98 \times 9 + 6 = 888$$
$$987 \times 9 + 5 = 8888$$
$$9876 \times 9 + 4 = 88888$$
$$98765 \times 9 + 3 = 888888$$
$$987654 \times 9 + 2 = 8888888$$
$$9876543 \times 9 + 1 = 88888888$$
$$98765432 \times 9 + 0 = 888888888$$

$$1 \times 1 = 1$$
$$11 \times 11 = 121$$
$$111 \times 111 = 12321$$
$$1111 \times 1111 = 1234321$$
$$11111 \times 11111 = 123454321$$
$$111111 \times 111111 = 12345654321$$
$$1111111 \times 1111111 = 1234567654321$$
$$11111111 \times 11111111 = 123456787654321$$
$$111111111 \times 111111111 = 12345678987654321$$
$$111,111,111 \times 111,111,111 = 12,345,678,987,654,321$$

$$1 \times 8 + 1 = 9$$
$$12 \times 8 + 2 = 98$$
$$123 \times 8 + 3 = 987$$
$$1234 \times 8 + 4 = 9876$$
$$12345 \times 8 + 5 = 98765$$
$$123456 \times 8 + 6 = 987654$$
$$1234567 \times 8 + 7 = 9876543$$
$$12345678 \times 8 + 8 = 98765432$$
$$123456789 \times 8 + 9 = 987654321$$

$$1 \times 9 + 2 = 11$$
$$12 \times 9 + 3 = 111$$
$$123 \times 9 + 4 = 1111$$
$$1234 \times 9 + 5 = 11111$$
$$12345 \times 9 + 6 = 111111$$
$$123456 \times 9 + 7 = 1111111$$
$$1234567 \times 9 + 8 = 11111111$$
$$12345678 \times 9 + 9 = 111111111$$
$$123456789 \times 9 + 10 = 1111111111$$

$$9 \times 9 + 7 = 88$$
$$98 \times 9 + 6 = 888$$
$$987 \times 9 + 5 = 8888$$
$$9876 \times 9 + 4 = 88888$$
$$98765 \times 9 + 3 = 888888$$
$$987654 \times 9 + 2 = 8888888$$
$$9876543 \times 9 + 1 = 88888888$$
$$98765432 \times 9 + 0 = 888888888$$

$$1 \times 1 = 1$$
$$11 \times 11 = 121$$
$$111 \times 111 = 12321$$
$$1111 \times 1111 = 1234321$$
$$11111 \times 11111 = 123454321$$
$$111111 \times 111111 = 12345654321$$
$$1111111 \times 1111111 = 1234567654321$$
$$11111111 \times 11111111 = 123456787654321$$
$$111111111 \times 111111111 = 12345678987654321$$

Closed-Form Mathematical Expressions

An expression is said to be a closed-form expression if, and only if, it can be expressed analytically in terms of a bounded number of certain well-known functions. Typically, these well-known functions are defined to be elementary functions: constants, one variable x, elementary operations of arithmetic ($+, -, \times, \div$), nth roots, exponent, and logarithm (which thus also include trigonometric functions and inverse trigonometric functions).

Here we will see some common forms of closed-form expressions:

$$\sum_{n=0}^{\infty} \frac{x^n}{n!} = e^x,$$

$$\sum_{n=0}^{\infty} \frac{x^n}{n} = \ln\left(\frac{1}{1-x}\right),$$

$$\sum_{n=0}^{k} x^n = \frac{x^{k+1} - 1}{x - 1}, \quad x \neq 1,$$

$$\sum_{n=1}^{k} x^n = \frac{x - x^{k+1}}{1 - x}, \quad x \neq 1,$$

$$\sum_{n=2}^{k} x^n = \frac{x^2 - x^{k+1}}{1 - x}, \quad x \neq 1,$$

$$\sum_{n=0}^{\infty} p^n = \frac{1}{1-p}, \quad \text{if } |p| < 1,$$

$$\sum_{n=0}^{\infty} n x^n = \frac{x}{(1-x)^2}, \quad x \neq 1,$$

$$\sum_{n=0}^{\infty} n^2 x^n = \frac{2x^2}{(1-x)^3} + \frac{x}{(1-x)^2} = \frac{x(1+x)}{(1-x)^3}, \quad |x| < 1,$$

$$\sum_{n=0}^{\infty} n^3 x^n = \frac{6x^3}{(1-x)^4} + \frac{6x^2}{(1-x)^3} + \frac{x}{(1-x)^2}, \quad |x| < 1,$$

$$\sum_{n=0}^{M} n x^n = \frac{x[1 - (M+1)x^M + Mx^{M+1}]}{(1-x)^2}, \quad |x| < 1,$$

$$\sum_{x=0}^{\infty} \binom{r+x-1}{x} u^x = (1-u)^{-r}, \quad \text{if } |u| < 1,$$

$$\sum_{k=1}^{\infty} (-1)^{k+1} \frac{1}{k} = 1 - \frac{1}{2} + \frac{1}{3} - \frac{1}{4} + \frac{1}{5} - \frac{1}{6} + \cdots = \ln 2,$$

$$\sum_{k=1}^{\infty} (-1)^{k+1} \frac{1}{2k-1} = 1 - \frac{1}{3} + \frac{1}{5} - \frac{1}{7} + \frac{1}{9} - \cdots = \frac{\pi}{4},$$

$$\sum_{k=0}^{\infty} (-1)^k x^k = \frac{1}{1+x}, \quad -1 < x < 1,$$

$$\sum_{k=1}^{n} (-1)^k \binom{n}{k} = 1, \quad \text{for } n \geqslant 2,$$

$$\sum_{k=0}^{n}\binom{n}{k}^{2}=\binom{2n}{n},$$

$$\sum_{k=1}^{n}k=1+2+3+\cdots+n=\frac{n(n+1)}{2},$$

$$\sum_{k=1}^{n}k^{2}=1+4+9+\cdots+n^{2}=\frac{n(n+1)(2n+1)}{6},$$

$$\sum_{k=0}^{n-1}k^{2}x^{k}=\frac{(x-1)^{2}n^{2}x^{n}-2(x-1)nx^{n+1}+x^{n+2}-x^{2}+x^{n+1}-x}{(x-1)^{3}},$$

$$\sum_{k=1}^{n}k^{3}=1+8+27+\cdots+n^{3}=\left(\frac{n(n+1)}{2}\right)^{2},$$

$$\sum_{k=1}^{n}(2k)=2+4+6+\cdots+2n=n(n-1),$$

$$\sum_{k=1}^{n}(2k-1)=1+3+5+\cdots+(2n-1)=n^{2},$$

$$\sum_{k=0}^{\infty}(a+kd)r^{k}=a+(a+d)r+(a+2d)r^{2}+\cdots=\frac{a}{1-r}+\frac{rd}{(1-r)^{2}},$$

$$\sum_{k=1}^{n}k^{3}=1+8+27+\cdots+n^{3}=\frac{n^{2}(n+1)^{2}}{4}=\left[\frac{n(n+1)^{2}}{2}\right]=\left[\sum_{k=1}^{n}k\right]^{2},$$

$$\sum_{x=1}^{\infty}\frac{1}{x}=1+\frac{1}{2}+\frac{1}{3}+\cdots\quad\text{(does not converge)},$$

$$\sum_{m=0}^{k}ma^{m}=\frac{a}{(1-a)^{2}}[1-(k+1)a^{k}+ka^{k+1}]=\sum_{m=1}^{k}ma^{m},$$

$$\sum_{k=0}^{n}(1)=n,$$

$$\sum_{k=0}^{n}\binom{n}{k}=2^{n},$$

$$(a+b)^n = \sum_{k=0}^{n} \binom{n}{k} a^k b^{n-k},$$

$$\prod_{n=1}^{\infty} a_n = e^{\left(\sum_{n=1}^{\infty} \ln(a_n)\right)},$$

$$\ln\left(\prod_{n=1}^{\infty} a_n\right) = \sum_{n=1}^{\infty} \ln a_n,$$

$$\ln(x) = \sum_{k=1}^{\infty} \frac{1}{k}\left(\frac{x-1}{x}\right)^k, \quad x \geqslant \frac{1}{2},$$

$$\lim_{h \to \infty} (1+h)^{1/h} = e,$$

$$\lim_{n \to \infty} \left(1 + \frac{x}{n}\right)^n = e^{-x},$$

$$\lim_{n \to \infty} \sum_{k=0}^{n} \frac{e^{-n} n^r}{K!} = \frac{1}{2},$$

$$\lim_{k \to \infty} \left(\frac{x^k}{k!}\right) = 0,$$

$$|x + y| \leqslant |x| + |y|,$$

$$|x - y| \geqslant |x| - |y|,$$

$$\ln(1+x) = \sum_{k=1}^{\infty} (-1)^{k+1}\left(\frac{x^k}{k}\right), \quad \text{if } -1 < x \leqslant 1,$$

$$\Gamma\left(\frac{1}{2}\right) = \sqrt{\pi},$$

$$\Gamma(\alpha + 1) = \alpha\Gamma(\alpha),$$

$$\Gamma\left(\frac{n}{2}\right) = \frac{\sqrt{\pi}(n-1)!}{2^{n-1}(\frac{1}{2}(n-1))!}, \quad \text{for odd } n,$$

$$\Gamma(n) = \int_{0}^{\infty} e^{-x} x^{n-1}\, dx,$$

$$\binom{n}{2} = \frac{1}{2}(n^2 - n) = \sum_{k=1}^{n-1} k,$$

$$\binom{n+1}{2} = \binom{n}{2} + n,$$

$$2 \cdot 4 \cdot 6 \cdot 8 \cdots 2n = \prod_{k=1}^{n} 2k = 2^n n!,$$

$$1 \cdot 3 \cdot 5 \cdot 7 \cdots (2n-1) = \frac{(2n-1)!}{2^{2n-2}(2n-2)!} = \frac{2n-1}{2^{2n-2}},$$

Derivation of closed-form expression for $\sum_{k=1}^{n} kx^k$:

$$\sum_{k=1}^{n} kx^k = x \sum_{k=1}^{n} kx^{k-1}$$

$$= x \sum_{k=1}^{n} \frac{d}{dx}[x^k]$$

$$= x \frac{d}{dx}\left[\sum_{k=1}^{n} x^k\right]$$

$$= x \frac{d}{dx}\left[\frac{x(1-x^n)}{1-x}\right]$$

$$= x\left[\frac{(1-(n+1)x^n)(1-x) - x(1-x^n)(-1)}{(1-x)^2}\right]$$

$$= \frac{x\left[1-(n+1)x^n + nx^{n+1}\right]}{(1-x)^2}, \quad x \neq 1,$$

Derivation of the Quadratic Formula

The roots of the quadratic equation

$$ax^2 + bx + c = 0$$

is given by the quadratic formula

$$x = \frac{-b \pm \sqrt{b^2 - 4ac}}{2a}.$$

The roots are complex if $b^2 - 4ac < 0$, the roots are real if $b^2 - 4ac > 0$, and the roots are real and repeated if $b^2 - 4ac = 0$. Formula:

$$ax^2 + bx + c = 0.$$

Dividing both sides by a, where $a \neq 0$, we obtain

$$x^2 + \frac{b}{a}x + \frac{c}{a} = 0.$$

Note if $a = 0$, the solution to $ax^2 + bx + c = 0$ is $x = -c/b$.

Rewrite

$$x^2 + \frac{b}{a}x + \frac{c}{a} = 0$$

as

$$\left(x + \frac{b}{2a}\right)^2 - \frac{b^2}{4a^2} + \frac{c}{a} = 0,$$

$$\left(x + \frac{b}{2a}\right)^2 = \frac{b^2}{4a^2} - \frac{c}{a} = \frac{b^2 - 4ac}{4a^2},$$

$$x + \frac{b}{2a} = \pm\sqrt{\frac{b^2 - 4ac}{4a^2}} = \pm\frac{\sqrt{b^2 - 4ac}}{2a},$$

$$x = -\frac{b}{2a} \pm \sqrt{\frac{b^2 - 4ac}{2a}},$$

$$x = \frac{-b \pm \sqrt{b^2 - 4ac}}{2a}.$$

Appendix B: Measurement Units, Notation, and Constants

Common Notation

Scientific notation uses exponents to express numerical figures. Here is an explanation of what scientific notation is, and examples of how to write numbers and perform addition, subtraction, multiplication, and division problems using scientific notation.

Measurement	Notation	Description
meter	m	length
hectare	ha	area
tonne	t	mass
kilogram	kg	mass
nautical mile	M	distance (navigation)
knot	kn	speed (navigation)
liter	L	volume or capacity
second	s	time
hertz	Hz	frequency
candela	cd	luminous intensity
degree Celsius	°C	temperature
kelvin	K	thermodynamic temperature
pascal	Pa	pressure, stress
joule	J	energy, work
newton	N	force
watt	W	power, radiant flux
ampere	A	electric current
volt	V	electric potential
ohm	Ω	electric resistance
coulomb	C	electric charge

Acre: An area of 43,560 square feet.
Agate: 1/14 inch (used in printing for measuring column length).
Ampere: Unit of electric current.

Astronomical unit (A.U.): 93,000,000 miles; the average distance of the earth from the sun (used in astronomy).

Bale: A large bundle of goods. In United States, approximate weight of a bale of cotton is 500 pounds. Weight of a bale may vary from country to country.

Board foot: 144 cubic inches (12 × 12 × 1 used for lumber).

Bolt: 40 yards (used for measuring cloth).

Btu: British thermal unit; amount of heat needed to increase the temperature of one pound of water by one degree Fahrenheit (252 calories).

Carat: 200 milligrams or 3086 troy; used for weighing precious stones (originally the weight of a seed of the carob tree in the Mediterranean region). *See also* Karat.

Chain: 66 feet; used in surveying (one mile = 80 chains).

Cubit: 18 inches (derived from distance between elbow and tip of middle finger).

Decibel: Unit of relative loudness.

Freight ton: 40 cubic feet of merchandise (used for cargo freight).

Gross: 12 dozen (144 units).

Hertz: Unit of measurement of electromagnetic wave frequencies (measures cycles per second).

Hogshead: Two liquid barrels or 14,653 cubic inches.

Horsepower: The power needed to lift 33,000 pounds a distance of one foot in one minute (about $1\frac{1}{2}$ times the power an average horse can exert); used for measuring power of mechanical engines.

Karat: A measure of the purity of gold. It indicates how many parts out of 24 are pure. 18 karat gold is $\frac{3}{4}$ pure gold.

Knot: Rate of speed of 1 nautical mile per hour; used for measuring speed of ships (not distance).

League: Approximately 3 miles.

Light year: 5,880,000,000,000 miles; distance traveled by light in one year at the rate of 186,281.7 miles per second; used for measurement of interstellar space.

Magnum: Two-quart bottle; used for measuring wine.

Ohm: Unit of electrical resistance.

Parsec: Approximately 3.26 light-years of 19.2 trillion miles; used for measuring interstellar distances.

Pi (π): 3.14159265+; the ratio of the circumference of a circle to its diameter.

Pica: 1/6 inch or 12 points; used in printing for measuring column width.

Pipe: 2 hogsheads; used for measuring wine and other liquids.

Point: 0.013837 (approximately 1/72 inch or 1/12 pica); used in printing for measuring type size.

Quintal: 100,000 grams or 220.46 pounds avoirdupois.

Quire: 24 or 25 sheets; used for measuring paper (20 quires is one ream).

Ream: 480 or 500 sheets; used for measuring paper.

Roentgen: Dosage unit of radiation exposure produced by x-rays.

Score: 20 units.

Span: 9 inches or 22.86 cm; derived from the distance between the end of the thumb and the end of the little finger when both are outstretched.

Square: 100 square feet; used in building.
Stone: 14 pounds avoirdupois in Great Britain.
Therm: 100,000 Btu.
Township: U.S. land measurement of almost 36 square miles; used in surveying.
Tun: 252 gallons (sometimes larger); used for measuring wine and other liquids.
Watt: Unit of power.

Scientific Constants

Speed of light	2.997925×10^{10} cm/s
	983.6×10^6 ft/s
	$186,284$ miles/s
Velocity of sound	340.3 m/s
	1116 ft/s
Gravity (acceleration)	9.80665 m/s^2
	32.174 ft/s^2
	386.089 in./s^2

Numbers and Prefixes

yotta (10^{24}):	1 000 000 000 000 000 000 000 000
zetta (10^{21}):	1 000 000 000 000 000 000 000
exa (10^{18}):	1 000 000 000 000 000 000
peta (10^{15}):	1 000 000 000 000 000
tera (10^{12}):	1 000 000 000 000
giga (10^{9}):	1 000 000 000
mega (10^{6}):	1 000 000
kilo (10^{3}):	1 000
hecto (10^{2}):	100
deca (10^{1}):	10
deci (10^{-1}):	0.1
centi (10^{-2}):	0.01
milli (10^{-3}):	0.001
micro (10^{-6}):	0.000 001
nano (10^{-9}):	0.000 000 001
pico (10^{-12}):	0.000 000 000 001
femto (10^{-15}):	0.000 000 000 000 001
atto (10^{-18}):	0.000 000 000 000 000 001
zepto (10^{-21}):	0.000 000 000 000 000 000 001
yacto (10^{-24}):	0.000 000 000 000 000 000 000 001
Stringo (10^{-35}):	0.000 000 000 000 000 000 000 000 000 000 000 01

Appendix C: Conversion Factors

A conversion factor changes something to a different version or form. A factor is something that brings results or a cause, whereas conversion is an action of changing the "version" of a thing.

Area Conversion Factors

Multiply	By	To Obtain
acres	43560	sq feet
	4047	sq meters
	4840	sq yards
	0.405	hectare
sq cm	0.155	sq inches
sq feet	144	sq inches
	0.0929	sq meters
	0.1111	sq yards
sq inches	645.16	sq millimeters
sq kilometers	0.3861	sq miles
sq meters	10.764	sq feet
	1.196	sq yards
sq miles	640	acres
	2.59	sq kilometers

Volume Conversion Factors

Multiply	By	To Obtain
acre-foot	1233.5	cubic meters
cubic cm	0.06102	cubic inches
cubic feet	1728	cubic inches
	7.48	gallons (U.S.)
	0.02832	cubic meters
	0.03704	cubic yards
liter	1.057	liquid quarts
	0.908	dry quarts
	61.024	cubic inches
gallons (U.S.)	231	cubic inches
	3.7854	liters
	4	quarts
	0.833	British gallons
	128	U.S. fluid ounces

Energy Conversion Factors

Multiply	By	To Obtain
BTU	1055.9	joules
	0.252	kg-calories
watt-hour	3600	joules
	3.409	BTU
HP (electric)	746	watts
BTU/second	1055.9	watts
watt-second	1	joules

Mass Conversion Factors

Multiply	By	To Obtain
carat	0.2	cubic grams
grams	0.03527	ounces
kilograms	2.2046	pounds
ounces	28.35	grams
pound	16	ounces
	453.6	grams
stone (UK)	6.35	kilograms
	14	pounds
ton (net)	907.2	kilograms
	2000	pounds
	0.893	gross ton
	0.907	metric ton
ton (gross)	2240	pounds
	1.12	net tons
	1.016	metric tons
tonne (metric)	2204.62	pounds
	0.984	gross pound
	1000	kilograms

Temperature Conversion Factors

TABLE C.1   Conversion Formulas

Celsius to Kelvin	$K = C + 273.15$
Celsius to Fahrenheit	$F = (9/5)C + 32$
Fahrenheit to Celsius	$C = (5/9)(F - 32)$
Fahrenheit to Kelvin	$K = (5/9)(F + 459.67)$
Fahrenheit to Rankin	$R = F + 459.67$
Rankin to Kelvin	$K = (5/9)R$

Velocity Conversion Factors

Multiply	By	To Obtain
feet/minute	5.08	mm/second
feet/second	0.3048	meters/second
inches/second	0.0254	meters/second
km/hour	0.6214	miles/hour
meters/second	3.2808	feet/second
	2.237	miles/hour
miles/hour	88	feet/minute
	0.44704	meters/second
	1.6093	km/hour
	0.8684	knots
knot	1.151	miles/hour

Pressure Conversion Factors

Multiply	By	To Obtain
atmospheres	1.01325	bars
	33.9	feet of water
	29.92	inches of mercury
	760	mm of mercury
bar	75.01	cm of mercury
	14.5	pounds/sq inch
dyne/sq cm	0.1	N/sq meter
newtons/sq cm	1.45	pounds/sq inch
pounds/sq inch	0.06805	atmospheres
	2.036	inches of mercury
	27.708	inches of water
	68.948	millibars
	51.72	mm of mercury

Distance Conversion Factors

Multiply	By	To Obtain
angstrom	10^{-10}	meters
feet	0.3048	meters
	12	inches
inches	25.4	millimeters
	0.0254	meters
	0.08333	feet
kilometers	3280.8	feet
	0.6214	miles
	1094	yards
meters	39.37	inches
	3.2808	feet
	1.094	yards
miles	5280	feet
	1.6093	kilometers
	0.8694	nautical miles
millimeters	0.03937	inches
nautical miles	6076	feet
	1.852	kilometers
yards	0.9144	meters
	3	feet
	36	inches

Physical Science Equations

$$D = \frac{m}{V}$$

$$P = \frac{W}{t}$$

$$d = vt$$

$$K.E. = \frac{1}{2}mv^2$$

$$a = \frac{v_f - v_i}{t}$$

$$F_e = \frac{kQ_1Q_2}{d^2}$$

$$d = v_i t + \frac{1}{2}at^2$$

$$V = \frac{W}{Q}$$

$$F = ma$$

$$I = \frac{Q}{t}$$

$$F_g = \frac{Gm_1 m_2}{d^2}$$

$$W = VIt$$

$$p = mv$$

$$p = VI$$

$$W = Fd$$

$$H = cm\Delta T$$

Note: D is the density (g/cm^3 = kg/m^3), m is the mass (kg), V is the volume, d is the distance (m), v is the velocity (m/s), t is the time (s), a is the acceleration (m/s^2), v_f is the final velocity (m/s), v_i is the initial velocity (m/s), F_g is the force of gravity (N), G is the universal gravitational constant ($G = 6.67 \times 10^{-11}$Nm^2/kg^2), m_1 and m_2 are the masses of the two objects (kg), p is the momentum (kg m/s), W is the work or electrical energy (J), P is the power (W), K.E. is the kinetic energy (J), F_e is the electric force (N), k Coulomb's constant ($k = 9 \times 10^9$Nm^2/C^2), Q, Q$_1$, and Q$_2$ are electric charges (C), V is the electric potential difference (V), I is the electric current (A), H is the heat energy (J), ΔT is the change in temperature (°C), and c is the specific heat (J/kg°C).

English and Metric Systems

	English System
1 foot (ft)	= 12 inches (in.); $1' = 12''$
1 yard (yd)	= 3 feet
1 mile (mi)	= 1760 yards
1 sq. foot	= 144 sq. inches
1 sq. yard	= 9 sq. feet
1 acre	= 4840 sq. yards = 43,560 ft^2
1 sq. mile	= 640 acres

	Metric System
millimeter (mm)	0.001 m
centimeter (cm)	0.01 m
decimeter (dm)	0.1 m
meter (m)	1 m
decameter (dam)	10 m
hectometer (hm)	100 m
kilometer (km)	1000 m

Note: Prefixes also apply to L (liter) and g (gram).

Household Measurement Conversion

A pinch	1/8 tsp. or less
3 tsp.	1 tbsp.
2 tbsp.	1/8 c.
4 tbsp.	1/4 c.
16 tbsp.	1 c.
5 tbsp. + 1 tsp.	1/3 c.
4 oz.	1/2 c.
8 oz.	1 c.
16 oz.	1 lbs.
1 oz.	2 tbsp. fat or liquid
1 c. of liquid	1/2 pt.
2 c.	1 pt.
2 pt.	1 qt.
4 c. of liquid	1 qt.
4 qts.	1 gallon
8 qts.	1 peck (such as apples, pears, etc.)
1 jigger	$1\frac{1}{2}$ fl. oz.
1 jigger	3 tbsp.

Appendix D: Interest Factors and Tables

Generally speaking, the interest factor denotes a numeral that is applied to work out the interest element of any mortgage payment. In fact, the interest factor is founded on how frequently mortgage payments are made—monthly, semimonthly, biweekly, or weekly. Having paid the interest component to the mortgagee, any surplus amount paid by the mortgagor goes towards lessening the principal loan amount. In other words, the money paid in excess of the fixed interest amount is considered as the principal component of that particular payment. The principal component or factor of any mortgage payment lessens the outstanding amount of the loan owed by the mortgagor to the mortgagee. Hence, the interest factor is also the number on which all future mortgage payment computations are rooted.

Formulas for Interest Factor

Name of Factor	Formula	Table Notation
Compound amount (single payment)	$(1+i)^N$	$(F/P, i, N)$
Present worth (single payment)	$(1+i)^{-N}$	$(P/F, i, N)$
Sinking fund	$\dfrac{i}{(1+i)^N - 1}$	$(A/F, i, N)$
Capital recovery	$\dfrac{i(1+i)^N}{(1+i)^N - 1}$	$(A/P, i, N)$
Compound amount (uniform series)	$\dfrac{(1+i)^N - 1}{i}$	$(F/A, i, N)$
Present worth (uniform series)	$\dfrac{(1+i)^N - 1}{i(1+i)^N}$	$(P/A, i, N)$

(continued)

Name of Factor	Formula	Table Notation
Arithmetic gradient to uniform series	$\dfrac{(1+i)^N - iN - 1}{i(1+i)^N - i}$	$(A/G, i, N)$
Arithmetic gradient to present worth	$\dfrac{(1+i)^N - iN - 1}{i^2(1+i)^N}$	$(P/G, i, N)$
Geometric gradient to present worth (for $1 \neq g$)	$\dfrac{1 - (1+g)^N(1+i)^{-N}}{i - g}$	$(P/A, g, i, N)$
Continuous compounding compound amount (single payment)	e^{rN}	$(F/P, r, N)$
Continuous compounding present worth (single payment)	e^{-rN}	$(P/F, r, N)$
Continuous compounding present worth (uniform series)	$\dfrac{e^{rN} - 1}{e^{rN}(e^r - 1)}$	$(P/A, r, N)$
Continuous compounding sinking fund	$\dfrac{e^r - 1}{e^{rN} - 1}$	$(A/F, r, N)$
Continuous compounding capital recovery	$\dfrac{e^{rN}(e^r - 1)}{e^{rN} - 1}$	$(A/P, r, N)$
Continuous compounding compound amount (uniform series)	$\dfrac{e^{rN} - 1}{e^r - 1}$	$(F/A, r, N)$
Continuous compounding present worth (single, continuous payment)	$\dfrac{i(1+i)^{-N}}{\ln(1+i)}$	$(P/\breve{F}, i, N)$
Continuous compounding compound amount (single, continuous payment)	$\dfrac{i(1+i)^{N-1}}{\ln(1+i)}$	$(F/\breve{P}, i, N)$
Continuous compounding sinking fund (continuous, uniform payments)	$\dfrac{\ln(1+i)}{(1+i)^N - 1}$	$(\bar{A}/F, i, N)$
Continuous compounding capital recovery (continuous, uniform payments)	$\dfrac{(1+i)^N \ln(1+i)}{(1+i)^N - 1}$	$(\bar{A}/P, i, N)$
Continuous compounding compound amount (continuous, uniform payments)	$\dfrac{(1+i)^N - 1}{\ln(1+i)}$	$(F/\bar{A}, i, N)$
Continuous compounding present worth (continuous, uniform payments)	$\dfrac{(1+i)^N - 1}{(1+i)^N \ln(1+i)}$	$(P/\bar{A}, i, N)$

Summation Formulas for Closed-Form Expressions

$$\sum_{t=0}^{n} x^t = \frac{1 - x^{n+1}}{1 - x},$$

$$\sum_{t=0}^{n-1} x^t = \frac{1 - x^n}{1 - x},$$

$$\sum_{t=1}^{n} x^t = \frac{x - x^{n+1}}{1 - x},$$

$$\sum_{t=1}^{n-1} x^t = \frac{x - x^n}{1 - x},$$

$$\sum_{t=2}^{n} x^t = \frac{x^2 - x^{n+1}}{1 - x},$$

$$\sum_{t=0}^{n} x^t = \frac{x[1 - (n+1)x^n + nx^{n+1}]}{(1 - x)^2}$$

$$= \frac{x - (1 - x)(n+1)x^{n+1} - x^{n+2}}{(1 - x)^2},$$

$$\sum_{t=1}^{n} tx^t = \sum_{t=0}^{n} tx^t - 0x^0 = \sum_{t=0}^{n} tx^t - 0 = \sum_{t=0}^{n} tx^t$$

$$= \frac{x[1 - (n+1)x^n + nx^{n+1}]}{(1 - x)^2} = \frac{x - (1 - x)(n+1)x^{n+1} - x^{n+2}}{(1 - x)^2}.$$

Interest Tables

0.25%	Compound Interest Factors								0.25%
Period	Single Payment		Uniform Payment Series				Arithmetic Gradient		Period
	Compound Amount Factor	Present Value Factor	Sinking Fund Factor	Capital Recovery Factor	Compound Amount Factor	Present Value Factor	Gradient Uniform Series	Gradient Present Value	
	Find F Given P	Find P Given F	Find A Given F	Find A Given P	Find F Given A	Find P Given A	Find A Given G	Find P Given G	
n	F/P	P/F	A/F	A/P	F/A	P/A	A/G	P/G	n
1	1.003	0.9975	1.0000	1.0025	1.000	0.998	0.000	0.000	1
2	1.005	0.9950	0.4994	0.5019	2.002	1.993	0.499	0.995	2
3	1.008	0.9925	0.3325	0.3350	3.008	2.985	0.998	2.980	3
4	1.010	0.9901	0.2491	0.2516	4.015	3.975	1.497	5.950	4
5	1.013	0.9876	0.1990	0.2015	5.025	4.963	1.995	9.901	5
6	1.015	0.9851	0.1656	0.1681	6.038	5.948	2.493	14.826	6
7	1.018	0.9827	0.1418	0.1443	7.053	6.931	2.990	20.722	7
8	1.020	0.9802	0.1239	0.1264	8.070	7.911	3.487	27.584	8

(continued)

0.25%					Compound Interest Factors				0.25%
Period	Single Payment		Uniform Payment Series				Arithmetic Gradient		Period
	Compound Amount Factor	Present Value Factor	Sinking Fund Factor	Capital Recovery Factor	Compound Amount Factor	Present Value Factor	Gradient Uniform Series	Gradient Present Value	
	Find F Given P F/P	Find P Given F P/F	Find A Given F A/F	Find A Given P A/P	Find F Given A F/A	Find P Given A P/A	Find A Given G A/G	Find P Given G P/G	
n									n
9	1.023	0.9778	0.1100	0.1125	9.091	8.889	3.983	35.406	9
10	1.025	0.9753	0.0989	0.1014	10.113	9.864	4.479	44.184	10
11	1.028	0.9729	0.0898	0.0923	11.139	10.837	4.975	53.913	11
12	1.030	0.9705	0.0822	0.0847	12.166	11.807	5.470	64.589	12
13	1.033	0.9681	0.0758	0.0783	13.197	12.775	5.965	76.205	13
14	1.036	0.9656	0.0703	0.0728	14.230	13.741	6.459	88.759	14
15	1.038	0.9632	0.0655	0.0680	15.265	14.704	6.953	102.244	15
16	1.041	0.9608	0.0613	0.0638	16.304	15.665	7.447	116.657	16
17	1.043	0.9584	0.0577	0.0602	17.344	16.623	7.940	131.992	17
18	1.046	0.9561	0.0544	0.0569	18.388	17.580	8.433	148.245	18
19	1.049	0.9537	0.0515	0.0540	19.434	18.533	8.925	165.411	19
20	1.051	0.9513	0.0488	0.0513	20.482	19.484	9.417	183.485	20
21	1.054	0.9489	0.0464	0.0489	21.533	20.433	9.908	202.463	21
22	1.056	0.9466	0.0443	0.0468	22.587	21.380	10.400	222.341	22
23	1.059	0.9442	0.0423	0.0448	23.644	22.324	10.890	243.113	23
24	1.062	0.9418	0.0405	0.0430	24.703	23.266	11.380	264.775	24
25	1.064	0.9395	0.0388	0.0413	25.765	24.205	11.870	287.323	25
26	1.067	0.9371	0.0373	0.0398	26.829	25.143	12.360	310.752	26
27	1.070	0.9348	0.0358	0.0383	27.896	26.077	12.849	335.057	27
28	1.072	0.9325	0.0345	0.0370	28.966	27.010	13.337	360.233	28
29	1.075	0.9301	0.0333	0.0358	30.038	27.940	13.825	386.278	29
30	1.078	0.9278	0.0321	0.0346	31.113	28.868	14.313	413.185	30
31	1.080	0.9255	0.0311	0.0336	32.191	29.793	14.800	440.950	31
32	1.083	0.9232	0.0301	0.0326	33.272	30.717	15.287	469.570	32
33	1.086	0.9209	0.0291	0.0316	34.355	31.638	15.774	499.039	33
34	1.089	0.9186	0.0282	0.0307	35.441	32.556	16.260	529.353	34
35	1.091	0.9163	0.0274	0.0299	36.529	33.472	16.745	560.508	35
36	1.094	0.9140	0.0266	0.0291	37.621	34.386	17.231	592.499	36
37	1.097	0.9118	0.0258	0.0283	38.715	35.298	17.715	625.322	37
38	1.100	0.9095	0.0251	0.0276	39.811	36.208	18.200	658.973	38
39	1.102	0.9072	0.0244	0.0269	40.911	37.115	18.684	693.447	39
40	1.105	0.9050	0.0238	0.0263	42.013	38.020	19.167	728.740	40
41	1.108	0.9027	0.0232	0.0257	43.118	38.923	19.650	764.848	41
42	1.111	0.9004	0.0226	0.0251	44.226	39.823	20.133	801.766	42
43	1.113	0.8982	0.0221	0.0246	45.337	40.721	20.616	839.490	43
44	1.116	0.8960	0.0215	0.0240	46.450	41.617	21.097	878.016	44
45	1.119	0.8937	0.0210	0.0235	47.566	42.511	21.579	917.340	45
46	1.122	0.8915	0.0205	0.0230	48.685	43.402	22.060	957.457	46
47	1.125	0.8893	0.0201	0.0226	49.807	44.292	22.541	998.364	47
48	1.127	0.8871	0.0196	0.0221	50.931	45.179	23.021	1040.055	48
49	1.130	0.8848	0.0192	0.0217	52.059	46.064	23.501	1082.528	49
50	1.133	0.8826	0.0188	0.0213	53.189	46.946	23.980	1125.777	50

0.5%				Compound Interest Factors					0.5%
Period	Single Payment		Uniform Payment Series				Arithmetic Gradient		Period
	Compound Amount Factor	Present Value Factor	Sinking Fund Factor	Capital Recovery Factor	Compound Amount Factor	Present Value Factor	Gradient Uniform Series	Gradient Present Value	
	Find F Given P F/P	Find P Given F P/F	Find A Given F A/F	Find A Given P A/P	Find F Given A F/A	Find P Given A P/A	Find A Given G A/G	Find P Given G P/G	
n									n
1	1.005	0.9950	1.0000	1.0050	1.000	0.995	0.000	0.000	1
2	1.010	0.9901	0.4988	0.5038	2.005	1.985	0.499	0.990	2
3	1.015	0.9851	0.3317	0.3367	3.015	2.970	0.997	2.960	3
4	1.020	0.9802	0.2481	0.2531	4.030	3.950	1.494	5.901	4
5	1.025	0.9754	0.1980	0.2030	5.050	4.926	1.990	9.803	5
6	1.030	0.9705	0.1646	0.1696	6.076	5.896	2.485	14.655	6
7	1.036	0.9657	0.1407	0.1457	7.106	6.862	2.980	20.449	7
8	1.041	0.9609	0.1228	0.1278	8.141	7.823	3.474	27.176	8
9	1.046	0.9561	0.1089	0.1139	9.182	8.779	3.967	34.824	9
10	1.051	0.9513	0.0978	0.1028	10.228	9.730	4.459	43.386	10
11	1.056	0.9466	0.0887	0.0937	11.279	10.677	4.950	52.853	11
12	1.062	0.9419	0.0811	0.0861	12.336	11.619	5.441	63.214	12
13	1.067	0.9372	0.0746	0.0796	13.397	12.556	5.930	74.460	13
14	1.072	0.9326	0.0691	0.0741	14.464	13.489	6.419	86.583	14
15	1.078	0.9279	0.0644	0.0694	15.537	14.417	6.907	99.574	15
16	1.083	0.9233	0.0602	0.0652	16.614	15.340	7.394	113.424	16
17	1.088	0.9187	0.0565	0.0615	17.697	16.259	7.880	128.123	17
18	1.094	0.9141	0.0532	0.0582	18.786	17.173	8.366	143.663	18
19	1.099	0.9096	0.0503	0.0553	19.880	18.082	8.850	160.036	19
20	1.105	0.9051	0.0477	0.0527	20.979	18.987	9.334	177.232	20
21	1.110	0.9006	0.0453	0.0503	22.084	19.888	9.817	195.243	21
22	1.116	0.8961	0.0431	0.0481	23.194	20.784	10.299	214.061	22
23	1.122	0.8916	0.0411	0.0461	24.310	21.676	10.781	233.677	23
24	1.127	0.8872	0.0393	0.0443	25.432	22.563	11.261	254.082	24
25	1.133	0.8828	0.0377	0.0427	26.559	23.446	11.741	275.269	25
26	1.138	0.8784	0.0361	0.0411	27.692	24.324	12.220	297.228	26
27	1.144	0.8740	0.0347	0.0397	28.830	25.198	12.698	319.952	27
28	1.150	0.8697	0.0334	0.0384	29.975	26.068	13.175	343.433	28
29	1.156	0.8653	0.0321	0.0371	31.124	26.933	13.651	367.663	29
30	1.161	0.8610	0.0310	0.0360	32.280	27.794	14.126	392.632	30
31	1.167	0.8567	0.0299	0.0349	33.441	28.651	14.601	418.335	31
32	1.173	0.8525	0.0289	0.0339	34.609	29.503	15.075	444.762	32
33	1.179	0.8482	0.0279	0.0329	35.782	30.352	15.548	471.906	33
34	1.185	0.8440	0.0271	0.0321	36.961	31.196	16.020	499.758	34
35	1.191	0.8398	0.0262	0.0312	38.145	32.035	16.492	528.312	35
36	1.197	0.8356	0.0254	0.0304	39.336	32.871	16.962	557.560	36
37	1.203	0.8315	0.0247	0.0297	40.533	33.703	17.432	587.493	37
38	1.209	0.8274	0.0240	0.0290	41.735	34.530	17.901	618.105	38
39	1.215	0.8232	0.0233	0.0283	42.944	35.353	18.369	649.388	39
40	1.221	0.8191	0.0226	0.0276	44.159	36.172	18.836	681.335	40
41	1.227	0.8151	0.0220	0.0270	45.380	36.987	19.302	713.937	41
42	1.233	0.8110	0.0215	0.0265	46.607	37.798	19.768	747.189	42
43	1.239	0.8070	0.0209	0.0259	47.840	38.605	20.233	781.081	43
44	1.245	0.8030	0.0204	0.0254	49.079	39.408	20.696	815.609	44
45	1.252	0.7990	0.0199	0.0249	50.324	40.207	21.159	850.763	45
46	1.258	0.7950	0.0194	0.0244	51.576	41.002	21.622	886.538	46
47	1.264	0.7910	0.0189	0.0239	52.834	41.793	22.083	922.925	47
48	1.270	0.7871	0.0185	0.0235	54.098	42.580	22.544	959.919	48

(continued)

0.5%				Compound Interest Factors						0.5%
Period	Single Payment		Uniform Payment Series					Arithmetic Gradient		Period
	Compound Amount Factor	Present Value Factor	Sinking Fund Factor	Capital Recovery Factor	Compound Amount Factor	Present Value Factor	Gradient Uniform Series	Gradient Present Value		
	Find F Given P F/P	Find P Given F P/F	Find A Given F A/F	Find A Given P A/P	Find F Given A F/A	Find P Given A P/A	Find A Given G A/G	Find P Given G P/G		
n										n
49	1.277	0.7832	0.0181	0.0231	55.368	43.364	23.003	997.512		49
50	1.283	0.7793	0.0177	0.0227	56.645	44.143	23.462	1035.697		50

0.75%				Compound Interest Factors						0.75%
Period	Single Payment		Uniform Payment Series					Arithmetic Gradient		Period
	Compound Amount Factor	Present Value Factor	Sinking Fund Factor	Capital Recovery Factor	Compound Amount Factor	Present Value Factor	Gradient Uniform Series	Gradient Present Value		
	Find F Given P F/P	Find P Given F P/F	Find A Given F A/F	Find A Given P A/P	Find F Given A F/A	Find P Given A P/A	Find A Given G A/G	Find P Given G P/G		
n										n
1	1.008	0.9926	1.0000	1.0075	1.000	0.993	0.000	0.000		1
2	1.015	0.9852	0.4981	0.5056	2.008	1.978	0.498	0.985		2
3	1.023	0.9778	0.3308	0.3383	3.023	2.956	0.995	2.941		3
4	1.030	0.9706	0.2472	0.2547	4.045	3.926	1.491	5.852		4
5	1.038	0.9633	0.1970	0.2045	5.076	4.889	1.985	9.706		5
6	1.046	0.9562	0.1636	0.1711	6.114	5.846	2.478	14.487		6
7	1.054	0.9490	0.1397	0.1472	7.159	6.795	2.970	20.181		7
8	1.062	0.9420	0.1218	0.1293	8.213	7.737	3.461	26.775		8
9	1.070	0.9350	0.1078	0.1153	9.275	8.672	3.950	34.254		9
10	1.078	0.9280	0.0967	0.1042	10.344	9.600	4.438	42.606		10
11	1.086	0.9211	0.0876	0.0951	11.422	10.521	4.925	51.817		11
12	1.094	0.9142	0.0800	0.0875	12.508	11.435	5.411	61.874		12
13	1.102	0.9074	0.0735	0.0810	13.601	12.342	5.895	72.763		13
14	1.110	0.9007	0.0680	0.0755	14.703	13.243	6.379	84.472		14
15	1.119	0.8940	0.0632	0.0707	15.814	14.137	6.861	96.988		15
16	1.127	0.8873	0.0591	0.0666	16.932	15.024	7.341	110.297		16
17	1.135	0.8807	0.0554	0.0629	18.059	15.905	7.821	124.389		17
18	1.144	0.8742	0.0521	0.0596	19.195	16.779	8.299	139.249		18
19	1.153	0.8676	0.0492	0.0567	20.339	17.647	8.776	154.867		19
20	1.161	0.8612	0.0465	0.0540	21.491	18.508	9.252	171.230		20
21	1.170	0.8548	0.0441	0.0516	22.652	19.363	9.726	188.325		21
22	1.179	0.8484	0.0420	0.0495	23.822	20.211	10.199	206.142		22
23	1.188	0.8421	0.0400	0.0475	25.001	21.053	10.671	224.668		23
24	1.196	0.8358	0.0382	0.0457	26.188	21.889	11.142	243.892		24
25	1.205	0.8296	0.0365	0.0440	27.385	22.719	11.612	263.803		25
26	1.214	0.8234	0.0350	0.0425	28.590	23.542	12.080	284.389		26
27	1.224	0.8173	0.0336	0.0411	29.805	24.359	12.547	305.639		27
28	1.233	0.8112	0.0322	0.0397	31.028	25.171	13.013	327.542		28
29	1.242	0.8052	0.0310	0.0385	32.261	25.976	13.477	350.087		29
30	1.251	0.7992	0.0298	0.0373	33.503	26.775	13.941	373.263		30
31	1.261	0.7932	0.0288	0.0363	34.754	27.568	14.403	397.060		31
32	1.270	0.7873	0.0278	0.0353	36.015	28.356	14.864	421.468		32
33	1.280	0.7815	0.0268	0.0343	37.285	29.137	15.323	446.475		33
34	1.289	0.7757	0.0259	0.0334	38.565	29.913	15.782	472.071		34

(continued)

0.75%	Compound Interest Factors								0.75%
Period	Single Payment		Uniform Payment Series				Arithmetic Gradient		Period
	Compound Amount Factor	Present Value Factor	Sinking Fund Factor	Capital Recovery Factor	Compound Amount Factor	Present Value Factor	Gradient Uniform Series	Gradient Present Value	
n	Find *F* Given *P* F/P	Find *P* Given *F* P/F	Find *A* Given *F* A/F	Find *A* Given *P* A/P	Find *F* Given *A* F/A	Find *P* Given *A* P/A	Find *A* Given *G* A/G	Find *P* Given *G* P/G	*n*
35	1.299	0.7699	0.0251	0.0326	39.854	30.683	16.239	498.247	35
36	1.309	0.7641	0.0243	0.0318	41.153	31.447	16.695	524.992	36
37	1.318	0.7585	0.0236	0.0311	42.461	32.205	17.149	552.297	37
38	1.328	0.7528	0.0228	0.0303	43.780	32.958	17.603	580.151	38
39	1.338	0.7472	0.0222	0.0297	45.108	33.705	18.055	608.545	39
40	1.348	0.7416	0.0215	0.0290	46.446	34.447	18.506	637.469	40
41	1.358	0.7361	0.0209	0.0284	47.795	35.183	18.956	666.914	41
42	1.369	0.7306	0.0203	0.0278	49.153	35.914	19.404	696.871	42
43	1.379	0.7252	0.0198	0.0273	50.522	36.639	19.851	727.330	43
44	1.389	0.7198	0.0193	0.0268	51.901	37.359	20.297	758.281	44
45	1.400	0.7145	0.0188	0.0263	53.290	38.073	20.742	789.717	45
46	1.410	0.7091	0.0183	0.0258	54.690	38.782	21.186	821.628	46
47	1.421	0.7039	0.0178	0.0253	56.100	39.486	21.628	854.006	47
48	1.431	0.6986	0.0174	0.0249	57.521	40.185	22.069	886.840	48
49	1.442	0.6934	0.0170	0.0245	58.952	40.878	22.509	920.124	49
50	1.453	0.6883	0.0166	0.0241	60.394	41.566	22.948	953.849	50

1%	Compound Interest Factors								1%
Period	Single Payment		Uniform Payment Series				Arithmetic Gradient		Period
	Compound Amount Factor	Present Value Factor	Sinking Fund Factor	Capital Recovery Factor	Compound Amount Factor	Present Value Factor	Gradient Uniform Series	Gradient Present Value	
n	Find *F* Given *P* F/P	Find *P* Given *F* P/F	Find *A* Given *F* A/F	Find *A* Given *P* A/P	Find *F* Given *A* F/A	Find *P* Given *A* P/A	Find *A* Given *G* A/G	Find *P* Given *G* P/G	*n*
1	1.010	0.9901	1.0000	1.0100	1.000	0.990	0.000	0.000	1
2	1.020	0.9803	0.4975	0.5075	2.010	1.970	0.498	0.980	2
3	1.030	0.9706	0.3300	0.3400	3.030	2.941	0.993	2.921	3
4	1.041	0.9610	0.2463	0.2563	4.060	3.902	1.488	5.804	4
5	1.051	0.9515	0.1960	0.2060	5.101	4.853	1.980	9.610	5
6	1.062	0.9420	0.1625	0.1725	6.152	5.795	2.471	14.321	6
7	1.072	0.9327	0.1386	0.1486	7.214	6.728	2.960	19.917	7
8	1.083	0.9235	0.1207	0.1307	8.286	7.652	3.448	26.381	8
9	1.094	0.9143	0.1067	0.1167	9.369	8.566	3.934	33.696	9
10	1.105	0.9053	0.0956	0.1056	10.462	9.471	4.418	41.843	10
11	1.116	0.8963	0.0865	0.0965	11.567	10.368	4.901	50.807	11
12	1.127	0.8874	0.0788	0.0888	12.683	11.255	5.381	60.569	12
13	1.138	0.8787	0.0724	0.0824	13.809	12.134	5.861	71.113	13
14	1.149	0.8700	0.0669	0.0769	14.947	13.004	6.338	82.422	14
15	1.161	0.8613	0.0621	0.0721	16.097	13.865	6.814	94.481	15
16	1.173	0.8528	0.0579	0.0679	17.258	14.718	7.289	107.273	16
17	1.184	0.8444	0.0543	0.0643	18.430	15.562	7.761	120.783	17
18	1.196	0.8360	0.0510	0.0610	19.615	16.398	8.232	134.996	18
19	1.208	0.8277	0.0481	0.0581	20.811	17.226	8.702	149.895	19
20	1.220	0.8195	0.0454	0.0554	22.019	18.046	9.169	165.466	20

(continued)

1%	Compound Interest Factors								1%
Period	Single Payment		Uniform Payment Series				Arithmetic Gradient		Period
	Compound Amount Factor	Present Value Factor	Sinking Fund Factor	Capital Recovery Factor	Compound Amount Factor	Present Value Factor	Gradient Uniform Series	Gradient Present Value	
	Find F Given P F/P	Find P Given F P/F	Find A Given F A/F	Find A Given P A/P	Find F Given A F/A	Find P Given A P/A	Find A Given G A/G	Find P Given G P/G	
n									n
21	1.232	0.8114	0.0430	0.0530	23.239	18.857	9.635	181.695	21
22	1.245	0.8034	0.0409	0.0509	24.472	19.660	10.100	198.566	22
23	1.257	0.7954	0.0389	0.0489	25.716	20.456	10.563	216.066	23
24	1.270	0.7876	0.0371	0.0471	26.973	21.243	11.024	234.180	24
25	1.282	0.7798	0.0354	0.0454	28.243	22.023	11.483	252.894	25
26	1.295	0.7720	0.0339	0.0439	29.526	22.795	11.941	272.196	26
27	1.308	0.7644	0.0324	0.0424	30.821	23.560	12.397	292.070	27
28	1.321	0.7568	0.0311	0.0411	32.129	24.316	12.852	312.505	28
29	1.335	0.7493	0.0299	0.0399	33.450	25.066	13.304	333.486	29
30	1.348	0.7419	0.0287	0.0387	34.785	25.808	13.756	355.002	30
31	1.361	0.7346	0.0277	0.0377	36.133	26.542	14.205	377.039	31
32	1.375	0.7273	0.0267	0.0367	37.494	27.270	14.653	399.586	32
33	1.389	0.7201	0.0257	0.0357	38.869	27.990	15.099	422.629	33
34	1.403	0.7130	0.0248	0.0348	40.258	28.703	15.544	446.157	34
35	1.417	0.7059	0.0240	0.0340	41.660	29.409	15.987	470.158	35
36	1.431	0.6989	0.0232	0.0332	43.077	30.108	16.428	494.621	36
37	1.445	0.6920	0.0225	0.0325	44.508	30.800	16.868	519.533	37
38	1.460	0.6852	0.0218	0.0318	45.953	31.485	17.306	544.884	38
39	1.474	0.6784	0.0211	0.0311	47.412	32.163	17.743	570.662	39
40	1.489	0.6717	0.0205	0.0305	48.886	32.835	18.178	596.856	40
41	1.504	0.6650	0.0199	0.0299	50.375	33.500	18.611	623.456	41
42	1.519	0.6584	0.0193	0.0293	51.879	34.158	19.042	650.451	42
43	1.534	0.6519	0.0187	0.0287	53.398	34.810	19.472	677.831	43
44	1.549	0.6454	0.0182	0.0282	54.932	35.455	19.901	705.585	44
45	1.565	0.6391	0.0177	0.0277	56.481	36.095	20.327	733.704	45
46	1.580	0.6327	0.0172	0.0272	58.046	36.727	20.752	762.176	46
47	1.596	0.6265	0.0168	0.0268	59.626	37.354	21.176	790.994	47
48	1.612	0.6203	0.0163	0.0263	61.223	37.974	21.598	820.146	48
49	1.628	0.6141	0.0159	0.0259	62.835	38.588	22.018	849.624	49
50	1.645	0.6080	0.0155	0.0255	64.463	39.196	22.436	879.418	50

1.25%	Compound Interest Factors								1.25%
Period	Single Payment		Uniform Payment Series				Arithmetic Gradient		Period
	Compound Amount Factor	Present Value Factor	Sinking Fund Factor	Capital Recovery Factor	Compound Amount Factor	Present Value Factor	Gradient Uniform Series	Gradient Present Value	
	Find F Given P F/P	Find P Given F P/F	Find A Given F A/F	Find A Given P A/P	Find F Given A F/A	Find P Given A P/A	Find A Given G A/G	Find P Given G P/G	
n									n
1	1.013	0.9877	1.0000	1.0125	1.000	0.988	0.000	0.000	1
2	1.025	0.9755	0.4969	0.5094	2.013	1.963	0.497	0.975	2
3	1.038	0.9634	0.3292	0.3417	3.038	2.927	0.992	2.902	3
4	1.051	0.9515	0.2454	0.2579	4.076	3.878	1.484	5.757	4
5	1.064	0.9398	0.1951	0.2076	5.127	4.818	1.975	9.516	5
6	1.077	0.9282	0.1615	0.1740	6.191	5.746	2.464	14.157	6

(continued)

1.25%			Compound Interest Factors						1.25%
Period	Single Payment		Uniform Payment Series				Arithmetic Gradient		Period
	Compound Amount Factor	Present Value Factor	Sinking Fund Factor	Capital Recovery Factor	Compound Amount Factor	Present Value Factor	Gradient Uniform Series	Gradient Present Value	
	Find F Given P F/P	Find P Given F P/F	Find A Given F A/F	Find A Given P A/P	Find F Given A F/A	Find P Given A P/A	Find A Given G A/G	Find P Given G P/G	
n									n
7	1.091	0.9167	0.1376	0.1501	7.268	6.663	2.950	19.657	7
8	1.104	0.9054	0.1196	0.1321	8.359	7.568	3.435	25.995	8
9	1.118	0.8942	0.1057	0.1182	9.463	8.462	3.917	33.149	9
10	1.132	0.8832	0.0945	0.1070	10.582	9.346	4.398	41.097	10
11	1.146	0.8723	0.0854	0.0979	11.714	10.218	4.876	49.820	11
12	1.161	0.8615	0.0778	0.0903	12.860	11.079	5.352	59.297	12
13	1.175	0.8509	0.0713	0.0838	14.021	11.930	5.826	69.507	13
14	1.190	0.8404	0.0658	0.0783	15.196	12.771	6.298	80.432	14
15	1.205	0.8300	0.0610	0.0735	16.386	13.601	6.768	92.052	15
16	1.220	0.8197	0.0568	0.0693	17.591	14.420	7.236	104.348	16
17	1.235	0.8096	0.0532	0.0657	18.811	15.230	7.702	117.302	17
18	1.251	0.7996	0.0499	0.0624	20.046	16.030	8.166	130.896	18
19	1.266	0.7898	0.0470	0.0595	21.297	16.819	8.628	145.111	19
20	1.282	0.7800	0.0443	0.0568	22.563	17.599	9.087	159.932	20
21	1.298	0.7704	0.0419	0.0544	23.845	18.370	9.545	175.339	21
22	1.314	0.7609	0.0398	0.0523	25.143	19.131	10.001	191.317	22
23	1.331	0.7515	0.0378	0.0503	26.457	19.882	10.454	207.850	23
24	1.347	0.7422	0.0360	0.0485	27.788	20.624	10.906	224.920	24
25	1.364	0.7330	0.0343	0.0468	29.135	21.357	11.355	242.513	25
26	1.381	0.7240	0.0328	0.0453	30.500	22.081	11.802	260.613	26
27	1.399	0.7150	0.0314	0.0439	31.881	22.796	12.248	279.204	27
28	1.416	0.7062	0.0300	0.0425	33.279	23.503	12.691	298.272	28
29	1.434	0.6975	0.0288	0.0413	34.695	24.200	13.132	317.802	29
30	1.452	0.6889	0.0277	0.0402	36.129	24.889	13.571	337.780	30
31	1.470	0.6804	0.0266	0.0391	37.581	25.569	14.009	358.191	31
32	1.488	0.6720	0.0256	0.0381	39.050	26.241	14.444	379.023	32
33	1.507	0.6637	0.0247	0.0372	40.539	26.905	14.877	400.261	33
34	1.526	0.6555	0.0238	0.0363	42.045	27.560	15.308	421.892	34
35	1.545	0.6474	0.0230	0.0355	43.571	28.208	15.737	443.904	35
36	1.564	0.6394	0.0222	0.0347	45.116	28.847	16.164	466.283	36
37	1.583	0.6315	0.0214	0.0339	46.679	29.479	16.589	489.018	37
38	1.603	0.6237	0.0207	0.0332	48.263	30.103	17.012	512.095	38
39	1.623	0.6160	0.0201	0.0326	49.866	30.719	17.433	535.504	39
40	1.644	0.6084	0.0194	0.0319	51.490	31.327	17.851	559.232	40
41	1.664	0.6009	0.0188	0.0313	53.133	31.928	18.268	583.268	41
42	1.685	0.5935	0.0182	0.0307	54.797	32.521	18.683	607.601	42
43	1.706	0.5862	0.0177	0.0302	56.482	33.107	19.096	632.219	43
44	1.727	0.5789	0.0172	0.0297	58.188	33.686	19.507	657.113	44
45	1.749	0.5718	0.0167	0.0292	59.916	34.258	19.916	682.271	45
46	1.771	0.5647	0.0162	0.0287	61.665	34.823	20.322	707.683	46
47	1.793	0.5577	0.0158	0.0283	63.435	35.381	20.727	733.339	47
48	1.815	0.5509	0.0153	0.0278	65.228	35.931	21.130	759.230	48
49	1.838	0.5441	0.0149	0.0274	67.044	36.476	21.531	785.344	49
50	1.861	0.5373	0.0145	0.0270	68.882	37.013	21.929	811.674	50

1.5%	Compound Interest Factors							1.5%
Period	Single Payment		Uniform Payment Series				Arithmetic Gradient	Period
	Compound Amount Factor	Present Value Factor	Sinking Fund Factor	Capital Recovery Factor	Compound Amount Factor	Present Value Factor	Gradient Uniform Series	Gradient Present Value
	Find F Given P F/P	Find P Given F P/F	Find A Given F A/F	Find A Given P A/P	Find F Given A F/A	Find P Given A P/A	Find A Given G A/G	Find P Given G P/G
n									n
1	1.015	0.9852	1.0000	1.0150	1.000	0.985	0.000	0.000	1
2	1.030	0.9707	0.4963	0.5113	2.015	1.956	0.496	0.971	2
3	1.046	0.9563	0.3284	0.3434	3.045	2.912	0.990	2.883	3
4	1.061	0.9422	0.2444	0.2594	4.091	3.854	1.481	5.710	4
5	1.077	0.9283	0.1941	0.2091	5.152	4.783	1.970	9.423	5
6	1.093	0.9145	0.1605	0.1755	6.230	5.697	2.457	13.996	6
7	1.110	0.9010	0.1366	0.1516	7.323	6.598	2.940	19.402	7
8	1.126	0.8877	0.1186	0.1336	8.433	7.486	3.422	25.616	8
9	1.143	0.8746	0.1046	0.1196	9.559	8.361	3.901	32.612	9
10	1.161	0.8617	0.0934	0.1084	10.703	9.222	4.377	40.367	10
11	1.178	0.8489	0.0843	0.0993	11.863	10.071	4.851	48.857	11
12	1.196	0.8364	0.0767	0.0917	13.041	10.908	5.323	58.057	12
13	1.214	0.8240	0.0702	0.0852	14.237	11.732	5.792	67.945	13
14	1.232	0.8118	0.0647	0.0797	15.450	12.543	6.258	78.499	14
15	1.250	0.7999	0.0599	0.0749	16.682	13.343	6.722	89.697	15
16	1.269	0.7880	0.0558	0.0708	17.932	14.131	7.184	101.518	16
17	1.288	0.7764	0.0521	0.0671	19.201	14.908	7.643	113.940	17
18	1.307	0.7649	0.0488	0.0638	20.489	15.673	8.100	126.943	18
19	1.327	0.7536	0.0459	0.0609	21.797	16.426	8.554	140.508	19
20	1.347	0.7425	0.0432	0.0582	23.124	17.169	9.006	154.615	20
21	1.367	0.7315	0.0409	0.0559	24.471	17.900	9.455	169.245	21
22	1.388	0.7207	0.0387	0.0537	25.838	18.621	9.902	184.380	22
23	1.408	0.7100	0.0367	0.0517	27.225	19.331	10.346	200.001	23
24	1.430	0.6995	0.0349	0.0499	28.634	20.030	10.788	216.090	24
25	1.451	0.6892	0.0333	0.0483	30.063	20.720	11.228	232.631	25
26	1.473	0.6790	0.0317	0.0467	31.514	21.399	11.665	249.607	26
27	1.495	0.6690	0.0303	0.0453	32.987	22.068	12.099	267.000	27
28	1.517	0.6591	0.0290	0.0440	34.481	22.727	12.531	284.796	28
29	1.540	0.6494	0.0278	0.0428	35.999	23.376	12.961	302.978	29
30	1.563	0.6398	0.0266	0.0416	37.539	24.016	13.388	321.531	30
31	1.587	0.6303	0.0256	0.0406	39.102	24.646	13.813	340.440	31
32	1.610	0.6210	0.0246	0.0396	40.688	25.267	14.236	359.691	32
33	1.634	0.6118	0.0236	0.0386	42.299	25.879	14.656	379.269	33
34	1.659	0.6028	0.0228	0.0378	43.933	26.482	15.073	399.161	34
35	1.684	0.5939	0.0219	0.0369	45.592	27.076	15.488	419.352	35
36	1.709	0.5851	0.0212	0.0362	47.276	27.661	15.901	439.830	36
37	1.735	0.5764	0.0204	0.0354	48.985	28.237	16.311	460.582	37
38	1.761	0.5679	0.0197	0.0347	50.720	28.805	16.719	481.595	38
39	1.787	0.5595	0.0191	0.0341	52.481	29.365	17.125	502.858	39
40	1.814	0.5513	0.0184	0.0334	54.268	29.916	17.528	524.357	40
41	1.841	0.5431	0.0178	0.0328	56.082	30.459	17.928	546.081	41
42	1.869	0.5351	0.0173	0.0323	57.923	30.994	18.327	568.020	42
43	1.897	0.5272	0.0167	0.0317	59.792	31.521	18.723	590.162	43
44	1.925	0.5194	0.0162	0.0312	61.689	32.041	19.116	612.496	44
45	1.954	0.5117	0.0157	0.0307	63.614	32.552	19.507	635.011	45
46	1.984	0.5042	0.0153	0.0303	65.568	33.056	19.896	657.698	46
47	2.013	0.4967	0.0148	0.0298	67.552	33.553	20.283	680.546	47
48	2.043	0.4894	0.0144	0.0294	69.565	34.043	20.667	703.546	48
49	2.074	0.4821	0.0140	0.0290	71.609	34.525	21.048	726.688	49
50	2.105	0.4750	0.0136	0.0286	73.683	35.000	21.428	749.964	50

1.75%				Compound Interest Factors						1.75%
Period	Single Payment		Uniform Payment Series				Arithmetic Gradient			Period
	Compound Amount Factor	Present Value Factor	Sinking Fund Factor	Capital Recovery Factor	Compound Amount Factor	Present Value Factor	Gradient Uniform Series	Gradient Present Value		
	Find F Given P F/P	Find P Given F P/F	Find A Given F A/F	Find A Given P A/P	Find F Given A F/A	Find P Given A P/A	Find A Given G A/G	Find P Given G P/G		
n										n
1	1.018	0.9828	1.0000	1.0175	1.000	0.983	0.000	0.000		1
2	1.035	0.9659	0.4957	0.5132	2.018	1.949	0.496	0.966		2
3	1.053	0.9493	0.3276	0.3451	3.053	2.898	0.988	2.864		3
4	1.072	0.9330	0.2435	0.2610	4.106	3.831	1.478	5.663		4
5	1.091	0.9169	0.1931	0.2106	5.178	4.748	1.965	9.331		5
6	1.110	0.9011	0.1595	0.1770	6.269	5.649	2.449	13.837		6
7	1.129	0.8856	0.1355	0.1530	7.378	6.535	2.931	19.151		7
8	1.149	0.8704	0.1175	0.1350	8.508	7.405	3.409	25.243		8
9	1.169	0.8554	0.1036	0.1211	9.656	8.260	3.884	32.087		9
10	1.189	0.8407	0.0924	0.1099	10.825	9.101	4.357	39.654		10
11	1.210	0.8263	0.0832	0.1007	12.015	9.927	4.827	47.916		11
12	1.231	0.8121	0.0756	0.0931	13.225	10.740	5.293	56.849		12
13	1.253	0.7981	0.0692	0.0867	14.457	11.538	5.757	66.426		13
14	1.275	0.7844	0.0637	0.0812	15.710	12.322	6.218	76.623		14
15	1.297	0.7709	0.0589	0.0764	16.984	13.093	6.677	87.415		15
16	1.320	0.7576	0.0547	0.0722	18.282	13.850	7.132	98.779		16
17	1.343	0.7446	0.0510	0.0685	19.602	14.595	7.584	110.693		17
18	1.367	0.7318	0.0477	0.0652	20.945	15.327	8.034	123.133		18
19	1.390	0.7192	0.0448	0.0623	22.311	16.046	8.480	136.078		19
20	1.415	0.7068	0.0422	0.0597	23.702	16.753	8.924	149.508		20
21	1.440	0.6947	0.0398	0.0573	25.116	17.448	9.365	163.401		21
22	1.465	0.6827	0.0377	0.0552	26.556	18.130	9.803	177.738		22
23	1.490	0.6710	0.0357	0.0532	28.021	18.801	10.239	192.500		23
24	1.516	0.6594	0.0339	0.0514	29.511	19.461	10.671	207.667		24
25	1.543	0.6481	0.0322	0.0497	31.027	20.109	11.101	223.221		25
26	1.570	0.6369	0.0307	0.0482	32.570	20.746	11.527	239.145		26
27	1.597	0.6260	0.0293	0.0468	34.140	21.372	11.951	255.421		27
28	1.625	0.6152	0.0280	0.0455	35.738	21.987	12.372	272.032		28
29	1.654	0.6046	0.0268	0.0443	37.363	22.592	12.791	288.962		29
30	1.683	0.5942	0.0256	0.0431	39.017	23.186	13.206	306.195		30
31	1.712	0.5840	0.0246	0.0421	40.700	23.770	13.619	323.716		31
32	1.742	0.5740	0.0236	0.0411	42.412	24.344	14.029	341.510		32
33	1.773	0.5641	0.0226	0.0401	44.154	24.908	14.436	359.561		33
34	1.804	0.5544	0.0218	0.0393	45.927	25.462	14.840	377.857		34
35	1.835	0.5449	0.0210	0.0385	47.731	26.007	15.241	396.382		35
36	1.867	0.5355	0.0202	0.0377	49.566	26.543	15.640	415.125		36
37	1.900	0.5263	0.0194	0.0369	51.434	27.069	16.036	434.071		37
38	1.933	0.5172	0.0187	0.0362	53.334	27.586	16.429	453.209		38
39	1.967	0.5083	0.0181	0.0356	55.267	28.095	16.819	472.526		39
40	2.002	0.4996	0.0175	0.0350	57.234	28.594	17.207	492.011		40
41	2.037	0.4910	0.0169	0.0344	59.236	29.085	17.591	511.651		41
42	2.072	0.4826	0.0163	0.0338	61.272	29.568	17.973	531.436		42
43	2.109	0.4743	0.0158	0.0333	63.345	30.042	18.353	551.355		43
44	2.145	0.4661	0.0153	0.0328	65.453	30.508	18.729	571.398		44
45	2.183	0.4581	0.0148	0.0323	67.599	30.966	19.103	591.554		45
46	2.221	0.4502	0.0143	0.0318	69.782	31.416	19.474	611.813		46
47	2.260	0.4425	0.0139	0.0314	72.003	31.859	19.843	632.167		47
48	2.300	0.4349	0.0135	0.0310	74.263	32.294	20.208	652.605		48
49	2.340	0.4274	0.0131	0.0306	76.562	32.721	20.571	673.120		49
50	2.381	0.4200	0.0127	0.0302	78.902	33.141	20.932	693.701		50

2%					Compound Interest Factors					2%
Period	Single Payment		Uniform Payment Series				Arithmetic Gradient		Period
	Compound Amount Factor	Present Value Factor	Sinking Fund Factor	Capital Recovery Factor	Compound Amount Factor	Present Value Factor	Gradient Uniform Series	Gradient Present Value	
	Find F Given P F/P	Find P Given F P/F	Find A Given F A/F	Find A Given P A/P	Find F Given A F/A	Find P Given A P/A	Find A Given G A/G	Find P Given G P/G	
n									n
1	1.020	0.9804	1.0000	1.0200	1.000	0.980	0.000	0.000	1
2	1.040	0.9612	0.4950	0.5150	2.020	1.942	0.495	0.961	2
3	1.061	0.9423	0.3268	0.3468	3.060	2.884	0.987	2.846	3
4	1.082	0.9238	0.2426	0.2626	4.122	3.808	1.475	5.617	4
5	1.104	0.9057	0.1922	0.2122	5.204	4.713	1.960	9.240	5
6	1.126	0.8880	0.1585	0.1785	6.308	5.601	2.442	13.680	6
7	1.149	0.8706	0.1345	0.1545	7.434	6.472	2.921	18.903	7
8	1.172	0.8535	0.1165	0.1365	8.583	7.325	3.396	24.878	8
9	1.195	0.8368	0.1025	0.1225	9.755	8.162	3.868	31.572	9
10	1.219	0.8203	0.0913	0.1113	10.950	8.983	4.337	38.955	10
11	1.243	0.8043	0.0822	0.1022	12.169	9.787	4.802	46.998	11
12	1.268	0.7885	0.0746	0.0946	13.412	10.575	5.264	55.671	12
13	1.294	0.7730	0.0681	0.0881	14.680	11.348	5.723	64.948	13
14	1.319	0.7579	0.0626	0.0826	15.974	12.106	6.179	74.800	14
15	1.346	0.7430	0.0578	0.0778	17.293	12.849	6.631	85.202	15
16	1.373	0.7284	0.0537	0.0737	18.639	13.578	7.080	96.129	16
17	1.400	0.7142	0.0500	0.0700	20.012	14.292	7.526	107.555	17
18	1.428	0.7002	0.0467	0.0667	21.412	14.992	7.968	119.458	18
19	1.457	0.6864	0.0438	0.0638	22.841	15.678	8.407	131.814	19
20	1.486	0.6730	0.0412	0.0612	24.297	16.351	8.843	144.600	20
21	1.516	0.6598	0.0388	0.0588	25.783	17.011	9.276	157.796	21
22	1.546	0.6468	0.0366	0.0566	27.299	17.658	9.705	171.379	22
23	1.577	0.6342	0.0347	0.0547	28.845	18.292	10.132	185.331	23
24	1.608	0.6217	0.0329	0.0529	30.422	18.914	10.555	199.630	24
25	1.641	0.6095	0.0312	0.0512	32.030	19.523	10.974	214.259	25
26	1.673	0.5976	0.0297	0.0497	33.671	20.121	11.391	229.199	26
27	1.707	0.5859	0.0283	0.0483	35.344	20.707	11.804	244.431	27
28	1.741	0.5744	0.0270	0.0470	37.051	21.281	12.214	259.939	28
29	1.776	0.5631	0.0258	0.0458	38.792	21.844	12.621	275.706	29
30	1.811	0.5521	0.0246	0.0446	40.568	22.396	13.025	291.716	30
31	1.848	0.5412	0.0236	0.0436	42.379	22.938	13.426	307.954	31
32	1.885	0.5306	0.0226	0.0426	44.227	23.468	13.823	324.403	32
33	1.922	0.5202	0.0217	0.0417	46.112	23.989	14.217	341.051	33
34	1.961	0.5100	0.0208	0.0408	48.034	24.499	14.608	357.882	34
35	2.000	0.5000	0.0200	0.0400	49.994	24.999	14.996	374.883	35
36	2.040	0.4902	0.0192	0.0392	51.994	25.489	15.381	392.040	36
37	2.081	0.4806	0.0185	0.0385	54.034	25.969	15.762	409.342	37
38	2.122	0.4712	0.0178	0.0378	56.115	26.441	16.141	426.776	38
39	2.165	0.4619	0.0172	0.0372	58.237	26.903	16.516	444.330	39
40	2.208	0.4529	0.0166	0.0366	60.402	27.355	16.889	461.993	40
41	2.252	0.4440	0.0160	0.0360	62.610	27.799	17.258	479.754	41
42	2.297	0.4353	0.0154	0.0354	64.862	28.235	17.624	497.601	42
43	2.343	0.4268	0.0149	0.0349	67.159	28.662	17.987	515.525	43
44	2.390	0.4184	0.0144	0.0344	69.503	29.080	18.347	533.517	44
45	2.438	0.4102	0.0139	0.0339	71.893	29.490	18.703	551.565	45
46	2.487	0.4022	0.0135	0.0335	74.331	29.892	19.057	569.662	46
47	2.536	0.3943	0.0130	0.0330	76.817	30.287	19.408	587.798	47
48	2.587	0.3865	0.0126	0.0326	79.354	30.673	19.756	605.966	48
49	2.639	0.3790	0.0122	0.0322	81.941	31.052	20.100	624.156	49
50	2.692	0.3715	0.0118	0.0318	84.579	31.424	20.442	642.361	50

2.5%			Compound Interest Factors						2.5%
Period	Single Payment		Uniform Payment Series				Arithmetic Gradient		Period
	Compound Amount Factor	Present Value Factor	Sinking Fund Factor	Capital Recovery Factor	Compound Amount Factor	Present Value Factor	Gradient Uniform Series	Gradient Present Value	
	Find F Given P F/P	Find P Given F P/F	Find A Given F A/F	Find A Given P A/P	Find F Given A F/A	Find P Given A P/A	Find A Given G A/G	Find P Given G P/G	
n									n
1	1.025	0.9756	1.0000	1.0250	1.000	0.976	0.000	0.000	1
2	1.051	0.9518	0.4938	0.5188	2.025	1.927	0.494	0.952	2
3	1.077	0.9286	0.3251	0.3501	3.076	2.856	0.984	2.809	3
4	1.104	0.9060	0.2408	0.2658	4.153	3.762	1.469	5.527	4
5	1.131	0.8839	0.1902	0.2152	5.256	4.646	1.951	9.062	5
6	1.160	0.8623	0.1565	0.1815	6.388	5.508	2.428	13.374	6
7	1.189	0.8413	0.1325	0.1575	7.547	6.349	2.901	18.421	7
8	1.218	0.8207	0.1145	0.1395	8.736	7.170	3.370	24.167	8
9	1.249	0.8007	0.1005	0.1255	9.955	7.971	3.836	30.572	9
10	1.280	0.7812	0.0893	0.1143	11.203	8.752	4.296	37.603	10
11	1.312	0.7621	0.0801	0.1051	12.483	9.514	4.753	45.225	11
12	1.345	0.7436	0.0725	0.0975	13.796	10.258	5.206	53.404	12
13	1.379	0.7254	0.0660	0.0910	15.140	10.983	5.655	62.109	13
14	1.413	0.7077	0.0605	0.0855	16.519	11.691	6.100	71.309	14
15	1.448	0.6905	0.0558	0.0808	17.932	12.381	6.540	80.976	15
16	1.485	0.6736	0.0516	0.0766	19.380	13.055	6.977	91.080	16
17	1.522	0.6572	0.0479	0.0729	20.865	13.712	7.409	101.595	17
18	1.560	0.6412	0.0447	0.0697	22.386	14.353	7.838	112.495	18
19	1.599	0.6255	0.0418	0.0668	23.946	14.979	8.262	123.755	19
20	1.639	0.6103	0.0391	0.0641	25.545	15.589	8.682	135.350	20
21	1.680	0.5954	0.0368	0.0618	27.183	16.185	9.099	147.257	21
22	1.722	0.5809	0.0346	0.0596	28.863	16.765	9.511	159.456	22
23	1.765	0.5667	0.0327	0.0577	30.584	17.332	9.919	171.923	23
24	1.809	0.5529	0.0309	0.0559	32.349	17.885	10.324	184.639	24
25	1.854	0.5394	0.0293	0.0543	34.158	18.424	10.724	197.584	25
26	1.900	0.5262	0.0278	0.0528	36.012	18.951	11.121	210.740	26
27	1.948	0.5134	0.0264	0.0514	37.912	19.464	11.513	224.089	27
28	1.996	0.5009	0.0251	0.0501	39.860	19.965	11.902	237.612	28
29	2.046	0.4887	0.0239	0.0489	41.856	20.454	12.286	251.295	29
30	2.098	0.4767	0.0228	0.0478	43.903	20.930	12.667	265.120	30
31	2.150	0.4651	0.0217	0.0467	46.000	21.395	13.044	279.074	31
32	2.204	0.4538	0.0208	0.0458	48.150	21.849	13.417	293.141	32
33	2.259	0.4427	0.0199	0.0449	50.354	22.292	13.786	307.307	33
34	2.315	0.4319	0.0190	0.0440	52.613	22.724	14.151	321.560	34
35	2.373	0.4214	0.0182	0.0432	54.928	23.145	14.512	335.887	35
36	2.433	0.4111	0.0175	0.0425	57.301	23.556	14.870	350.275	36
37	2.493	0.4011	0.0167	0.0417	59.734	23.957	15.223	364.713	37
38	2.556	0.3913	0.0161	0.0411	62.227	24.349	15.573	379.191	38
39	2.620	0.3817	0.0154	0.0404	64.783	24.730	15.920	393.697	39
40	2.685	0.3724	0.0148	0.0398	67.403	25.103	16.262	408.222	40
41	2.752	0.3633	0.0143	0.0393	70.088	25.466	16.601	422.756	41
42	2.821	0.3545	0.0137	0.0387	72.840	25.821	16.936	437.290	42
43	2.892	0.3458	0.0132	0.0382	75.661	26.166	17.267	451.815	43
44	2.964	0.3374	0.0127	0.0377	78.552	26.504	17.595	466.323	44
45	3.038	0.3292	0.0123	0.0373	81.516	26.833	17.918	480.807	45
46	3.114	0.3211	0.0118	0.0368	84.554	27.154	18.239	495.259	46
47	3.192	0.3133	0.0114	0.0364	87.668	27.467	18.555	509.671	47
48	3.271	0.3057	0.0110	0.0360	90.860	27.773	18.868	524.038	48
49	3.353	0.2982	0.0106	0.0356	94.131	28.071	19.178	538.352	49
50	3.437	0.2909	0.0103	0.0353	97.484	28.362	19.484	552.608	50

3%					Compound Interest Factors				3%
Period	Single Payment		Uniform Payment Series				Arithmetic Gradient		Period
	Compound Amount Factor	Present Value Factor	Sinking Fund Factor	Capital Recovery Factor	Compound Amount Factor	Present Value Factor	Gradient Uniform Series	Gradient Present Value	
	Find F Given P F/P	Find P Given F P/F	Find A Given F A/F	Find A Given P A/P	Find F Given A F/A	Find P Given A P/A	Find A Given G A/G	Find P Given G P/G	
n									n
1	1.030	0.9709	1.0000	1.0300	1.000	0.971	0.000	0.000	1
2	1.061	0.9426	0.4926	0.5226	2.030	1.913	0.493	0.943	2
3	1.093	0.9151	0.3235	0.3535	3.091	2.829	0.980	2.773	3
4	1.126	0.8885	0.2390	0.2690	4.184	3.717	1.463	5.438	4
5	1.159	0.8626	0.1884	0.2184	5.309	4.580	1.941	8.889	5
6	1.194	0.8375	0.1546	0.1846	6.468	5.417	2.414	13.076	6
7	1.230	0.8131	0.1305	0.1605	7.662	6.230	2.882	17.955	7
8	1.267	0.7894	0.1125	0.1425	8.892	7.020	3.345	23.481	8
9	1.305	0.7664	0.0984	0.1284	10.159	7.786	3.803	29.612	9
10	1.344	0.7441	0.0872	0.1172	11.464	8.530	4.256	36.309	10
11	1.384	0.7224	0.0781	0.1081	12.808	9.253	4.705	43.533	11
12	1.426	0.7014	0.0705	0.1005	14.192	9.954	5.148	51.248	12
13	1.469	0.6810	0.0640	0.0940	15.618	10.635	5.587	59.420	13
14	1.513	0.6611	0.0585	0.0885	17.086	11.296	6.021	68.014	14
15	1.558	0.6419	0.0538	0.0838	18.599	11.938	6.450	77.000	15
16	1.605	0.6232	0.0496	0.0796	20.157	12.561	6.874	86.348	16
17	1.653	0.6050	0.0460	0.0760	21.762	13.166	7.294	96.028	17
18	1.702	0.5874	0.0427	0.0727	23.414	13.754	7.708	106.014	18
19	1.754	0.5703	0.0398	0.0698	25.117	14.324	8.118	116.279	19
20	1.806	0.5537	0.0372	0.0672	26.870	14.877	8.523	126.799	20
21	1.860	0.5375	0.0349	0.0649	28.676	15.415	8.923	137.550	21
22	1.916	0.5219	0.0327	0.0627	30.537	15.937	9.319	148.509	22
23	1.974	0.5067	0.0308	0.0608	32.453	16.444	9.709	159.657	23
24	2.033	0.4919	0.0290	0.0590	34.426	16.936	10.095	170.971	24
25	2.094	0.4776	0.0274	0.0574	36.459	17.413	10.477	182.434	25
26	2.157	0.4637	0.0259	0.0559	38.553	17.877	10.853	194.026	26
27	2.221	0.4502	0.0246	0.0546	40.710	18.327	11.226	205.731	27
28	2.288	0.4371	0.0233	0.0533	42.931	18.764	11.593	217.532	28
29	2.357	0.4243	0.0221	0.0521	45.219	19.188	11.956	229.414	29
30	2.427	0.4120	0.0210	0.0510	47.575	19.600	12.314	241.361	30
31	2.500	0.4000	0.0200	0.0500	50.003	20.000	12.668	253.361	31
32	2.575	0.3883	0.0190	0.0490	52.503	20.389	13.017	265.399	32
33	2.652	0.3770	0.0182	0.0482	55.078	20.766	13.362	277.464	33
34	2.732	0.3660	0.0173	0.0473	57.730	21.132	13.702	289.544	34
35	2.814	0.3554	0.0165	0.0465	60.462	21.487	14.037	301.627	35
36	2.898	0.3450	0.0158	0.0458	63.276	21.832	14.369	313.703	36
37	2.985	0.3350	0.0151	0.0451	66.174	22.167	14.696	325.762	37
38	3.075	0.3252	0.0145	0.0445	69.159	22.492	15.018	337.796	38
39	3.167	0.3158	0.0138	0.0438	72.234	22.808	15.336	349.794	39
40	3.262	0.3066	0.0133	0.0433	75.401	23.115	15.650	361.750	40
41	3.360	0.2976	0.0127	0.0427	78.663	23.412	15.960	373.655	41
42	3.461	0.2890	0.0122	0.0422	82.023	23.701	16.265	385.502	42
43	3.565	0.2805	0.0117	0.0417	85.484	23.982	16.566	397.285	43
44	3.671	0.2724	0.0112	0.0412	89.048	24.254	16.863	408.997	44
45	3.782	0.2644	0.0108	0.0408	92.720	24.519	17.156	420.632	45
46	3.895	0.2567	0.0104	0.0404	96.501	24.775	17.444	432.186	46
47	4.012	0.2493	0.0100	0.0400	100.397	25.025	17.729	443.652	47
48	4.132	0.2420	0.0096	0.0396	104.408	25.267	18.009	455.025	48
49	4.256	0.2350	0.0092	0.0392	108.541	25.502	18.285	466.303	49
50	4.384	0.2281	0.0089	0.0389	112.797	25.730	18.558	477.480	50

3.5%			Compound Interest Factors						3.5%
Period	Single Payment		Uniform Payment Series				Arithmetic Gradient		Period
	Compound Amount Factor	Present Value Factor	Sinking Fund Factor	Capital Recovery Factor	Compound Amount Factor	Present Value Factor	Gradient Uniform Series	Gradient Present Value	
n	Find *F* Given *P* F/P	Find *P* Given *F* P/F	Find *A* Given *F* A/F	Find *A* Given *P* A/P	Find *F* Given *A* F/A	Find *P* Given *A* P/A	Find *A* Given *G* A/G	Find *P* Given *G* P/G	*n*
1	1.035	0.9662	1.0000	1.0350	1.000	0.966	0.000	0.000	1
2	1.071	0.9335	0.4914	0.5264	2.035	1.900	0.491	0.934	2
3	1.109	0.9019	0.3219	0.3569	3.106	2.802	0.977	2.737	3
4	1.148	0.8714	0.2373	0.2723	4.215	3.673	1.457	5.352	4
5	1.188	0.8420	0.1865	0.2215	5.362	4.515	1.931	8.720	5
6	1.229	0.8135	0.1527	0.1877	6.550	5.329	2.400	12.787	6
7	1.272	0.7860	0.1285	0.1635	7.779	6.115	2.863	17.503	7
8	1.317	0.7594	0.1105	0.1455	9.052	6.874	3.320	22.819	8
9	1.363	0.7337	0.0964	0.1314	10.368	7.608	3.771	28.689	9
10	1.411	0.7089	0.0852	0.1202	11.731	8.317	4.217	35.069	10
11	1.460	0.6849	0.0761	0.1111	13.142	9.002	4.657	41.919	11
12	1.511	0.6618	0.0685	0.1035	14.602	9.663	5.091	49.198	12
13	1.564	0.6394	0.0621	0.0971	16.113	10.303	5.520	56.871	13
14	1.619	0.6178	0.0566	0.0916	17.677	10.921	5.943	64.902	14
15	1.675	0.5969	0.0518	0.0868	19.296	11.517	6.361	73.259	15
16	1.734	0.5767	0.0477	0.0827	20.971	12.094	6.773	81.909	16
17	1.795	0.5572	0.0440	0.0790	22.705	12.651	7.179	90.824	17
18	1.857	0.5384	0.0408	0.0758	24.500	13.190	7.580	99.977	18
19	1.923	0.5202	0.0379	0.0729	26.357	13.710	7.975	109.339	19
20	1.990	0.5026	0.0354	0.0704	28.280	14.212	8.365	118.888	20
21	2.059	0.4856	0.0330	0.0680	30.269	14.698	8.749	128.600	21
22	2.132	0.4692	0.0309	0.0659	32.329	15.167	9.128	138.452	22
23	2.206	0.4533	0.0290	0.0640	34.460	15.620	9.502	148.424	23
24	2.283	0.4380	0.0273	0.0623	36.667	16.058	9.870	158.497	24
25	2.363	0.4231	0.0257	0.0607	38.950	16.482	10.233	168.653	25
26	2.446	0.4088	0.0242	0.0592	41.313	16.890	10.590	178.874	26
27	2.532	0.3950	0.0229	0.0579	43.759	17.285	10.942	189.144	27
28	2.620	0.3817	0.0216	0.0566	46.291	17.667	11.289	199.448	28
29	2.712	0.3687	0.0204	0.0554	48.911	18.036	11.631	209.773	29
30	2.807	0.3563	0.0194	0.0544	51.623	18.392	11.967	220.106	30
31	2.905	0.3442	0.0184	0.0534	54.429	18.736	12.299	230.432	31
32	3.007	0.3326	0.0174	0.0524	57.335	19.069	12.625	240.743	32
33	3.112	0.3213	0.0166	0.0516	60.341	19.390	12.946	251.026	33
34	3.221	0.3105	0.0158	0.0508	63.453	19.701	13.262	261.271	34
35	3.334	0.3000	0.0150	0.0500	66.674	20.001	13.573	271.471	35
36	3.450	0.2898	0.0143	0.0493	70.008	20.290	13.879	281.615	36
37	3.571	0.2800	0.0136	0.0486	73.458	20.571	14.180	291.696	37
38	3.696	0.2706	0.0130	0.0480	77.029	20.841	14.477	301.707	38
39	3.825	0.2614	0.0124	0.0474	80.725	21.102	14.768	311.640	39
40	3.959	0.2526	0.0118	0.0468	84.550	21.355	15.055	321.491	40
41	4.098	0.2440	0.0113	0.0463	88.510	21.599	15.336	331.252	41
42	4.241	0.2358	0.0108	0.0458	92.607	21.835	15.613	340.919	42
43	4.390	0.2278	0.0103	0.0453	96.849	22.063	15.886	350.487	43
44	4.543	0.2201	0.0099	0.0449	101.238	22.283	16.154	359.951	44
45	4.702	0.2127	0.0095	0.0445	105.782	22.495	16.417	369.308	45
46	4.867	0.2055	0.0091	0.0441	110.484	22.701	16.676	378.554	46
47	5.037	0.1985	0.0087	0.0437	115.351	22.899	16.930	387.686	47
48	5.214	0.1918	0.0083	0.0433	120.388	23.091	17.180	396.701	48
49	5.396	0.1853	0.0080	0.0430	125.602	23.277	17.425	405.596	49
50	5.585	0.1791	0.0076	0.0426	130.998	23.456	17.666	414.370	50

4%				Compound Interest Factors					4%
Period	Single Payment		Uniform Payment Series				Arithmetic Gradient		Period
	Compound Amount Factor	Present Value Factor	Sinking Fund Factor	Capital Recovery Factor	Compound Amount Factor	Present Value Factor	Gradient Uniform Series	Gradient Present Value	
	Find F Given P	Find P Given F	Find A Given F	Find A Given P	Find F Given A	Find P Given A	Find A Given G	Find P Given G	
n	F/P	P/F	A/F	A/P	F/A	P/A	A/G	P/G	n
1	1.040	0.9615	1.0000	1.0400	1.000	0.962	0.000	0.000	1
2	1.082	0.9246	0.4902	0.5302	2.040	1.886	0.490	0.925	2
3	1.125	0.8890	0.3203	0.3603	3.122	2.775	0.974	2.703	3
4	1.170	0.8548	0.2355	0.2755	4.246	3.630	1.451	5.267	4
5	1.217	0.8219	0.1846	0.2246	5.416	4.452	1.922	8.555	5
6	1.265	0.7903	0.1508	0.1908	6.633	5.242	2.386	12.506	6
7	1.316	0.7599	0.1266	0.1666	7.898	6.002	2.843	17.066	7
8	1.369	0.7307	0.1085	0.1485	9.214	6.733	3.294	22.181	8
9	1.423	0.7026	0.0945	0.1345	10.583	7.435	3.739	27.801	9
10	1.480	0.6756	0.0833	0.1233	12.006	8.111	4.177	33.881	10
11	1.539	0.6496	0.0741	0.1141	13.486	8.760	4.609	40.377	11
12	1.601	0.6246	0.0666	0.1066	15.026	9.385	5.034	47.248	12
13	1.665	0.6006	0.0601	0.1001	16.627	9.986	5.453	54.455	13
14	1.732	0.5775	0.0547	0.0947	18.292	10.563	5.866	61.962	14
15	1.801	0.5553	0.0499	0.0899	20.024	11.118	6.272	69.735	15
16	1.873	0.5339	0.0458	0.0858	21.825	11.652	6.672	77.744	16
17	1.948	0.5134	0.0422	0.0822	23.698	12.166	7.066	85.958	17
18	2.026	0.4936	0.0390	0.0790	25.645	12.659	7.453	94.350	18
19	2.107	0.4746	0.0361	0.0761	27.671	13.134	7.834	102.893	19
20	2.191	0.4564	0.0336	0.0736	29.778	13.590	8.209	111.565	20
21	2.279	0.4388	0.0313	0.0713	31.969	14.029	8.578	120.341	21
22	2.370	0.4220	0.0292	0.0692	34.248	14.451	8.941	129.202	22
23	2.465	0.4057	0.0273	0.0673	36.618	14.857	9.297	138.128	23
24	2.563	0.3901	0.0256	0.0656	39.083	15.247	9.648	147.101	24
25	2.666	0.3751	0.0240	0.0640	41.646	15.622	9.993	156.104	25
26	2.772	0.3607	0.0226	0.0626	44.312	15.983	10.331	165.121	26
27	2.883	0.3468	0.0212	0.0612	47.084	16.330	10.664	174.138	27
28	2.999	0.3335	0.0200	0.0600	49.968	16.663	10.991	183.142	28
29	3.119	0.3207	0.0189	0.0589	52.966	16.984	11.312	192.121	29
30	3.243	0.3083	0.0178	0.0578	56.085	17.292	11.627	201.062	30
31	3.373	0.2965	0.0169	0.0569	59.328	17.588	11.937	209.956	31
32	3.508	0.2851	0.0159	0.0559	62.701	17.874	12.241	218.792	32
33	3.648	0.2741	0.0151	0.0551	66.210	18.148	12.540	227.563	33
34	3.794	0.2636	0.0143	0.0543	69.858	18.411	12.832	236.261	34
35	3.946	0.2534	0.0136	0.0536	73.652	18.665	13.120	244.877	35
36	4.104	0.2437	0.0129	0.0529	77.598	18.908	13.402	253.405	36
37	4.268	0.2343	0.0122	0.0522	81.702	19.143	13.678	261.840	37
38	4.439	0.2253	0.0116	0.0516	85.970	19.368	13.950	270.175	38
39	4.616	0.2166	0.0111	0.0511	90.409	19.584	14.216	278.407	39
40	4.801	0.2083	0.0105	0.0505	95.026	19.793	14.477	286.530	40
41	4.993	0.2003	0.0100	0.0500	99.827	19.993	14.732	294.541	41
42	5.193	0.1926	0.0095	0.0495	104.820	20.186	14.983	302.437	42
43	5.400	0.1852	0.0091	0.0491	110.012	20.371	15.228	310.214	43
44	5.617	0.1780	0.0087	0.0487	115.413	20.549	15.469	317.870	44
45	5.841	0.1712	0.0083	0.0483	121.029	20.720	15.705	325.403	45
46	6.075	0.1646	0.0079	0.0479	126.871	20.885	15.936	332.810	46
47	6.318	0.1583	0.0075	0.0475	132.945	21.043	16.162	340.091	47
48	6.571	0.1522	0.0072	0.0472	139.263	21.195	16.383	347.245	48
49	6.833	0.1463	0.0069	0.0469	145.834	21.341	16.600	354.269	49
50	7.107	0.1407	0.0066	0.0466	152.667	21.482	16.812	361.164	50

4.5%		Compound Interest Factors								4.5%
Period	Single Payment		Uniform Payment Series				Arithmetic Gradient		Period	
	Compound Amount Factor	Present Value Factor	Sinking Fund Factor	Capital Recovery Factor	Compound Amount Factor	Present Value Factor	Gradient Uniform Series	Gradient Present Value		
	Find F Given P F/P	Find P Given F P/F	Find A Given F A/F	Find A Given P A/P	Find F Given A F/A	Find P Given A P/A	Find A Given G A/G	Find P Given G P/G		
n									n	
1	1.045	0.9569	1.0000	1.0450	1.000	0.957	0.000	0.000	1	
2	1.092	0.9157	0.4890	0.5340	2.045	1.873	0.489	0.916	2	
3	1.141	0.8763	0.3188	0.3638	3.137	2.749	0.971	2.668	3	
4	1.193	0.8386	0.2337	0.2787	4.278	3.588	1.445	5.184	4	
5	1.246	0.8025	0.1828	0.2278	5.471	4.390	1.912	8.394	5	
6	1.302	0.7679	0.1489	0.1939	6.717	5.158	2.372	12.233	6	
7	1.361	0.7348	0.1247	0.1697	8.019	5.893	2.824	16.642	7	
8	1.422	0.7032	0.1066	0.1516	9.380	6.596	3.269	21.565	8	
9	1.486	0.6729	0.0926	0.1376	10.802	7.269	3.707	26.948	9	
10	1.553	0.6439	0.0814	0.1264	12.288	7.913	4.138	32.743	10	
11	1.623	0.6162	0.0722	0.1172	13.841	8.529	4.562	38.905	11	
12	1.696	0.5897	0.0647	0.1097	15.464	9.119	4.978	45.391	12	
13	1.772	0.5643	0.0583	0.1033	17.160	9.683	5.387	52.163	13	
14	1.852	0.5400	0.0528	0.0978	18.932	10.223	5.789	59.182	14	
15	1.935	0.5167	0.0481	0.0931	20.784	10.740	6.184	66.416	15	
16	2.022	0.4945	0.0440	0.0890	22.719	11.234	6.572	73.833	16	
17	2.113	0.4732	0.0404	0.0854	24.742	11.707	6.953	81.404	17	
18	2.208	0.4528	0.0372	0.0822	26.855	12.160	7.327	89.102	18	
19	2.308	0.4333	0.0344	0.0794	29.064	12.593	7.695	96.901	19	
20	2.412	0.4146	0.0319	0.0769	31.371	13.008	8.055	104.780	20	
21	2.520	0.3968	0.0296	0.0746	33.783	13.405	8.409	112.715	21	
22	2.634	0.3797	0.0275	0.0725	36.303	13.784	8.755	120.689	22	
23	2.752	0.3634	0.0257	0.0707	38.937	14.148	9.096	128.683	23	
24	2.876	0.3477	0.0240	0.0690	41.689	14.495	9.429	136.680	24	
25	3.005	0.3327	0.0224	0.0674	44.565	14.828	9.756	144.665	25	
26	3.141	0.3184	0.0210	0.0660	47.571	15.147	10.077	152.625	26	
27	3.282	0.3047	0.0197	0.0647	50.711	15.451	10.391	160.547	27	
28	3.430	0.2916	0.0185	0.0635	53.993	15.743	10.698	168.420	28	
29	3.584	0.2790	0.0174	0.0624	57.423	16.022	10.999	176.232	29	
30	3.745	0.2670	0.0164	0.0614	61.007	16.289	11.295	183.975	30	
31	3.914	0.2555	0.0154	0.0604	64.752	16.544	11.583	191.640	31	
32	4.090	0.2445	0.0146	0.0596	68.666	16.789	11.866	199.220	32	
33	4.274	0.2340	0.0137	0.0587	72.756	17.023	12.143	206.707	33	
34	4.466	0.2239	0.0130	0.0580	77.030	17.247	12.414	214.096	34	
35	4.667	0.2143	0.0123	0.0573	81.497	17.461	12.679	221.380	35	
36	4.877	0.2050	0.0116	0.0566	86.164	17.666	12.938	228.556	36	
37	5.097	0.1962	0.0110	0.0560	91.041	17.862	13.191	235.619	37	
38	5.326	0.1878	0.0104	0.0554	96.138	18.050	13.439	242.566	38	
39	5.566	0.1797	0.0099	0.0549	101.464	18.230	13.681	249.393	39	
40	5.816	0.1719	0.0093	0.0543	107.030	18.402	13.917	256.099	40	
41	6.078	0.1645	0.0089	0.0539	112.847	18.566	14.148	262.680	41	
42	6.352	0.1574	0.0084	0.0534	118.925	18.724	14.374	269.135	42	
43	6.637	0.1507	0.0080	0.0530	125.276	18.874	14.595	275.462	43	
44	6.936	0.1442	0.0076	0.0526	131.914	19.018	14.810	281.662	44	
45	7.248	0.1380	0.0072	0.0522	138.850	19.156	15.020	287.732	45	
46	7.574	0.1320	0.0068	0.0518	146.098	19.288	15.225	293.673	46	
47	7.915	0.1263	0.0065	0.0515	153.673	19.415	15.426	299.485	47	
48	8.271	0.1209	0.0062	0.0512	161.588	19.536	15.621	305.167	48	
49	8.644	0.1157	0.0059	0.0509	169.859	19.651	15.812	310.720	49	
50	9.033	0.1107	0.0056	0.0506	178.503	19.762	15.998	316.145	50	

5%	Compound Interest Factors								5%
Period	Single Payment		Uniform Payment Series				Arithmetic Gradient		Period
	Compound Amount Factor	Present Value Factor	Sinking Fund Factor	Capital Recovery Factor	Compound Amount Factor	Present Value Factor	Gradient Uniform Series	Gradient Present Value	
n	Find *F* Given *P* *F/P*	Find *P* Given *F* *P/F*	Find *A* Given *F* *A/F*	Find *A* Given *P* *A/P*	Find *F* Given *A* *F/A*	Find *P* Given *A* *P/A*	Find *A* Given *G* *A/G*	Find *P* Given *G* *P/G*	*n*
1	1.050	0.9524	1.0000	1.0500	1.000	0.952	0.000	0.000	1
2	1.103	0.9070	0.4878	0.5378	2.050	1.859	0.488	0.907	2
3	1.158	0.8638	0.3172	0.3672	3.153	2.723	0.967	2.635	3
4	1.216	0.8227	0.2320	0.2820	4.310	3.546	1.439	5.103	4
5	1.276	0.7835	0.1810	0.2310	5.526	4.329	1.903	8.237	5
6	1.340	0.7462	0.1470	0.1970	6.802	5.076	2.358	11.968	6
7	1.407	0.7107	0.1228	0.1728	8.142	5.786	2.805	16.232	7
8	1.477	0.6768	0.1047	0.1547	9.549	6.463	3.245	20.970	8
9	1.551	0.6446	0.0907	0.1407	11.027	7.108	3.676	26.127	9
10	1.629	0.6139	0.0795	0.1295	12.578	7.722	4.099	31.652	10
11	1.710	0.5847	0.0704	0.1204	14.207	8.306	4.514	37.499	11
12	1.796	0.5568	0.0628	0.1128	15.917	8.863	4.922	43.624	12
13	1.886	0.5303	0.0565	0.1065	17.713	9.394	5.322	49.988	13
14	1.980	0.5051	0.0510	0.1010	19.599	9.899	5.713	56.554	14
15	2.079	0.4810	0.0463	0.0963	21.579	10.380	6.097	63.288	15
16	2.183	0.4581	0.0423	0.0923	23.657	10.838	6.474	70.160	16
17	2.292	0.4363	0.0387	0.0887	25.840	11.274	6.842	77.140	17
18	2.407	0.4155	0.0355	0.0855	28.132	11.690	7.203	84.204	18
19	2.527	0.3957	0.0327	0.0827	30.539	12.085	7.557	91.328	19
20	2.653	0.3769	0.0302	0.0802	33.066	12.462	7.903	98.488	20
21	2.786	0.3589	0.0280	0.0780	35.719	12.821	8.242	105.667	21
22	2.925	0.3418	0.0260	0.0760	38.505	13.163	8.573	112.846	22
23	3.072	0.3256	0.0241	0.0741	41.430	13.489	8.897	120.009	23
24	3.225	0.3101	0.0225	0.0725	44.502	13.799	9.214	127.140	24
25	3.386	0.2953	0.0210	0.0710	47.727	14.094	9.524	134.228	25
26	3.556	0.2812	0.0196	0.0696	51.113	14.375	9.827	141.259	26
27	3.733	0.2678	0.0183	0.0683	54.669	14.643	10.122	148.223	27
28	3.920	0.2551	0.0171	0.0671	58.403	14.898	10.411	155.110	28
29	4.116	0.2429	0.0160	0.0660	62.323	15.141	10.694	161.913	29
30	4.322	0.2314	0.0151	0.0651	66.439	15.372	10.969	168.623	30
31	4.538	0.2204	0.0141	0.0641	70.761	15.593	11.238	175.233	31
32	4.765	0.2099	0.0133	0.0633	75.299	15.803	11.501	181.739	32
33	5.003	0.1999	0.0125	0.0625	80.064	16.003	11.757	188.135	33
34	5.253	0.1904	0.0118	0.0618	85.067	16.193	12.006	194.417	34
35	5.516	0.1813	0.0111	0.0611	90.320	16.374	12.250	200.581	35
36	5.792	0.1727	0.0104	0.0604	95.836	16.547	12.487	206.624	36
37	6.081	0.1644	0.0098	0.0598	101.628	16.711	12.719	212.543	37
38	6.385	0.1566	0.0093	0.0593	107.710	16.868	12.944	218.338	38
39	6.705	0.1491	0.0088	0.0588	114.095	17.017	13.164	224.005	39
40	7.040	0.1420	0.0083	0.0583	120.800	17.159	13.377	229.545	40
41	7.392	0.1353	0.0078	0.0578	127.840	17.294	13.586	234.956	41
42	7.762	0.1288	0.0074	0.0574	135.232	17.423	13.788	240.239	42
43	8.150	0.1227	0.0070	0.0570	142.993	17.546	13.986	245.392	43
44	8.557	0.1169	0.0066	0.0566	151.143	17.663	14.178	250.417	44
45	8.985	0.1113	0.0063	0.0563	159.700	17.774	14.364	255.315	45
46	9.434	0.1060	0.0059	0.0559	168.685	17.880	14.546	260.084	46
47	9.906	0.1009	0.0056	0.0556	178.119	17.981	14.723	264.728	47
48	10.401	0.0961	0.0053	0.0553	188.025	18.077	14.894	269.247	48
49	10.921	0.0916	0.0050	0.0550	198.427	18.169	15.061	273.642	49
50	11.467	0.0872	0.0048	0.0548	209.348	18.256	15.223	277.915	50

6%				Compound Interest Factors					6%
Period	Single Payment		Uniform Payment Series				Arithmetic Gradient		Period
	Compound Amount Factor	Present Value Factor	Sinking Fund Factor	Capital Recovery Factor	Compound Amount Factor	Present Value Factor	Gradient Uniform Series	Gradient Present Value	
	Find *F* Given *P* *F/P*	Find *P* Given *F* *P/F*	Find *A* Given *F* *A/F*	Find *A* Given *P* *A/P*	Find *F* Given *A* *F/A*	Find *P* Given *A* *P/A*	Find *A* Given *G* *A/G*	Find *P* Given *G* *P/G*	
n									*n*
1	1.060	0.9434	1.0000	1.0600	1.000	0.943	0.000	0.000	1
2	1.124	0.8900	0.4854	0.5454	2.060	1.833	0.485	0.890	2
3	1.191	0.8396	0.3141	0.3741	3.184	2.673	0.961	2.569	3
4	1.262	0.7921	0.2286	0.2886	4.375	3.465	1.427	4.946	4
5	1.338	0.7473	0.1774	0.2374	5.637	4.212	1.884	7.935	5
6	1.419	0.7050	0.1434	0.2034	6.975	4.917	2.330	11.459	6
7	1.504	0.6651	0.1191	0.1791	8.394	5.582	2.768	15.450	7
8	1.594	0.6274	0.1010	0.1610	9.897	6.210	3.195	19.842	8
9	1.689	0.5919	0.0870	0.1470	11.491	6.802	3.613	24.577	9
10	1.791	0.5584	0.0759	0.1359	13.181	7.360	4.022	29.602	10
11	1.898	0.5268	0.0668	0.1268	14.972	7.887	4.421	34.870	11
12	2.012	0.4970	0.0593	0.1193	16.870	8.384	4.811	40.337	12
13	2.133	0.4688	0.0530	0.1130	18.882	8.853	5.192	45.963	13
14	2.261	0.4423	0.0476	0.1076	21.015	9.295	5.564	51.713	14
15	2.397	0.4173	0.0430	0.1030	23.276	9.712	5.926	57.555	15
16	2.540	0.3936	0.0390	0.0990	25.673	10.106	6.279	63.459	16
17	2.693	0.3714	0.0354	0.0954	28.213	10.477	6.624	69.401	17
18	2.854	0.3503	0.0324	0.0924	30.906	10.828	6.960	75.357	18
19	3.026	0.3305	0.0296	0.0896	33.760	11.158	7.287	81.306	19
20	3.207	0.3118	0.0272	0.0872	36.786	11.470	7.605	87.230	20
21	3.400	0.2942	0.0250	0.0850	39.993	11.764	7.915	93.114	21
22	3.604	0.2775	0.0230	0.0830	43.392	12.042	8.217	98.941	22
23	3.820	0.2618	0.0213	0.0813	46.996	12.303	8.510	104.701	23
24	4.049	0.2470	0.0197	0.0797	50.816	12.550	8.795	110.381	24
25	4.292	0.2330	0.0182	0.0782	54.865	12.783	9.072	115.973	25
26	4.549	0.2198	0.0169	0.0769	59.156	13.003	9.341	121.468	26
27	4.822	0.2074	0.0157	0.0757	63.706	13.211	9.603	126.860	27
28	5.112	0.1956	0.0146	0.0746	68.528	13.406	9.857	132.142	28
29	5.418	0.1846	0.0136	0.0736	73.640	13.591	10.103	137.310	29
30	5.743	0.1741	0.0126	0.0726	79.058	13.765	10.342	142.359	30
31	6.088	0.1643	0.0118	0.0718	84.802	13.929	10.574	147.286	31
32	6.453	0.1550	0.0110	0.0710	90.890	14.084	10.799	152.090	32
33	6.841	0.1462	0.0103	0.0703	97.343	14.230	11.017	156.768	33
34	7.251	0.1379	0.0096	0.0696	104.184	14.368	11.228	161.319	34
35	7.686	0.1301	0.0090	0.0690	111.435	14.498	11.432	165.743	35
36	8.147	0.1227	0.0084	0.0684	119.121	14.621	11.630	170.039	36
37	8.636	0.1158	0.0079	0.0679	127.268	14.737	11.821	174.207	37
38	9.154	0.1092	0.0074	0.0674	135.904	14.846	12.007	178.249	38
39	9.704	0.1031	0.0069	0.0669	145.058	14.949	12.186	182.165	39
40	10.286	0.0972	0.0065	0.0665	154.762	15.046	12.359	185.957	40
41	10.903	0.0917	0.0061	0.0661	165.048	15.138	12.526	189.626	41
42	11.557	0.0865	0.0057	0.0657	175.951	15.225	12.688	193.173	42
43	12.250	0.0816	0.0053	0.0653	187.508	15.306	12.845	196.602	43
44	12.985	0.0770	0.0050	0.0650	199.758	15.383	12.996	199.913	44
45	13.765	0.0727	0.0047	0.0647	212.744	15.456	13.141	203.110	45
46	14.590	0.0685	0.0044	0.0644	226.508	15.524	13.282	206.194	46
47	15.466	0.0647	0.0041	0.0641	241.099	15.589	13.418	209.168	47
48	16.394	0.0610	0.0039	0.0639	256.565	15.650	13.549	212.035	48
49	17.378	0.0575	0.0037	0.0637	272.958	15.708	13.675	214.797	49
50	18.420	0.0543	0.0034	0.0634	290.336	15.762	13.796	217.457	50

7%				Compound Interest Factors					7%
Period	Single Payment		Uniform Payment Series				Arithmetic Gradient		Period
	Compound Amount Factor	Present Value Factor	Sinking Fund Factor	Capital Recovery Factor	Compound Amount Factor	Present Value Factor	Gradient Uniform Series	Gradient Present Value	
	Find F Given P F/P	Find P Given F P/F	Find A Given F A/F	Find A Given P A/P	Find F Given A F/A	Find P Given A P/A	Find A Given G A/G	Find P Given G P/G	
n									n
1	1.070	0.9346	1.0000	1.0700	1.000	0.935	0.000	0.000	1
2	1.145	0.8734	0.4831	0.5531	2.070	1.808	0.483	0.873	2
3	1.225	0.8163	0.3111	0.3811	3.215	2.624	0.955	2.506	3
4	1.311	0.7629	0.2252	0.2952	4.440	3.387	1.416	4.795	4
5	1.403	0.7130	0.1739	0.2439	5.751	4.100	1.865	7.647	5
6	1.501	0.6663	0.1398	0.2098	7.153	4.767	2.303	10.978	6
7	1.606	0.6227	0.1156	0.1856	8.654	5.389	2.730	14.715	7
8	1.718	0.5820	0.0975	0.1675	10.260	5.971	3.147	18.789	8
9	1.838	0.5439	0.0835	0.1535	11.978	6.515	3.552	23.140	9
10	1.967	0.5083	0.0724	0.1424	13.816	7.024	3.946	27.716	10
11	2.105	0.4751	0.0634	0.1334	15.784	7.499	4.330	32.466	11
12	2.252	0.4440	0.0559	0.1259	17.888	7.943	4.703	37.351	12
13	2.410	0.4150	0.0497	0.1197	20.141	8.358	5.065	42.330	13
14	2.579	0.3878	0.0443	0.1143	22.550	8.745	5.417	47.372	14
15	2.759	0.3624	0.0398	0.1098	25.129	9.108	5.758	52.446	15
16	2.952	0.3387	0.0359	0.1059	27.888	9.447	6.090	57.527	16
17	3.159	0.3166	0.0324	0.1024	30.840	9.763	6.411	62.592	17
18	3.380	0.2959	0.0294	0.0994	33.999	10.059	6.722	67.622	18
19	3.617	0.2765	0.0268	0.0968	37.379	10.336	7.024	72.599	19
20	3.870	0.2584	0.0244	0.0944	40.995	10.594	7.316	77.509	20
21	4.141	0.2415	0.0223	0.0923	44.865	10.836	7.599	82.339	21
22	4.430	0.2257	0.0204	0.0904	49.006	11.061	7.872	87.079	22
23	4.741	0.2109	0.0187	0.0887	53.436	11.272	8.137	91.720	23
24	5.072	0.1971	0.0172	0.0872	58.177	11.469	8.392	96.255	24
25	5.427	0.1842	0.0158	0.0858	63.249	11.654	8.639	100.676	25
26	5.807	0.1722	0.0146	0.0846	68.676	11.826	8.877	104.981	26
27	6.214	0.1609	0.0134	0.0834	74.484	11.987	9.107	109.166	27
28	6.649	0.1504	0.0124	0.0824	80.698	12.137	9.329	113.226	28
29	7.114	0.1406	0.0114	0.0814	87.347	12.278	9.543	117.162	29
30	7.612	0.1314	0.0106	0.0806	94.461	12.409	9.749	120.972	30
31	8.145	0.1228	0.0098	0.0798	102.073	12.532	9.947	124.655	31
32	8.715	0.1147	0.0091	0.0791	110.218	12.647	10.138	128.212	32
33	9.325	0.1072	0.0084	0.0784	118.933	12.754	10.322	131.643	33
34	9.978	0.1002	0.0078	0.0778	128.259	12.854	10.499	134.951	34
35	10.677	0.0937	0.0072	0.0772	138.237	12.948	10.669	138.135	35
36	11.424	0.0875	0.0067	0.0767	148.913	13.035	10.832	141.199	36
37	12.224	0.0818	0.0062	0.0762	160.337	13.117	10.989	144.144	37
38	13.079	0.0765	0.0058	0.0758	172.561	13.193	11.140	146.973	38
39	13.995	0.0715	0.0054	0.0754	185.640	13.265	11.285	149.688	39
40	14.974	0.0668	0.0050	0.0750	199.635	13.332	11.423	152.293	40
41	16.023	0.0624	0.0047	0.0747	214.610	13.394	11.557	154.789	41
42	17.144	0.0583	0.0043	0.0743	230.632	13.452	11.684	157.181	42
43	18.344	0.0545	0.0040	0.0740	247.776	13.507	11.807	159.470	43
44	19.628	0.0509	0.0038	0.0738	266.121	13.558	11.924	161.661	44
45	21.002	0.0476	0.0035	0.0735	285.749	13.606	12.036	163.756	45
46	22.473	0.0445	0.0033	0.0733	306.752	13.650	12.143	165.758	46
47	24.046	0.0416	0.0030	0.0730	329.224	13.692	12.246	167.671	47
48	25.729	0.0389	0.0028	0.0728	353.270	13.730	12.345	169.498	48
49	27.530	0.0363	0.0026	0.0726	378.999	13.767	12.439	171.242	49
50	29.457	0.0339	0.0025	0.0725	406.529	13.801	12.529	172.905	50

8%					Compound Interest Factors					8%
Period	Single Payment		Uniform Payment Series				Arithmetic Gradient		Period	
	Compound Amount Factor	Present Value Factor	Sinking Fund Factor	Capital Recovery Factor	Compound Amount Factor	Present Value Factor	Gradient Uniform Series	Gradient Present Value		
	Find F Given P F/P	Find P Given F P/F	Find A Given F A/F	Find A Given P A/P	Find F Given A F/A	Find P Given A P/A	Find A Given G A/G	Find P Given G P/G		
n									n	
1	1.080	0.9259	1.0000	1.0800	1.000	0.926	0.000	0.000	1	
2	1.166	0.8573	0.4808	0.5608	2.080	1.783	0.481	0.857	2	
3	1.260	0.7938	0.3080	0.3880	3.246	2.577	0.949	2.445	3	
4	1.360	0.7350	0.2219	0.3019	4.506	3.312	1.404	4.650	4	
5	1.469	0.6806	0.1705	0.2505	5.867	3.993	1.846	7.372	5	
6	1.587	0.6302	0.1363	0.2163	7.336	4.623	2.276	10.523	6	
7	1.714	0.5835	0.1121	0.1921	8.923	5.206	2.694	14.024	7	
8	1.851	0.5403	0.0940	0.1740	10.637	5.747	3.099	17.806	8	
9	1.999	0.5002	0.0801	0.1601	12.488	6.247	3.491	21.808	9	
10	2.159	0.4632	0.0690	0.1490	14.487	6.710	3.871	25.977	10	
11	2.332	0.4289	0.0601	0.1401	16.645	7.139	4.240	30.266	11	
12	2.518	0.3971	0.0527	0.1327	18.977	7.536	4.596	34.634	12	
13	2.720	0.3677	0.0465	0.1265	21.495	7.904	4.940	39.046	13	
14	2.937	0.3405	0.0413	0.1213	24.215	8.244	5.273	43.472	14	
15	3.172	0.3152	0.0368	0.1168	27.152	8.559	5.594	47.886	15	
16	3.426	0.2919	0.0330	0.1130	30.324	8.851	5.905	52.264	16	
17	3.700	0.2703	0.0296	0.1096	33.750	9.122	6.204	56.588	17	
18	3.996	0.2502	0.0267	0.1067	37.450	9.372	6.492	60.843	18	
19	4.316	0.2317	0.0241	0.1041	41.446	9.604	6.770	65.013	19	
20	4.661	0.2145	0.0219	0.1019	45.762	9.818	7.037	69.090	20	
21	5.034	0.1987	0.0198	0.0998	50.423	10.017	7.294	73.063	21	
22	5.437	0.1839	0.0180	0.0980	55.457	10.201	7.541	76.926	22	
23	5.871	0.1703	0.0164	0.0964	60.893	10.371	7.779	80.673	23	
24	6.341	0.1577	0.0150	0.0950	66.765	10.529	8.007	84.300	24	
25	6.848	0.1460	0.0137	0.0937	73.106	10.675	8.225	87.804	25	
26	7.396	0.1352	0.0125	0.0925	79.954	10.810	8.435	91.184	26	
27	7.988	0.1252	0.0114	0.0914	87.351	10.935	8.636	94.439	27	
28	8.627	0.1159	0.0105	0.0905	95.339	11.051	8.829	97.569	28	
29	9.317	0.1073	0.0096	0.0896	103.966	11.158	9.013	100.574	29	
30	10.063	0.0994	0.0088	0.0888	113.283	11.258	9.190	103.456	30	
31	10.868	0.0920	0.0081	0.0881	123.346	11.350	9.358	106.216	31	
32	11.737	0.0852	0.0075	0.0875	134.214	11.435	9.520	108.857	32	
33	12.676	0.0789	0.0069	0.0869	145.951	11.514	9.674	111.382	33	
34	13.690	0.0730	0.0063	0.0863	158.627	11.587	9.821	113.792	34	
35	14.785	0.0676	0.0058	0.0858	172.317	11.655	9.961	116.092	35	
36	15.968	0.0626	0.0053	0.0853	187.102	11.717	10.095	118.284	36	
37	17.246	0.0580	0.0049	0.0849	203.070	11.775	10.222	120.371	37	
38	18.625	0.0537	0.0045	0.0845	220.316	11.829	10.344	122.358	38	
39	20.115	0.0497	0.0042	0.0842	238.941	11.879	10.460	124.247	39	
40	21.725	0.0460	0.0039	0.0839	259.057	11.925	10.570	126.042	40	
41	23.462	0.0426	0.0036	0.0836	280.781	11.967	10.675	127.747	41	
42	25.339	0.0395	0.0033	0.0833	304.244	12.007	10.774	129.365	42	
43	27.367	0.0365	0.0030	0.0830	329.583	12.043	10.869	130.900	43	
44	29.556	0.0338	0.0028	0.0828	356.950	12.077	10.959	132.355	44	
45	31.920	0.0313	0.0026	0.0826	386.506	12.108	11.045	133.733	45	
46	34.474	0.0290	0.0024	0.0824	418.426	12.137	11.126	135.038	46	
47	37.232	0.0269	0.0022	0.0822	452.900	12.164	11.203	136.274	47	
48	40.211	0.0249	0.0020	0.0820	490.132	12.189	11.276	137.443	48	
49	43.427	0.0230	0.0019	0.0819	530.343	12.212	11.345	138.548	49	
50	46.902	0.0213	0.0017	0.0817	573.770	12.233	11.411	139.593	50	

9%				Compound Interest Factors					9%
Period	Single Payment		Uniform Payment Series				Arithmetic Gradient		Period
	Compound Amount Factor	Present Value Factor	Sinking Fund Factor	Capital Recovery Factor	Compound Amount Factor	Present Value Factor	Gradient Uniform Series	Gradient Present Value	
	Find F Given P F/P	Find P Given F P/F	Find A Given F A/F	Find A Given P A/P	Find F Given A F/A	Find P Given A P/A	Find A Given G A/G	Find P Given G P/G	
n									n
1	1.090	0.9174	1.0000	1.0900	1.000	0.917	0.000	0.000	1
2	1.188	0.8417	0.4785	0.5685	2.090	1.759	0.478	0.842	2
3	1.295	0.7722	0.3051	0.3951	3.278	2.531	0.943	2.386	3
4	1.412	0.7084	0.2187	0.3087	4.573	3.240	1.393	4.511	4
5	1.539	0.6499	0.1671	0.2571	5.985	3.890	1.828	7.111	5
6	1.677	0.5963	0.1329	0.2229	7.523	4.486	2.250	10.092	6
7	1.828	0.5470	0.1087	0.1987	9.200	5.033	2.657	13.375	7
8	1.993	0.5019	0.0907	0.1807	11.028	5.535	3.051	16.888	8
9	2.172	0.4604	0.0768	0.1668	13.021	5.995	3.431	20.571	9
10	2.367	0.4224	0.0658	0.1558	15.193	6.418	3.798	24.373	10
11	2.580	0.3875	0.0569	0.1469	17.560	6.805	4.151	28.248	11
12	2.813	0.3555	0.0497	0.1397	20.141	7.161	4.491	32.159	12
13	3.066	0.3262	0.0436	0.1336	22.953	7.487	4.818	36.073	13
14	3.342	0.2992	0.0384	0.1284	26.019	7.786	5.133	39.963	14
15	3.642	0.2745	0.0341	0.1241	29.361	8.061	5.435	43.807	15
16	3.970	0.2519	0.0303	0.1203	33.003	8.313	5.724	47.585	16
17	4.328	0.2311	0.0270	0.1170	36.974	8.544	6.002	51.282	17
18	4.717	0.2120	0.0242	0.1142	41.301	8.756	6.269	54.886	18
19	5.142	0.1945	0.0217	0.1117	46.018	8.950	6.524	58.387	19
20	5.604	0.1784	0.0195	0.1095	51.160	9.129	6.767	61.777	20
21	6.109	0.1637	0.0176	0.1076	56.765	9.292	7.001	65.051	21
22	6.659	0.1502	0.0159	0.1059	62.873	9.442	7.223	68.205	22
23	7.258	0.1378	0.0144	0.1044	69.532	9.580	7.436	71.236	23
24	7.911	0.1264	0.0130	0.1030	76.790	9.707	7.638	74.143	24
25	8.623	0.1160	0.0118	0.1018	84.701	9.823	7.832	76.926	25
26	9.399	0.1064	0.0107	0.1007	93.324	9.929	8.016	79.586	26
27	10.245	0.0976	0.0097	0.0997	102.723	10.027	8.191	82.124	27
28	11.167	0.0895	0.0089	0.0989	112.968	10.116	8.357	84.542	28
29	12.172	0.0822	0.0081	0.0981	124.135	10.198	8.515	86.842	29
30	13.268	0.0754	0.0073	0.0973	136.308	10.274	8.666	89.028	30
31	14.462	0.0691	0.0067	0.0967	149.575	10.343	8.808	91.102	31
32	15.763	0.0634	0.0061	0.0961	164.037	10.406	8.944	93.069	32
33	17.182	0.0582	0.0056	0.0956	179.800	10.464	9.072	94.931	33
34	18.728	0.0534	0.0051	0.0951	196.982	10.518	9.193	96.693	34
35	20.414	0.0490	0.0046	0.0946	215.711	10.567	9.308	98.359	35
36	22.251	0.0449	0.0042	0.0942	236.125	10.612	9.417	99.932	36
37	24.254	0.0412	0.0039	0.0939	258.376	10.653	9.520	101.416	37
38	26.437	0.0378	0.0035	0.0935	282.630	10.691	9.617	102.816	38
39	28.816	0.0347	0.0032	0.0932	309.066	10.726	9.709	104.135	39
40	31.409	0.0318	0.0030	0.0930	337.882	10.757	9.796	105.376	40
41	34.236	0.0292	0.0027	0.0927	369.292	10.787	9.878	106.545	41
42	37.318	0.0268	0.0025	0.0925	403.528	10.813	9.955	107.643	42
43	40.676	0.0246	0.0023	0.0923	440.846	10.838	10.027	108.676	43
44	44.337	0.0226	0.0021	0.0921	481.522	10.861	10.096	109.646	44
45	48.327	0.0207	0.0019	0.0919	525.859	10.881	10.160	110.556	45
46	52.677	0.0190	0.0017	0.0917	574.186	10.900	10.221	111.410	46
47	57.418	0.0174	0.0016	0.0916	626.863	10.918	10.278	112.211	47
48	62.585	0.0160	0.0015	0.0915	684.280	10.934	10.332	112.962	48
49	68.218	0.0147	0.0013	0.0913	746.866	10.948	10.382	113.666	49
50	74.358	0.0134	0.0012	0.0912	815.084	10.962	10.430	114.325	50

10%	Compound Interest Factors								10%
Period	Single Payment		Uniform Payment Series				Arithmetic Gradient		Period
	Compound Amount Factor	Present Value Factor	Sinking Fund Factor	Capital Recovery Factor	Compound Amount Factor	Present Value Factor	Gradient Uniform Series	Gradient Present Value	
n	Find *F* Given *P* *F/P*	Find *P* Given *F* *P/F*	Find *A* Given *F* *A/F*	Find *A* Given *P* *A/P*	Find *F* Given *A* *F/A*	Find *P* Given *A* *P/A*	Find *A* Given *G* *A/G*	Find *P* Given *G* *P/G*	*n*
1	1.100	0.9091	1.0000	1.1000	1.000	0.909	0.000	0.000	1
2	1.210	0.8264	0.4762	0.5762	2.100	1.736	0.476	0.826	2
3	1.331	0.7513	0.3021	0.4021	3.310	2.487	0.937	2.329	3
4	1.464	0.6830	0.2155	0.3155	4.641	3.170	1.381	4.378	4
5	1.611	0.6209	0.1638	0.2638	6.105	3.791	1.810	6.862	5
6	1.772	0.5645	0.1296	0.2296	7.716	4.355	2.224	9.684	6
7	1.949	0.5132	0.1054	0.2054	9.487	4.868	2.622	12.763	7
8	2.144	0.4665	0.0874	0.1874	11.436	5.335	3.004	16.029	8
9	2.358	0.4241	0.0736	0.1736	13.579	5.759	3.372	19.421	9
10	2.594	0.3855	0.0627	0.1627	15.937	6.145	3.725	22.891	10
11	2.853	0.3505	0.0540	0.1540	18.531	6.495	4.064	26.396	11
12	3.138	0.3186	0.0468	0.1468	21.384	6.814	4.388	29.901	12
13	3.452	0.2897	0.0408	0.1408	24.523	7.103	4.699	33.377	13
14	3.797	0.2633	0.0357	0.1357	27.975	7.367	4.996	36.800	14
15	4.177	0.2394	0.0315	0.1315	31.772	7.606	5.279	40.152	15
16	4.595	0.2176	0.0278	0.1278	35.950	7.824	5.549	43.416	16
17	5.054	0.1978	0.0247	0.1247	40.545	8.022	5.807	46.582	17
18	5.560	0.1799	0.0219	0.1219	45.599	8.201	6.053	49.640	18
19	6.116	0.1635	0.0195	0.1195	51.159	8.365	6.286	52.583	19
20	6.727	0.1486	0.0175	0.1175	57.275	8.514	6.508	55.407	20
21	7.400	0.1351	0.0156	0.1156	64.002	8.649	6.719	58.110	21
22	8.140	0.1228	0.0140	0.1140	71.403	8.772	6.919	60.689	22
23	8.954	0.1117	0.0126	0.1126	79.543	8.883	7.108	63.146	23
24	9.850	0.1015	0.0113	0.1113	88.497	8.985	7.288	65.481	24
25	10.835	0.0923	0.0102	0.1102	98.347	9.077	7.458	67.696	25
26	11.918	0.0839	0.0092	0.1092	109.182	9.161	7.619	69.794	26
27	13.110	0.0763	0.0083	0.1083	121.100	9.237	7.770	71.777	27
28	14.421	0.0693	0.0075	0.1075	134.210	9.307	7.914	73.650	28
29	15.863	0.0630	0.0067	0.1067	148.631	9.370	8.049	75.415	29
30	17.449	0.0573	0.0061	0.1061	164.494	9.427	8.176	77.077	30
31	19.194	0.0521	0.0055	0.1055	181.943	9.479	8.296	78.640	31
32	21.114	0.0474	0.0050	0.1050	201.138	9.526	8.409	80.108	32
33	23.225	0.0431	0.0045	0.1045	222.252	9.569	8.515	81.486	33
34	25.548	0.0391	0.0041	0.1041	245.477	9.609	8.615	82.777	34
35	28.102	0.0356	0.0037	0.1037	271.024	9.644	8.709	83.987	35
36	30.913	0.0323	0.0033	0.1033	299.127	9.677	8.796	85.119	36
37	34.004	0.0294	0.0030	0.1030	330.039	9.706	8.879	86.178	37
38	37.404	0.0267	0.0027	0.1027	364.043	9.733	8.956	87.167	38
39	41.145	0.0243	0.0025	0.1025	401.448	9.757	9.029	88.091	39
40	45.259	0.0221	0.0023	0.1023	442.593	9.779	9.096	88.953	40
41	49.785	0.0201	0.0020	0.1020	487.852	9.799	9.160	89.756	41
42	54.764	0.0183	0.0019	0.1019	537.637	9.817	9.219	90.505	42
43	60.240	0.0166	0.0017	0.1017	592.401	9.834	9.274	91.202	43
44	66.264	0.0151	0.0015	0.1015	652.641	9.849	9.326	91.851	44
45	72.890	0.0137	0.0014	0.1014	718.905	9.863	9.374	92.454	45
46	80.180	0.0125	0.0013	0.1013	791.795	9.875	9.419	93.016	46
47	88.197	0.0113	0.0011	0.1011	871.975	9.887	9.461	93.537	47
48	97.017	0.0103	0.0010	0.1010	960.172	9.897	9.500	94.102	48
49	106.719	0.0094	0.0009	0.1009	1057.190	9.906	9.537	94.471	49
50	117.391	0.0085	0.0009	0.1009	1163.909	9.915	9.570	94.889	50

12%	Compound Interest Factors								12%
Period	Single Payment		Uniform Payment Series				Arithmetic Gradient		Period
	Compound Amount Factor	Present Value Factor	Sinking Fund Factor	Capital Recovery Factor	Compound Amount Factor	Present Value Factor	Gradient Uniform Series	Gradient Present Value	
	Find F Given P F/P	Find P Given F P/F	Find A Given F A/F	Find A Given P A/P	Find F Given A F/A	Find P Given A P/A	Find A Given G A/G	Find P Given G P/G	
n									n
1	1.120	0.8929	1.0000	1.1200	1.000	0.893	0.000	0.000	1
2	1.254	0.7972	0.4717	0.5917	2.120	1.690	0.472	0.797	2
3	1.405	0.7118	0.2963	0.4163	3.374	2.402	0.925	2.221	3
4	1.574	0.6355	0.2092	0.3292	4.779	3.037	1.359	4.127	4
5	1.762	0.5674	0.1574	0.2774	6.353	3.605	1.775	6.397	5
6	1.974	0.5066	0.1232	0.2432	8.115	4.111	2.172	8.930	6
7	2.211	0.4523	0.0991	0.2191	10.089	4.564	2.551	11.644	7
8	2.476	0.4039	0.0813	0.2013	12.300	4.968	2.913	14.471	8
9	2.773	0.3606	0.0677	0.1877	14.776	5.328	3.257	17.356	9
10	3.106	0.3220	0.0570	0.1770	17.549	5.650	3.585	20.254	10
11	3.479	0.2875	0.0484	0.1684	20.655	5.938	3.895	23.129	11
12	3.896	0.2567	0.0414	0.1614	24.133	6.194	4.190	25.952	12
13	4.363	0.2292	0.0357	0.1557	28.029	6.424	4.468	28.702	13
14	4.887	0.2046	0.0309	0.1509	32.393	6.628	4.732	31.362	14
15	5.474	0.1827	0.0268	0.1468	37.280	6.811	4.980	33.920	15
16	6.130	0.1631	0.0234	0.1434	42.753	6.974	5.215	36.367	16
17	6.866	0.1456	0.0205	0.1405	48.884	7.120	5.435	38.697	17
18	7.690	0.1300	0.0179	0.1379	55.750	7.250	5.643	40.908	18
19	8.613	0.1161	0.0158	0.1358	63.440	7.366	5.838	42.998	19
20	9.646	0.1037	0.0139	0.1339	72.052	7.469	6.020	44.968	20
21	10.804	0.0926	0.0122	0.1322	81.699	7.562	6.191	46.819	21
22	12.100	0.0826	0.0108	0.1308	92.503	7.645	6.351	48.554	22
23	13.552	0.0738	0.0096	0.1296	104.603	7.718	6.501	50.178	23
24	15.179	0.0659	0.0085	0.1285	118.155	7.784	6.641	51.693	24
25	17.000	0.0588	0.0075	0.1275	133.334	7.843	6.771	53.105	25
26	19.040	0.0525	0.0067	0.1267	150.334	7.896	6.892	54.418	26
27	21.325	0.0469	0.0059	0.1259	169.374	7.943	7.005	55.637	27
28	23.884	0.0419	0.0052	0.1252	190.699	7.984	7.110	56.767	28
29	26.750	0.0374	0.0047	0.1247	214.583	8.022	7.207	57.814	29
30	29.960	0.0334	0.0041	0.1241	241.333	8.055	7.297	58.782	30
31	33.555	0.0298	0.0037	0.1237	271.293	8.085	7.381	59.676	31
32	37.582	0.0266	0.0033	0.1233	304.848	8.112	7.459	60.501	32
33	42.092	0.0238	0.0029	0.1229	342.429	8.135	7.530	61.261	33
34	47.143	0.0212	0.0026	0.1226	384.521	8.157	7.596	61.961	34
35	52.800	0.0189	0.0023	0.1223	431.663	8.176	7.658	62.605	35
36	59.136	0.0169	0.0021	0.1221	484.463	8.192	7.714	63.197	36
37	66.232	0.0151	0.0018	0.1218	543.599	8.208	7.766	63.741	37
38	74.180	0.0135	0.0016	0.1216	609.831	8.221	7.814	64.239	38
39	83.081	0.0120	0.0015	0.1215	684.010	8.233	7.858	64.697	39
40	93.051	0.0107	0.0013	0.1213	767.091	8.244	7.899	65.116	40
41	104.217	0.0096	0.0012	0.1212	860.142	8.253	7.936	65.500	41
42	116.723	0.0086	0.0010	0.1210	964.359	8.262	7.970	65.851	42
43	130.730	0.0076	0.0009	0.1209	1081.083	8.270	8.002	66.172	43
44	146.418	0.0068	0.0008	0.1208	1211.813	8.276	8.031	66.466	44
45	163.988	0.0061	0.0007	0.1207	1358.230	8.283	8.057	66.734	45
46	183.666	0.0054	0.0007	0.1207	1522.218	8.288	8.082	66.979	46
47	205.706	0.0049	0.0006	0.1206	1705.884	8.293	8.104	67.203	47
48	230.391	0.0043	0.0005	0.1205	1911.590	8.297	8.124	67.407	48
49	258.038	0.0039	0.0005	0.1205	2141.981	8.301	8.143	67.593	49
50	289.002	0.0035	0.0004	0.1204	2400.018	8.304	8.160	67.762	50

14%			Compound Interest Factors						14%
Period	Single Payment		Uniform Payment Series				Arithmetic Gradient		Period
	Compound Amount Factor	Present Value Factor	Sinking Fund Factor	Capital Recovery Factor	Compound Amount Factor	Present Value Factor	Gradient Uniform Series	Gradient Present Value	
	Find F Given P F/P	Find P Given F P/F	Find A Given F A/F	Find A Given P A/P	Find F Given A F/A	Find P Given A P/A	Find A Given G A/G	Find P Given G P/G	
n									n
1	1.140	0.8772	1.0000	1.1400	1.000	0.877	0.000	0.000	1
2	1.300	0.7695	0.4673	0.6073	2.140	1.647	0.467	0.769	2
3	1.482	0.6750	0.2907	0.4307	3.440	2.322	0.913	2.119	3
4	1.689	0.5921	0.2032	0.3432	4.921	2.914	1.337	3.896	4
5	1.925	0.5194	0.1513	0.2913	6.610	3.433	1.740	5.973	5
6	2.195	0.4556	0.1172	0.2572	8.536	3.889	2.122	8.251	6
7	2.502	0.3996	0.0932	0.2332	10.730	4.288	2.483	10.649	7
8	2.853	0.3506	0.0756	0.2156	13.233	4.639	2.825	13.103	8
9	3.252	0.3075	0.0622	0.2022	16.085	4.946	3.146	15.563	9
10	3.707	0.2697	0.0517	0.1917	19.337	5.216	3.449	17.991	10
11	4.226	0.2366	0.0434	0.1834	23.045	5.453	3.733	20.357	11
12	4.818	0.2076	0.0367	0.1767	27.271	5.660	4.000	22.640	12
13	5.492	0.1821	0.0312	0.1712	32.089	5.842	4.249	24.825	13
14	6.261	0.1597	0.0266	0.1666	37.581	6.002	4.482	26.901	14
15	7.138	0.1401	0.0228	0.1628	43.842	6.142	4.699	28.862	15
16	8.137	0.1229	0.0196	0.1596	50.980	6.265	4.901	30.706	16
17	9.276	0.1078	0.0169	0.1569	59.118	6.373	5.089	32.430	17
18	10.575	0.0946	0.0146	0.1546	68.394	6.467	5.263	34.038	18
19	12.056	0.0829	0.0127	0.1527	78.969	6.550	5.424	35.531	19
20	13.743	0.0728	0.0110	0.1510	91.025	6.623	5.573	36.914	20
21	15.668	0.0638	0.0095	0.1495	104.768	6.687	5.711	38.190	21
22	17.861	0.0560	0.0083	0.1483	120.436	6.743	5.838	39.366	22
23	20.362	0.0491	0.0072	0.1472	138.297	6.792	5.955	40.446	23
24	23.212	0.0431	0.0063	0.1463	158.659	6.835	6.062	41.437	24
25	26.462	0.0378	0.0055	0.1455	181.871	6.873	6.161	42.344	25
26	30.167	0.0331	0.0048	0.1448	208.333	6.906	6.251	43.173	26
27	34.390	0.0291	0.0042	0.1442	238.499	6.935	6.334	43.929	27
28	39.204	0.0255	0.0037	0.1437	272.889	6.961	6.410	44.618	28
29	44.693	0.0224	0.0032	0.1432	312.094	6.983	6.479	45.244	29
30	50.950	0.0196	0.0028	0.1428	356.787	7.003	6.542	45.813	30
31	58.083	0.0172	0.0025	0.1425	407.737	7.020	6.600	46.330	31
32	66.215	0.0151	0.0021	0.1421	465.820	7.035	6.652	46.798	32
33	75.485	0.0132	0.0019	0.1419	532.035	7.048	6.700	47.222	33
34	86.053	0.0116	0.0016	0.1416	607.520	7.060	6.743	47.605	34
35	98.100	0.0102	0.0014	0.1414	693.573	7.070	6.782	47.952	35
36	111.834	0.0089	0.0013	0.1413	791.673	7.079	6.818	48.265	36
37	127.491	0.0078	0.0011	0.1411	903.507	7.087	6.850	48.547	37
38	145.340	0.0069	0.0010	0.1410	1030.998	7.094	6.880	48.802	38
39	165.687	0.0060	0.0009	0.1409	1176.338	7.100	6.906	49.031	39
40	188.884	0.0053	0.0007	0.1407	1342.025	7.105	6.930	49.238	40
41	215.327	0.0046	0.0007	0.1407	1530.909	7.110	6.952	49.423	41
42	245.473	0.0041	0.0006	0.1406	1746.236	7.114	6.971	49.590	42
43	279.839	0.0036	0.0005	0.1405	1991.709	7.117	6.989	49.741	43
44	319.017	0.0031	0.0004	0.1404	2271.548	7.120	7.004	49.875	44
45	363.679	0.0027	0.0004	0.1404	2590.565	7.123	7.019	49.996	45
46	414.594	0.0024	0.0003	0.1403	2954.244	7.126	7.032	50.105	46
47	472.637	0.0021	0.0003	0.1403	3368.838	7.128	7.043	50.202	47
48	538.807	0.0019	0.0003	0.1403	3841.475	7.130	7.054	50.289	48
49	614.239	0.0016	0.0002	0.1402	4380.282	7.131	7.063	50.368	49
50	700.233	0.0014	0.0002	0.1402	4994.521	7.133	7.071	50.438	50

16%	Compound Interest Factors						Arithmetic Gradient		16%
Period	Single Payment		Uniform Payment Series				Arithmetic Gradient		Period
	Compound Amount Factor	Present Value Factor	Sinking Fund Factor	Capital Recovery Factor	Compound Amount Factor	Present Value Factor	Gradient Uniform Series	Gradient Present Value	
	Find F Given P F/P	Find P Given F P/F	Find A Given F A/F	Find A Given P A/P	Find F Given A F/A	Find P Given A P/A	Find A Given G A/G	Find P Given G P/G	
n									n
1	1.160	0.8621	1.0000	1.1600	1.000	0.862	0.000	0.000	1
2	1.346	0.7432	0.4630	0.6230	2.160	1.605	0.463	0.743	2
3	1.561	0.6407	0.2853	0.4453	3.506	2.246	0.901	2.024	3
4	1.811	0.5523	0.1974	0.3574	5.066	2.798	1.316	3.681	4
5	2.100	0.4761	0.1454	0.3054	6.877	3.274	1.706	5.586	5
6	2.436	0.4104	0.1114	0.2714	8.977	3.685	2.073	7.638	6
7	2.826	0.3538	0.0876	0.2476	11.414	4.039	2.417	9.761	7
8	3.278	0.3050	0.0702	0.2302	14.240	4.344	2.739	11.896	8
9	3.803	0.2630	0.0571	0.2171	17.519	4.607	3.039	14.000	9
10	4.411	0.2267	0.0469	0.2069	21.321	4.833	3.319	16.040	10
11	5.117	0.1954	0.0389	0.1989	25.733	5.029	3.578	17.994	11
12	5.936	0.1685	0.0324	0.1924	30.850	5.197	3.819	19.847	12
13	6.886	0.1452	0.0272	0.1872	36.786	5.342	4.041	21.590	13
14	7.988	0.1252	0.0229	0.1829	43.672	5.468	4.246	23.217	14
15	9.266	0.1079	0.0194	0.1794	51.660	5.575	4.435	24.728	15
16	10.748	0.0930	0.0164	0.1764	60.925	5.668	4.609	26.124	16
17	12.468	0.0802	0.0140	0.1740	71.673	5.749	4.768	27.407	17
18	14.463	0.0691	0.0119	0.1719	84.141	5.818	4.913	28.583	18
19	16.777	0.0596	0.0101	0.1701	98.603	5.877	5.046	29.656	19
20	19.461	0.0514	0.0087	0.1687	115.380	5.929	5.167	30.632	20
21	22.574	0.0443	0.0074	0.1674	134.841	5.973	5.277	31.518	21
22	26.186	0.0382	0.0064	0.1664	157.415	6.011	5.377	32.320	22
23	30.376	0.0329	0.0054	0.1654	183.601	6.044	5.467	33.044	23
24	35.236	0.0284	0.0047	0.1647	213.978	6.073	5.549	33.697	24
25	40.874	0.0245	0.0040	0.1640	249.214	6.097	5.623	34.284	25
26	47.414	0.0211	0.0034	0.1634	290.088	6.118	5.690	34.811	26
27	55.000	0.0182	0.0030	0.1630	337.502	6.136	5.750	35.284	27
28	63.800	0.0157	0.0025	0.1625	392.503	6.152	5.804	35.707	28
29	74.009	0.0135	0.0022	0.1622	456.303	6.166	5.853	36.086	29
30	85.850	0.0116	0.0019	0.1619	530.312	6.177	5.896	36.423	30
31	99.586	0.0100	0.0016	0.1616	616.162	6.187	5.936	36.725	31
32	115.520	0.0087	0.0014	0.1614	715.747	6.196	5.971	36.993	32
33	134.003	0.0075	0.0012	0.1612	831.267	6.203	6.002	37.232	33
34	155.443	0.0064	0.0010	0.1610	965.270	6.210	6.030	37.444	34
35	180.314	0.0055	0.0009	0.1609	1120.713	6.215	6.055	37.633	35
36	209.164	0.0048	0.0008	0.1608	1301.027	6.220	6.077	37.800	36
37	242.631	0.0041	0.0007	0.1607	1510.191	6.224	6.097	37.948	37
38	281.452	0.0036	0.0006	0.1606	1752.822	6.228	6.115	38.080	38
39	326.484	0.0031	0.0005	0.1605	2034.273	6.231	6.130	38.196	39
40	378.721	0.0026	0.0004	0.1604	2360.757	6.233	6.144	38.299	40
41	439.317	0.0023	0.0004	0.1604	2739.478	6.236	6.156	38.390	41
42	509.607	0.0020	0.0003	0.1603	3178.795	6.238	6.167	38.471	42
43	591.144	0.0017	0.0003	0.1603	3688.402	6.239	6.177	38.542	43
44	685.727	0.0015	0.0002	0.1602	4279.546	6.241	6.186	38.605	44
45	795.444	0.0013	0.0002	0.1602	4965.274	6.242	6.193	38.660	45
46	922.715	0.0011	0.0002	0.1602	5760.718	6.243	6.200	38.709	46
47	1070.349	0.0009	0.0001	0.1601	6683.433	6.244	6.206	38.752	47
48	1241.605	0.0008	0.0001	0.1601	7753.782	6.245	6.211	38.789	48
49	1440.262	0.0007	0.0001	0.1601	8995.387	6.246	6.216	38.823	49
50	1670.704	0.0006	0.0001	0.1601	10435.649	6.246	6.220	38.852	50

18%				Compound Interest Factors					18%
Period	Single Payment		Uniform Payment Series				Arithmetic Gradient		Period
	Compound Amount Factor	Present Value Factor	Sinking Fund Factor	Capital Recovery Factor	Compound Amount Factor	Present Value Factor	Gradient Uniform Series	Gradient Present Value	
	Find F Given P F/P	Find P Given F P/F	Find A Given F A/F	Find A Given P A/P	Find F Given A F/A	Find P Given A P/A	Find A Given G A/G	Find P Given G P/G	
n									n
1	1.180	0.8475	1.0000	1.1800	1.000	0.847	0.000	0.000	1
2	1.392	0.7182	0.4587	0.6387	2.180	1.566	0.459	0.718	2
3	1.643	0.6086	0.2799	0.4599	3.572	2.174	0.890	1.935	3
4	1.939	0.5158	0.1917	0.3717	5.215	2.690	1.295	3.483	4
5	2.288	0.4371	0.1398	0.3198	7.154	3.127	1.673	5.231	5
6	2.700	0.3704	0.1059	0.2859	9.442	3.498	2.025	7.083	6
7	3.185	0.3139	0.0824	0.2624	12.142	3.812	2.353	8.967	7
8	3.759	0.2660	0.0652	0.2452	15.327	4.078	2.656	10.829	8
9	4.435	0.2255	0.0524	0.2324	19.086	4.303	2.936	12.633	9
10	5.234	0.1911	0.0425	0.2225	23.521	4.494	3.194	14.352	10
11	6.176	0.1619	0.0348	0.2148	28.755	4.656	3.430	15.972	11
12	7.288	0.1372	0.0286	0.2086	34.931	4.793	3.647	17.481	12
13	8.599	0.1163	0.0237	0.2037	42.219	4.910	3.845	18.877	13
14	10.147	0.0985	0.0197	0.1997	50.818	5.008	4.025	20.158	14
15	11.974	0.0835	0.0164	0.1964	60.965	5.092	4.189	21.327	15
16	14.129	0.0708	0.0137	0.1937	72.939	5.162	4.337	22.389	16
17	16.672	0.0600	0.0115	0.1915	87.068	5.222	4.471	23.348	17
18	19.673	0.0508	0.0096	0.1896	103.740	5.273	4.592	24.212	18
19	23.214	0.0431	0.0081	0.1881	123.414	5.316	4.700	24.988	19
20	27.393	0.0365	0.0068	0.1868	146.628	5.353	4.798	25.681	20
21	32.324	0.0309	0.0057	0.1857	174.021	5.384	4.885	26.300	21
22	38.142	0.0262	0.0048	0.1848	206.345	5.410	4.963	26.851	22
23	45.008	0.0222	0.0041	0.1841	244.487	5.432	5.033	27.339	23
24	53.109	0.0188	0.0035	0.1835	289.494	5.451	5.095	27.772	24
25	62.669	0.0160	0.0029	0.1829	342.603	5.467	5.150	28.155	25
26	73.949	0.0135	0.0025	0.1825	405.272	5.480	5.199	28.494	26
27	87.260	0.0115	0.0021	0.1821	479.221	5.492	5.243	28.791	27
28	102.967	0.0097	0.0018	0.1818	566.481	5.502	5.281	29.054	28
29	121.501	0.0082	0.0015	0.1815	669.447	5.510	5.315	29.284	29
30	143.371	0.0070	0.0013	0.1813	790.948	5.517	5.345	29.486	30
31	169.177	0.0059	0.0011	0.1811	934.319	5.523	5.371	29.664	31
32	199.629	0.0050	0.0009	0.1809	1103.496	5.528	5.394	29.819	32
33	235.563	0.0042	0.0008	0.1808	1303.125	5.532	5.415	29.955	33
34	277.964	0.0036	0.0006	0.1806	1538.688	5.536	5.433	30.074	34
35	327.997	0.0030	0.0006	0.1806	1816.652	5.539	5.449	30.177	35
36	387.037	0.0026	0.0005	0.1805	2144.649	5.541	5.462	30.268	36
37	456.703	0.0022	0.0004	0.1804	2531.686	5.543	5.474	30.347	37
38	538.910	0.0019	0.0003	0.1803	2988.389	5.545	5.485	30.415	38
39	635.914	0.0016	0.0003	0.1803	3527.299	5.547	5.494	30.475	39
40	750.378	0.0013	0.0002	0.1802	4163.213	5.548	5.502	30.527	40
41	885.446	0.0011	0.0002	0.1802	4913.591	5.549	5.509	30.572	41
42	1044.827	0.0010	0.0002	0.1802	5799.038	5.550	5.515	30.611	42
43	1232.896	0.0008	0.0001	0.1801	6843.865	5.551	5.521	30.645	43
44	1454.817	0.0007	0.0001	0.1801	8076.760	5.552	5.525	30.675	44
45	1716.684	0.0006	0.0001	0.1801	9531.577	5.552	5.529	30.701	45
46	2025.687	0.0005	0.0001	0.1801	11248.261	5.553	5.533	30.723	46
47	2390.311	0.0004	0.0001	0.1801	13273.948	5.553	5.536	30.742	47
48	2820.567	0.0004	0.0001	0.1801	15664.259	5.554	5.539	30.759	48
49	3328.269	0.0003	0.0001	0.1801	18484.825	5.554	5.541	30.773	49
50	3927.357	0.0003	0.0000	0.1800	21813.094	5.554	5.543	30.786	50

20%				Compound Interest Factors					20%
Period	Single Payment		Uniform Payment Series				Arithmetic Gradient		Period
	Compound Amount Factor	Present Value Factor	Sinking Fund Factor	Capital Recovery Factor	Compound Amount Factor	Present Value Factor	Gradient Uniform Series	Gradient Present Value	
	Find F Given P F/P	Find P Given F P/F	Find A Given F A/F	Find A Given P A/P	Find F Given A F/A	Find P Given A P/A	Find A Given G A/G	Find P Given G P/G	
n									n
1	1.200	0.8333	1.0000	1.2000	1.000	0.833	0.000	0.000	1
2	1.440	0.6944	0.4545	0.6545	2.200	1.528	0.455	0.694	2
3	1.728	0.5787	0.2747	0.4747	3.640	2.106	0.879	1.852	3
4	2.074	0.4823	0.1863	0.3863	5.368	2.589	1.274	3.299	4
5	2.488	0.4019	0.1344	0.3344	7.442	2.991	1.641	4.906	5
6	2.986	0.3349	0.1007	0.3007	9.930	3.326	1.979	6.581	6
7	3.583	0.2791	0.0774	0.2774	12.916	3.605	2.290	8.255	7
8	4.300	0.2326	0.0606	0.2606	16.499	3.837	2.576	9.883	8
9	5.160	0.1938	0.0481	0.2481	20.799	4.031	2.836	11.434	9
10	6.192	0.1615	0.0385	0.2385	25.959	4.192	3.074	12.887	10
11	7.430	0.1346	0.0311	0.2311	32.150	4.327	3.289	14.233	11
12	8.916	0.1122	0.0253	0.2253	39.581	4.439	3.484	15.467	12
13	10.699	0.0935	0.0206	0.2206	48.497	4.533	3.660	16.588	13
14	12.839	0.0779	0.0169	0.2169	59.196	4.611	3.817	17.601	14
15	15.407	0.0649	0.0139	0.2139	72.035	4.675	3.959	18.509	15
16	18.488	0.0541	0.0114	0.2114	87.442	4.730	4.085	19.321	16
17	22.186	0.0451	0.0094	0.2094	105.931	4.775	4.198	20.042	17
18	26.623	0.0376	0.0078	0.2078	128.117	4.812	4.298	20.680	18
19	31.948	0.0313	0.0065	0.2065	154.740	4.843	4.386	21.244	19
20	38.338	0.0261	0.0054	0.2054	186.688	4.870	4.464	21.739	20
21	46.005	0.0217	0.0044	0.2044	225.026	4.891	4.533	22.174	21
22	55.206	0.0181	0.0037	0.2037	271.031	4.909	4.594	22.555	22
23	66.247	0.0151	0.0031	0.2031	326.237	4.925	4.647	22.887	23
24	79.497	0.0126	0.0025	0.2025	392.484	4.937	4.694	23.176	24
25	95.396	0.0105	0.0021	0.2021	471.981	4.948	4.735	23.428	25
26	114.475	0.0087	0.0018	0.2018	567.377	4.956	4.771	23.646	26
27	137.371	0.0073	0.0015	0.2015	681.853	4.964	4.802	23.835	27
28	164.845	0.0061	0.0012	0.2012	819.223	4.970	4.829	23.999	28
29	197.814	0.0051	0.0010	0.2010	984.068	4.975	4.853	24.141	29
30	237.376	0.0042	0.0008	0.2008	1181.882	4.979	4.873	24.263	30
31	284.852	0.0035	0.0007	0.2007	1419.258	4.982	4.891	24.368	31
32	341.822	0.0029	0.0006	0.2006	1704.109	4.985	4.906	24.459	32
33	410.186	0.0024	0.0005	0.2005	2045.931	4.988	4.919	24.537	33
34	492.224	0.0020	0.0004	0.2004	2456.118	4.990	4.931	24.604	34
35	590.668	0.0017	0.0003	0.2003	2948.341	4.992	4.941	24.661	35
36	708.802	0.0014	0.0003	0.2003	3539.009	4.993	4.949	24.711	36
37	850.562	0.0012	0.0002	0.2002	4247.811	4.994	4.956	24.753	37
38	1020.675	0.0010	0.0002	0.2002	5098.373	4.995	4.963	24.789	38
39	1224.810	0.0008	0.0002	0.2002	6119.048	4.996	4.968	24.820	39
40	1469.772	0.0007	0.0001	0.2001	7343.858	4.997	4.973	24.847	40
41	1763.726	0.0006	0.0001	0.2001	8813.629	4.997	4.977	24.870	41
42	2116.471	0.0005	0.0001	0.2001	10577.355	4.998	4.980	24.889	42
43	2539.765	0.0004	0.0001	0.2001	12693.826	4.998	4.983	24.906	43
44	3047.718	0.0003	0.0001	0.2001	15233.592	4.998	4.986	24.920	44
45	3657.262	0.0003	0.0001	0.2001	18281.310	4.999	4.988	24.932	45
46	4388.714	0.0002	0.0000	0.2000	21938.572	4.999	4.990	24.942	46
47	5266.457	0.0002	0.0000	0.2000	26327.286	4.999	4.991	24.951	47
48	6319.749	0.0002	0.0000	0.2000	31593.744	4.999	4.992	24.958	48
49	7583.698	0.0001	0.0000	0.2000	37913.492	4.999	4.994	24.964	49
50	9100.438	0.0001	0.0000	0.2000	45497.191	4.999	4.995	24.970	50

25%				Compound Interest Factors					25%
Period	Single Payment		Uniform Payment Series				Arithmetic Gradient		Period
	Compound Amount Factor	Present Value Factor	Sinking Fund Factor	Capital Recovery Factor	Compound Amount Factor	Present Value Factor	Gradient Uniform Series	Gradient Present Value	
	Find F Given P F/P	Find P Given F P/F	Find A Given F A/F	Find A Given P A/P	Find F Given A F/A	Find P Given A P/A	Find A Given G A/G	Find P Given G P/G	
n									n
1	1.250	0.8000	1.0000	1.2500	1.000	0.800	0.000	0.000	1
2	1.563	0.6400	0.4444	0.6944	2.250	1.440	0.444	0.640	2
3	1.953	0.5120	0.2623	0.5123	3.813	1.952	0.852	1.664	3
4	2.441	0.4096	0.1734	0.4234	5.766	2.362	1.225	2.893	4
5	3.052	0.3277	0.1218	0.3718	8.207	2.689	1.563	4.204	5
6	3.815	0.2621	0.0888	0.3388	11.259	2.951	1.868	5.514	6
7	4.768	0.2097	0.0663	0.3163	15.073	3.161	2.142	6.773	7
8	5.960	0.1678	0.0504	0.3004	19.842	3.329	2.387	7.947	8
9	7.451	0.1342	0.0388	0.2888	25.802	3.463	2.605	9.021	9
10	9.313	0.1074	0.0301	0.2801	33.253	3.571	2.797	9.987	10
11	11.642	0.0859	0.0235	0.2735	42.566	3.656	2.966	10.846	11
12	14.552	0.0687	0.0184	0.2684	54.208	3.725	3.115	11.602	12
13	18.190	0.0550	0.0145	0.2645	68.760	3.780	3.244	12.262	13
14	22.737	0.0440	0.0115	0.2615	86.949	3.824	3.356	12.833	14
15	28.422	0.0352	0.0091	0.2591	109.687	3.859	3.453	13.326	15
16	35.527	0.0281	0.0072	0.2572	138.109	3.887	3.537	13.748	16
17	44.409	0.0225	0.0058	0.2558	173.636	3.910	3.608	14.108	17
18	55.511	0.0180	0.0046	0.2546	218.045	3.928	3.670	14.415	18
19	69.389	0.0144	0.0037	0.2537	273.556	3.942	3.722	14.674	19
20	86.736	0.0115	0.0029	0.2529	342.945	3.954	3.767	14.893	20
21	108.420	0.0092	0.0023	0.2523	429.681	3.963	3.805	15.078	21
22	135.525	0.0074	0.0019	0.2519	538.101	3.970	3.836	15.233	22
23	169.407	0.0059	0.0015	0.2515	673.626	3.976	3.863	15.362	23
24	211.758	0.0047	0.0012	0.2512	843.033	3.981	3.886	15.471	24
25	264.698	0.0038	0.0009	0.2509	1054.791	3.985	3.905	15.562	25
26	330.872	0.0030	0.0008	0.2508	1319.489	3.988	3.921	15.637	26
27	413.590	0.0024	0.0006	0.2506	1650.361	3.990	3.935	15.700	27
28	516.988	0.0019	0.0005	0.2505	2063.952	3.992	3.946	15.752	28
29	646.235	0.0015	0.0004	0.2504	2580.939	3.994	3.955	15.796	29
30	807.794	0.0012	0.0003	0.2503	3227.174	3.995	3.963	15.832	30
31	1009.742	0.0010	0.0002	0.2502	4034.968	3.996	3.969	15.861	31
32	1262.177	0.0008	0.0002	0.2502	5044.710	3.997	3.975	15.886	32
33	1577.722	0.0006	0.0002	0.2502	6306.887	3.997	3.979	15.906	33
34	1972.152	0.0005	0.0001	0.2501	7884.609	3.998	3.983	15.923	34
35	2465.190	0.0004	0.0001	0.2501	9856.761	3.998	3.986	15.937	35
36	3081.488	0.0003	0.0001	0.2501	12321.952	3.999	3.988	15.948	36
37	3851.860	0.0003	0.0001	0.2501	15403.440	3.999	3.990	15.957	37
38	4814.825	0.0002	0.0001	0.2501	19255.299	3.999	3.992	15.965	38
39	6018.531	0.0002	0.0000	0.2500	24070.124	3.999	3.994	15.971	39
40	7523.164	0.0001	0.0000	0.2500	30088.655	3.999	3.995	15.977	40
41	9403.955	0.0001	0.0000	0.2500	37611.819	4.000	3.996	15.981	41
42	11754.944	0.0001	0.0000	0.2500	47015.774	4.000	3.996	15.984	42
43	14693.679	0.0001	0.0000	0.2500	58770.718	4.000	3.997	15.987	43
44	18367.099	0.0001	0.0000	0.2500	73464.397	4.000	3.998	15.990	44
45	22958.874	0.0000	0.0000	0.2500	91831.496	4.000	3.998	15.991	45
46	28698.593	0.0000	0.0000	0.2500	114790.370	4.000	3.998	15.993	46
47	35873.241	0.0000	0.0000	0.2500	143488.963	4.000	3.999	15.994	47
48	44841.551	0.0000	0.0000	0.2500	179362.203	4.000	3.999	15.995	48
49	56051.939	0.0000	0.0000	0.2500	224203.754	4.000	3.999	15.996	49
50	70064.923	0.0000	0.0000	0.2500	280255.693	4.000	3.999	15.997	50

30%	Compound Interest Factors								30%
Period	Single Payment		Uniform Payment Series				Arithmetic Gradient		Period
	Compound Amount Factor	Present Value Factor	Sinking Fund Factor	Capital Recovery Factor	Compound Amount Factor	Present Value Factor	Gradient Uniform Series	Gradient Present Value	
	Find F Given P F/P	Find P Given F P/F	Find A Given F A/F	Find A Given P A/P	Find F Given A F/A	Find P Given A P/A	Find A Given G A/G	Find P Given G P/G	
n									n
1	1.300	0.7692	1.0000	1.3000	1.000	0.769	0.000	0.000	1
2	1.690	0.5917	0.4348	0.7348	2.300	1.361	0.435	0.592	2
3	2.197	0.4552	0.2506	0.5506	3.990	1.816	0.827	1.502	3
4	2.856	0.3501	0.1616	0.4616	6.187	2.166	1.178	2.552	4
5	3.713	0.2693	0.1106	0.4106	9.043	2.436	1.490	3.630	5
6	4.827	0.2072	0.0784	0.3784	12.756	2.643	1.765	4.666	6
7	6.275	0.1594	0.0569	0.3569	17.583	2.802	2.006	5.622	7
8	8.157	0.1226	0.0419	0.3419	23.858	2.925	2.216	6.480	8
9	10.604	0.0943	0.0312	0.3312	32.015	3.019	2.396	7.234	9
10	13.786	0.0725	0.0235	0.3235	42.619	3.092	2.551	7.887	10
11	17.922	0.0558	0.0177	0.3177	56.405	3.147	2.683	8.445	11
12	23.298	0.0429	0.0135	0.3135	74.327	3.190	2.795	8.917	12
13	30.288	0.0330	0.0102	0.3102	97.625	3.223	2.889	9.314	13
14	39.374	0.0254	0.0078	0.3078	127.913	3.249	2.969	9.644	14
15	51.186	0.0195	0.0060	0.3060	167.286	3.268	3.034	9.917	15
16	66.542	0.0150	0.0046	0.3046	218.472	3.283	3.089	10.143	16
17	86.504	0.0116	0.0035	0.3035	285.014	3.295	3.135	10.328	17
18	112.455	0.0089	0.0027	0.3027	371.518	3.304	3.172	10.479	18
19	146.192	0.0068	0.0021	0.3021	483.973	3.311	3.202	10.602	19
20	190.050	0.0053	0.0016	0.3016	630.165	3.316	3.228	10.702	20
21	247.065	0.0040	0.0012	0.3012	820.215	3.320	3.248	10.783	21
22	321.184	0.0031	0.0009	0.3009	1067.280	3.323	3.265	10.848	22
23	417.539	0.0024	0.0007	0.3007	1388.464	3.325	3.278	10.901	23
24	542.801	0.0018	0.0006	0.3006	1806.003	3.327	3.289	10.943	24
25	705.641	0.0014	0.0004	0.3004	2348.803	3.329	3.298	10.977	25
26	917.333	0.0011	0.0003	0.3003	3054.444	3.330	3.305	11.005	26
27	1192.533	0.0008	0.0003	0.3003	3971.778	3.331	3.311	11.026	27
28	1550.293	0.0006	0.0002	0.3002	5164.311	3.331	3.315	11.044	28
29	2015.381	0.0005	0.0001	0.3001	6714.604	3.332	3.319	11.058	29
30	2619.996	0.0004	0.0001	0.3001	8729.985	3.332	3.322	11.069	30
31	3405.994	0.0003	0.0001	0.3001	11349.981	3.332	3.324	11.078	31
32	4427.793	0.0002	0.0001	0.3001	14755.975	3.333	3.326	11.085	32
33	5756.130	0.0002	0.0001	0.3001	19183.768	3.333	3.328	11.090	33
34	7482.970	0.0001	0.0000	0.3000	24939.899	3.333	3.329	11.094	34
35	9727.860	0.0001	0.0000	0.3000	32422.868	3.333	3.330	11.098	35
36	12646.219	0.0001	0.0000	0.3000	42150.729	3.333	3.330	11.101	36
37	16440.084	0.0001	0.0000	0.3000	54796.947	3.333	3.331	11.103	37
38	21372.109	0.0000	0.0000	0.3000	71237.031	3.333	3.332	11.105	38
39	27783.742	0.0000	0.0000	0.3000	92609.141	3.333	3.332	11.106	39
40	36118.865	0.0000	0.0000	0.3000	120392.883	3.333	3.332	11.107	40
41	46954.524	0.0000	0.0000	0.3000	156511.748	3.333	3.332	11.108	41
42	61040.882	0.0000	0.0000	0.3000	203466.272	3.333	3.333	11.109	42
43	79353.146	0.0000	0.0000	0.3000	264507.153	3.333	3.333	11.109	43
44	103159.090	0.0000	0.0000	0.3000	343860.299	3.333	3.333	11.110	44
45	134106.817	0.0000	0.0000	0.3000	447019.389	3.333	3.333	11.110	45
46	174338.862	0.0000	0.0000	0.3000	581126.206	3.333	3.333	11.110	46
47	226640.520	0.0000	0.0000	0.3000	755465.067	3.333	3.333	11.110	47
48	294632.676	0.0000	0.0000	0.3000	982105.588	3.333	3.333	11.111	48
49	383022.479	0.0000	0.0000	0.3000	1276738.264	3.333	3.333	11.111	49
50	497929.223	0.0000	0.0000	0.3000	1659760.743	3.333	3.333	11.111	50

Appendix E:
Greek Symbols and
Roman Numerals

Greek Symbols

The Greek alphabet is a set of twenty-four letters that has been used to write the Greek language since the late-ninth- or early-eighth-century BCE. It is the first and oldest alphabet in the narrow sense that it notes each vowel and consonant with a separate symbol.

Capital	Lower Case	Greek Name	Pronunciation	English
A	α	Alpha	al-fah	a
B	β	Beta	bay-tah	b
Γ	γ	Gamma	gam-ah	g
Δ	δ	Delta	del-tah	d
E	ε	Epsilon	ep-si-lon	e
Z	ζ	Zeta	zat-tah	z
H	η	Eta	ay-tah	h
Θ	θ	Theta	thay-tah	th
I	ι	Iota	eye-oh-tah	i
K	κ	Kappa	cap-ah	k
Λ	λ	Lambda	lamb-da	l
M	μ	Mu	mew	m
N	ν	Nu	new	n
Ξ	ξ	Xi	sah-eye	x
O	o	Omicron	oh-mi-cron	o
Π	π	Pi	pie	p
P	ρ	Rho	roe	r
Σ	σ	Sigma	sig-mah	s
T	τ	Tau	tah-hoe	t
Y	υ	Upsilon	oop-si-lon	u
Φ	φ	Phi	fah-eye	Ph
X	χ	Chi	kigh	ch
Ψ	ψ	Psi	sigh	ps
Ω	ω	Omega	Oh-mega	o

Roman Numerals

Roman numerals are a system of numerical notations used by the Romans. They are an additive (and subtractive) system in which letters are used to denote certain "base" numbers, and arbitrary numbers are then denoted using combinations of symbols.

The following table gives the Latin letters used in Roman numerals and the corresponding numerical values they represent.

1	I	14	XIV	27	XXVII	150	CL
2	II	15	XV	28	XXVIII	200	CC
3	III	16	XVI	29	XXIX	300	CCC
4	IV	17	XVII	30	XXX	400	CD
5	V	18	XVIII	31	XXXI	500	D
6	VI	19	XIX	40	XL	600	DC
7	VII	20	XX	50	L	700	DCC
8	VIII	21	XXI	60	LX	800	DCCC
9	IX	22	XXII	70	LXX	900	CM
10	X	23	XXIII	80	LXXX	1000	M
11	XI	24	XXIV	90	XC	1600	MDC
12	XII	25	XXV	100	C	1700	MDCC
13	XIII	26	XXVI	101	CI	1900	MCM

Note: C stands for *centum*, the Latin word for 100. Derivatives: "centurion," "century," "cent."

Bibliography

Abramowitz, M. and Stegun, I. A., eds. 1972. *Handbook of Mathematical Functions with Formulas, Graphs, and Mathematical Tables.* New York: Dover Publications.

Badiru, A. B., ed. 2006. *Handbook of Industrial & Systems Engineering.* Boca Raton, FL: Taylor & Francis/CRC Press.

Beyer, W. H., ed. 2000a. *CRC Standard Mathematical Tables*, 26th edn. Boca Raton, FL: CRC Press.

Beyer, W. H., ed. 2000b. *CRC Probability and Statistics Tables and Formulae.* Boca Raton, FL: CRC Press.

Boljanovic, V. 2007. *Applied Mathematical & Physical Formulas.* New York: Industrial Press.

Brown, T. H. 2005. *Mark's Calculations for Machine Design.* McGraw-Hill, New York.

Dorf, R. C. 2004. *CRC Handbook of Engineering Tables.* Boca Raton, FL: CRC Press.

Evans, M., Hastings, N., and Peacock, B. 2000. *Statistical Distributions*, 3rd edn. New York: Wiley.

Fischbeck, H. J. and Fischbeck, K. H. 1987. *Formulas, Facts, and Constants for Students and Professionals in Engineering, Chemistry, and Physics.* New York: Springer.

Fogiel, M., ed. 2001. *Handbook of Mathematical, Scientific, and Engineering Formulas, Tables, Functions, Graph Transforms.* Piscataway, NJ: Research and Education Association.

Galer, C. K. 2008. Best-fit circle. *The Bent of Tau Beta Pi*, p. 37.

Ganic, E. H. and Hicks, T. G. 2003. *McGraw-Hill's Engineering Companion.* New York: McGraw-Hill.

Gieck, K. and Gieck, R. 2006. *Engineering Formulas*, 8th edn. McGraw-Hill.

Heisler, S. I. 1998. *Wiley Engineer's Desk Reference: A Concise Guide for the Professional Engineer*, 2nd edn. New York: Wiley.

Hicks, T. G., ed. 1995. *Standard Handbook of Engineering Calculations*, 3rd edn. New York: McGraw-Hill.

Hicks, T. G. 2003. *Mechanical Engineering Formulas.* New York: McGraw-Hill.

Hicks, T. G., ed. 2005. *Standard Handbook of Engineering Calculations*, 4th edn. New York: McGraw-Hill.

Hicks, T. G. 2006. *Handbook of Mechanical Engineering Calculations*, 2nd edn. New York: McGraw-Hill.

Kokoska, S. and Nevison, C. 1989. *Statistical Tables and Formulae.* New York: Springer.

Leemis, L. M. 1995. *Reliability: Probability Models and Statistical Methods.* Englewood Cliffs, NY: Prentice-Hall.

Leemis, L. M. and Park, S. K. 2006. *Discrete-Event Simulation: A First Course.* Upper Saddle River, NJ: Prentice-Hall.

Luderer, B., Volker, N., and Vetters, K. 2002. *Mathematical Formulas for Economists.* New York: Springer.

Index

Printed in the United States
by Baker & Taylor Publisher Services